THE ART OF RECKONING

analysis of performance criteria

this is a volume in the Decision Science series
edited by Ingram Olkin and Gerald Lieberman

THE ART OF RECKONING

analysis of performance criteria

Samuel Eilon

Department of Management Science
Imperial College of Science and Technology, London

1984

ACADEMIC PRESS
(Harcourt Brace Jovanovich, Publishers)

LONDON · ORLANDO · SAN DIEGO · SAN FRANCISCO · NEW YORK
TORONTO · MONTREAL · SYDNEY · TOKYO · SÃO PAULO

ACADEMIC PRESS INC. (LONDON) LTD.
24/28 Oval Road,
London NW1

United States Edition published by
ACADEMIC PRESS INC.
(Harcourt Brace Jovanovich, Inc.)
Orlando, Florida 32837

British Library Cataloguing in Publication Data

Eilon, Samuel
The art of reckoning.
1. Corporations—Finance—Decision making
I. Title
658.1′51 HG4026

ISBN 0-12-234080-9

Printed in Great Britain at
Thomson Litho Ltd., East Kilbride, Scotland

Preface

The business environment abounds with performance criteria, some real and many surrogate. In the pursuit of their objectives, managers make various assumptions about the relationships between these criteria. Consider for example, the following commonly held views:

(a) It is extremely difficult to increase profit without an increase in sales and market share.
(b) An increase in profit margin always leads to an increase in profit.
(c) If the profit per unit of output increases, the profit margin must increase.
(d) It is not possible to improve the net profit margin without increasing the gross margin.
(e) If labor productivity rises, production costs per unit are bound to fall.
(f) An increase in added value per employee is a sure way to achieve an increase in profit and in return on capital.

All these statements sound plausible enough, and it would be tempting to draw normative conclusions from them for managerial actions. And yet, as general statements, they can all be shown to be fallacious. They represent oversimplifications and extrapolations from narrowly defined circumstances, and as such they are typical of an inflated body of myths that has grown in the business world. Many examples in this volume demonstrate the general lack of compatibility between various performance criteria, thereby indicating the need for a methodology of reckoning, namely for a modeling approach that helps to establish relationships between the various measures in question.

An industrial or business enterprise is a complex system with many interrelated variables. Some of the variables are controlled by external forces: The general economic climate, intensity of competition, emergence of new products or processes likely to threaten existing ones, new government regulations that impinge on product specification or trading methods, changes in prices of raw materials and energy, fluctuations in interest rates and currency rates, new wage settlements agreed at the national level, and so on. Each of these factors, and many others, may have a profound impact on the performance of a business, and management needs to have an understanding of the cause and effect relationships and to anticipate, whenever possible, the effects of future changes in the external environment.

Apart from external factors, there are also those that are determined within the enterprise, as a consequence of prevailing operating practices and managerial decisions. Any attempt to change these practices may have an effect on performance, but because most operating variables interact, and because any change in one variable can trigger a chain reaction involving a sequence of changes, the eventual outcome is not always obvious or easy to determine quantitatively. As a result, many executives tend to oversimplify the interrelationships that are involved, and to ignore the possible effects of such chain reactions, whereas a systematic analysis, attempting to plot the changes that are likely to occur, may well reveal that the oversimplification is not justified and at times may even be misleading.

The task for management, as has been said often enough, is to make good use of available resources, namely to allocate resources to appropriate products and activities, as well as to ensure that resources are efficiently used. But how can managers know whether their resources are indeed well utilized? They may employ a negative set of measures, such as waste of materials, idle time of men and machines, percentage of defective products, frequency of breakdowns and stoppages, etc. Each of these is a reflection on the level of effective utilization, so that a great deal of managerial effort is often aimed at reducing each of the negative measures in order to contribute to an improvement in the use of resources.

In contrast, positive performance measures attempt to identify the objective for which the resources are managed, and numerous discourses in the literature are devoted to exploring the nature and meaning of such an objective. Many authors argue that the sole purpose of an industrial or business enterprise is to make money. Others argue that it has a social function and that several diverse interested parties need to be satisfied, such as the customers, the employees, and the shareholders. In that case it is not appropriate to define a single objective function, but to recognize that the resource allocation problem is subject to many constraints and to multidimensional objectives, which are often in conflict with each other.

Even within narrowly defined financial goals—such as "making money"—a single objective is often untenable. For many people in industry financial success is measured solely by return on capital, while many others are content to concentrate on profit. It is often possible to increase profit by a substantial injection of fresh capital into the business, but if the amount of new capital is excessive, return on total capital may decline. Nevertheless, profit has assumed a prominent position as a prime measure of performance. One reason for this is that profit is simpler to monitor (on a monthly basis) than return on capital; also, when assets are revalued, or when write-offs are made from time to time (for example, annually), the return on capital may be subject to a drastic change, which may not be in any way related to the day-to-day conduct of affairs at the operational level of the enterprise. This is why it is generally considered that, while return on capital is a sound measure of overall financial performance, profit is an indispensable criterion in its own right; indeed, in many allocation models in mathematical programming, profit is specified as the sole objective function to be optimized.

Now, profit is defined as the difference between revenue and costs, and these are measures which can be more readily identified with managerial responsibilities at the operating level. A production manager in charge of resources allocated to the production function cannot be made directly accountable for profit, but he* can be accountable for costs, because he is able to control at least some of the operating variables that affect costs. Since the output is prescribed by the production plan, and presumably by the quantities needed to meet customer demand and/or to replenish stock, his attention naturally turns to the control of cost per unit (or "unit cost" for short) and the various components that make up the unit cost: labor input, materials usage, processing times, utilization of labor and machines, handling times, amount of work in progress, etc. The sales manager, on the other hand, can do little to reduce manufacturing costs and his attention concentrates on revenue and on means by which it can be increased.

We thus see how the objective of increasing profit is converted into the dual objectives of increasing revenue and reducing costs. It might be said that these two are surrogate objectives, in the sense that they may be regarded as a means to an end. The development of the concept of accountability demands that echelons in the managerial hierarchy be given clear objectives that are meaningful in the context of the ability of each echelon to control specific tasks and activities. It follows that numerous surrogate objectives need to be defined, some in absolute terms, others in the form of ratios. This is why accountants and financial analysts resort to a plethora of ratios to describe many facets of the workings of an enterprise, some relating to the immediate state of health of the business, while others attempt to evaluate its long term prospects and its ability to withstand future shocks.

Managerial reality, therefore, requires that numerous objectives are included in the overall control system. For the individual manager, a defined objective, by which his own performance is judged, ceases to be a surrogate objective and becomes a real objective, to be pursued relentlessly and often in total disregard of objectives defined for others. When a sales manager is set the task of increasing sales revenue or market share, he will endeavour to do just that. These are his real objectives, and his achievements in this respect determine his financial reward, his career prospects, and the regard

* Throughout this volume the masculine embraces the feminine: He means he or she, his stands for his or her.

with which he is held by colleagues and superiors. Thus, what is meant by a real objective versus a surrogate objective is a matter of definition. Profit is often regarded as a real objective for the reasons discussed earlier, but in one sense it could be seen as a surrogate to the criterion of return on capital. While an attempt is made in Chapter 2 to describe various performance ratios as surrogate objectives, it should be recognized that there is an arbitrary element in all such selections. What needs to be understood, though, is that while surrogate objectives are constructed with the aim of enhancing the real or main objectives of an enterprise, total compatibility is not always achievable.

Most treatises in managerial economics are content with general descriptive expositions of the nature of variables in operating conditions, but these are usually inadequate even in providing a full explanation for changes in past performance, let alone in giving persuasive guidance for decision making purposes. For example, the manager may want to know what the expected outcome would be of a 10% increase in wage rates, coupled with a 5% improvement in labor productivity, a 10% increase in selling price, a 4% increase in advertising expenditure and a reduction of two points in the interest rate on bank loans. A qualitative discussion, however sophisticated, will not enlighten the manger as to whether unit cost and total profit are likely to improve as a result of this set of changes, and if so by how much. To answer such questions, a modeling approach is needed. This is where the methodology developed in this volume can be of assistance in the analysis of various plausible scenarios, so that the manger can develop a better understanding of the behavior of the system under his control and examine the likely outcome of alternative decisions.

The approach starts with the premise that *all performance evaluation is based on comparisons*. Of the many comparisons, three are most commonly used:

1. *A comparison with past performance*, which indicates whether an improvement over time has taken place with respect to some selected criteria.
2. *A comparison against budget performance* specified in a corporate plan, so that any shortfalls can be highlighted and action taken whenever appropriate.
3. *A comparison with other plants or enterprises* engaged in the same line of business, or with the average performance of the relevant industrial sector, to give an indication of relative competitiveness and strength.

Not included in this list is *a comparison with optimal performance, determined with hindsight*. In such an exercise, the discrepancy between the actual and the best possible outcome becomes a measure of efficiency, or the so-called measure of regret, that quantifies the opportunity missed by deviating from the optimal course. Although in principle such a comparison may be instructive, any unfavorable inference made on the quality of managerial decisions, based on hindsight, is naturally resisted, on the grounds that events that turned out to have a crucial impact on performance could not have been foreseen at the planning stage.

It is, therefore, often argued that a comparison with the budget, or planned performance, is a more realistic procedure. But what is the budget to be based on? A bottom up planning approach relies on targets supplied by the operating departments or divisions, reflecting what those responsible believe they can achieve. A top down approach signifies what the senior executives at headquarters expect, in terms of performance measures, and these expectations may be based on extrapolations from past performance, or on an assessment of what the competition is up to. Alternatively, the formulation of a budget may be the solution of an optimization model, namely the answer to the following question: Given the resources and constraints *at the time of drawing up the corporate plan*, and given the best forecasts then available of demand and price levels, what would be the optimum allocation of resources to achieve the best possible result in terms of certain pre-defined criteria? Such an approach avoids the criticism of employing hindsight, alluded to earlier, and attempts to replace some subjective (and often arbitrary) estimates of managers by an objective yardstick, against which subsequent actual performance can be measured. It is understandable, therefore, why forward budget plans, based on an optimal solution of a detailed structural model, have an intuitive appeal to corporate planners.

But I have long held the view that optimization is a questionable doctrine. If the optimal solution is very close to the current mode of operation, its value is necessarily limited. If the solution is substantially different from current operations, it relies on the assumptions that the functional relationships between variables will remain valid when substantial changes in the value of the variables take place. In practice, such an assumption rarely holds, and as soon as operating conditions are subject to change, so are the postulated functional relationships. For example, it has often been pointed out that while linear relationships are assumed in linear programming models, they hold for relatively narrow ranges of values for the variables concerned. This is, therefore, the dilemma facing the proponents of optimization models: A solution that advocates minor adjustments in managerial control variables holds promise for only a modest improvement in performance, while a solution that advocates drastic changes may be totally invalid. This is why optimization, which is widely heralded in the literature, is of limited consequence in practice, where—as Herbert Simon has postulated—managers tend to satisfice rather than optimize. Admittedly, there are in this book several examples for which optimal solutions are sought, but these should be viewed as exercises aimed at giving managers an indication as to the *direction* in which it would be desirable to move, rather than a definitive normative prescription for action.

Allied to the concept of optimization is the concern with the behavior of variables at the margin, which is the essence of the contribution made by differential calculus. If the derivative of a variable y can be determined with respect to another variable x, then we can draw conclusions about the functional relationship between the two variables and their behavioral characteristics, including the conditions that lead to the optimal value of y. As Tolstoy wrote in "War and Peace",

> "It is impossible for the human intellect to grasp the idea of continuity of motion. Laws of motion of any kind only become comprehensible to man when he can examine arbitrarily selected units of that motion. But at the same time it is this arbitrary division of continuous motion into discontinuous units which gives rise to a large proportion of human error . . . It is only by admitting infinitesimal quantities . . . that we can arrive at the same solution of the problem."

There is no denying that differential calculus is a powerful analytical tool in numerous branches of science (and there are examples of its application in this book). But differential calculus relies on the assumption that variables undergo continuous change, in other words, that functions are differentiable, so that their rate of change at the margin can be ascertained, whereas the problems that face business analysts do not conform to such a requirement: Wage rates, prices of raw materials, market demand, expenditure on promotion, the number of machines in a plant, currency exchange and interest rates—all these variables (and many others) do not change gradually but quite abruptly from one day (or week, or month) to the next. Instead of differential calculus, which is concerned with what happens at the margin, we need a calculus that can cope with *discrete changes* in variables, and where information on their magnitude is inferred from a series of snapshots of accumulated figures, instead of being given in the form of a continuous record. For example, the level of sales or costs is usually ascertained for a month, a quarter, a year, not for every single minute during the year, and this monitoring procedure is a reflection of the fact that a detailed record would be impossible or too expensive to compile, or that the manager would find little use for it.

It was this simple realization that led me to the development of *incremental calculus* (or, to be precise, *the calculus of relative increments*), where the major concern is with the magnitude of *relative changes* (or percentage changes) of key variables in the system. Such a concept accords with the way performance data are reported and discussed in the business environment:

Volume increased by 10% this year compared with last year, revenue fell by 5%, cost of materials rose by 2%, etc. Analysis of data for several industrial case studies in the early seventies led me to realize how useful such a methodology could be, both as a diagnosic tool and also for planning purposes, and I first formulated some basic ideas in two notes published in the mid-seventies*, leading to a series of publications since then, The basic rules of incremental calculus are quite simple, and the methodology is presented in Chapter 4, while its use is demonstrated in subsequent chapters dealing with cost, productivity, revenue, profit, return on capital, and many other performance measures employed in practice.

The main feature of the methodology expounded here is its flexibility. It can be used in simple models, involving only a handful of variables, or in large networks, where the sequence of changes and interactions between variables needs to be systematically traced before the total effect of these changes on key performance criteria can be ascertained. Furthermore, it allows analysis at the aggregate level as well as disaggregation to any desired level of detail, compatible with the availability of reliable data. It can be employed in conjunction with any chosen performance criteria to provide a suite of models to suit the needs of particular control and decision centers. It can help in sensitivity analyses to determine the type and accuracy of data to be monitored for future use.

This flexibility means that the methodology presented in this volume can be used to construct a wide variety of interrelated models, with the output of some becoming the input to others, both for the purposes of analysing past performance and also for developing guidelines for decision making. Some of the models that can be developed with the proposed methodology for common use are:

(1) *A cost model*—to describe the total cost incurred by the enterprise in terms of its constituent parts, as well as its dependence on volume and various external factors.

(2) *A unit cost model*—to determine the total cost per unit of output and to examine ways in which the unit cost can be maintained or reduced. The disaggregation of total cost and unit cost can take the form of identifying specific factor inputs, such as labor, materials, marketing and distribution, interest charges, and other overheads. Alternatively, the cost structure may follow the convention of segregating fixed cost elements from variable costs. In either case, the model can be decomposed to incorporate any desired level of detail, for example by breaking down labor into skilled, unskilled, managerial, supervisory, etc.; or materials costs into several categories of materials and bought-in components.

* Eilon, S. (1975): Changes in profitability components, *Omega*, vol. **3**, pp 353–4
Eilon, S. (1976): Sensitivity of the rate of return, *Omega*, vol. **4**, pp 350–2

(3) *A revenue model*—to express the total revenue in terms of global price and volume measures, or by taking account of the contribution to total revenue made by each of the individual products in the product-mix.
(4) *A profit model*—to incorporate the variables in the system that affect the level of profit and to highlight the effect of changes in these variables.
(5) *A profit margin model*—to show how the profit margin can be improved and the circumstances under which such improvements are compatible with increased profit.
(6) *An added value model*—to provide an analysis of how added value is obtained, how it is affected by its constituent parts, and how it can be improved.
(7) *A return on capital model*—to relate profit to total investment, to identify the reasons for change and to determine the degree of correlation between an improvement in return on capital and changes in profit levels.

The book is intended for students and practitioners alike (knowledge of College algebra and calculus should be adequate to follow the text). It can be used extensively in business schools and in courses of managerial and/or business economics, accountancy, finance, performance analysis, and operations research. In the business world the book will be useful to business analysts, accountants, operations research analysts, corporate planners, management scientists, and—last but not least—senior executives.

The list of contents is self-explanatory. The first two chapters discuss the enterprise as a system, highlighting the relationship between many variables, some identified as inputs, some as outputs, while others reflect operational practices. Various performance measures are described, including a wide range of management ratios to measure efficiency, productivity and profitability. The third chapter is devoted to index numbers and their use for measuring changes in volume, price and revenue. This chapter is not essential for the perusal of subsequent chapters and may well be omitted during a first reading or in a curtailed course of study. The chapter does, however, address itself to the important problem of measuring output for a multi-product system, and by realizing that more than one method may be applicable, the reader should get a better appreciation of the more general problem of measurement practices.

Chapter 4 gives an exposition of the methodology of incremental calculus, accompanied by many examples to illustrate its use in model building and in sensitivity analysis. The next three chapters are devoted to the analysis of costs (both total cost and unit cost), starting with definitions of various types of costs and a description of alternative cost models (Chapter 5), then analysing the factor inputs cost model in which labor and materials inputs play a prominent part (Chapter 6), followed by the ramifications of the fixed and variable cost model (Chapter 7).

Chapter 8 analyses the relationship between price and volume, gives formal definitions for three alternative measures of price elasticity of demand and shows the relationship between them. Chapter 9 deals with revenue and profit and specifies the conditions under which they are expected to rise or fall, with the change in unit price shown as a prominent decision variable. Attention is also drawn to circumstances that lead to a rise in profit accompanied by a fall in revenue. The net and gross profit margins are discussed in Chapter 10, and attention is particularly drawn to how they are affected by changes in price, profit and revenue. Chapter 11 is devoted to the analysis of return on investment and discusses how a change in working capital can have a profound effect on various performance criteria. Chapter 12 shows how the models developed earlier can be applied to the two product case and explores circumstances likely to lead to improved profit level and/or profit margin. Neither Chapter 11 nor 12 is essential for those wishing to proceed directly to the next chapter. Finally, Chapter 13 discusses ways in which the corporate planning function can benefit from the methodology presented earlier. Throughout this book numerous examples are presented to illustrated the model building process in relation to selected performance measures. In addition, a set of exercises is given at the end of each chapter, and a separate tutor's manual with answers to these exercises is available.

This volume is the outcome of extensive studies and applied research into profitability and productivity analysis over the past fifteen years. During this time I have greatly benefited from discussions with colleagues and research associates, with students, and with practitioners, as well as with senior executives in whose businesses a variety of models propounded here have been used at various levels of detail. The constructive comments of all these people have helped to clarify my own thinking about the various issues involved, and have greatly encouraged me to continue with the task of developing and refining the approach advocated here. It would be invidious to single out specific names, but I must mention Professor George Cosmetatos, who has collaborated with me in several publications exploring various theoretical aspects relating to the analysis of productivity and its potential role in negotiations between labor and management, as well as between executives with conflicting responsibilities. My secretary, Miss Jennifer Everett, has gallantly battled with a succession of revisions and rewrites, and I am most grateful for her efforts.

April 1984 *Samuel Eilon*

to

Amir

Ronit

Carmel

Daniel

Contents

1
An Introduction to Performance Analysis

Themes in this chapter

The purposes of performance analysis
Types of analyses
Ratios of:
- **financial output to financial input**
- **physical output to physical inputs**
- **financial output to physical inputs**
- **physical output to financial input**

The meaning of efficiency:
- **technical efficiency**
- **cost efficiency**
- **capacity utilization**
- **revenue efficiency**

THE NEED FOR PERFORMANCE ANALYSIS

An industrial enterprise is engaged in a series of activities aimed at providing products or services to meet active or latent market demand. To that end, a variety of resources need to be employed—manpower, materials, production facilities and, of course, money. These resources need to be managed so as to ensure that the output conforms to predetermined specifications and can be offered to the customer at a competitive price. It is immaterial, in this context, whether the customer is a member of the general public, a business corporation, a public institution or another department within the same firm. What we are concerned with here is a purposeful system with a series of inputs, which are converted into a series of outputs. An evaluation of the performance of the system is often called for in order to ascertain whether it meets the expectations or aspirations of various interested parties:

Investors— wish to have a reasonable return on their investment, while avoiding undue risks

Lenders—wish to lend to a healthy business, in order to safeguard their loans and payment of the interest due on the loans

Customers—want good quality and reliability at a reasonable price

Employees—wish to enjoy high remuneration and good working conditions, coupled with job satisfaction and job security

Suppliers—are interested in continuity of demand from healthy and cooperative debtors

Management—wishes to improve the use of resources under its control in order to maintain or expand the demand for its products or services and, thereby, ensure the economic viability of the system

Although all these interested parties seem to diverge in their objectives, they are all concerned with the *system's performance* and, although different people are bound to emphasize different goals, the need for analysis of performance is an inherent ingredient in any evaluation process. Since management is the custodian of the enterprise and the main instrument through which tighter control or a change of direction can be engineered, and since management needs in any event to take account of the interests of all the parties concerned, it is natural for the managerial viewpoint and managerial actions to be the subject of close scrutiny in any performance analysis.

Measures of performance constitute a focal point in the managerial control process [4]. Starting with the multiplicity and hierarchy of objectives, management needs to plan the means by which these objectives can best be achieved, including the acquisition of resources and their allocation to various activities against a desirable time-scale. Planning is followed by execution and the control function comes into operation (see Fig. 1.1), which consists of:

Monitoring and feedback—to determine progress, to compare with the plan, to note deviations and to report to those concerned.

Corrective action—to rearrange activities or the allocation of resources in order to reduce the gap between what is achieved and what is intended.

Post-operational evaluation—to find out how well a plant (or an enterprise as a whole) has performed in the past, to learn the appropriate lessons for the future and to feed these lessons into the planning function.

For the control function to be effective, it is necessary to define a wide range of performance measures: some relate to engineering aspects of the product, to the technology employed by the plant and to the methods of manufacture; some are solely concerned with financial performance such as cost, revenue, profit and return on investment; some concentrate on the organizational structure and the procedures used to institute effective responses when the

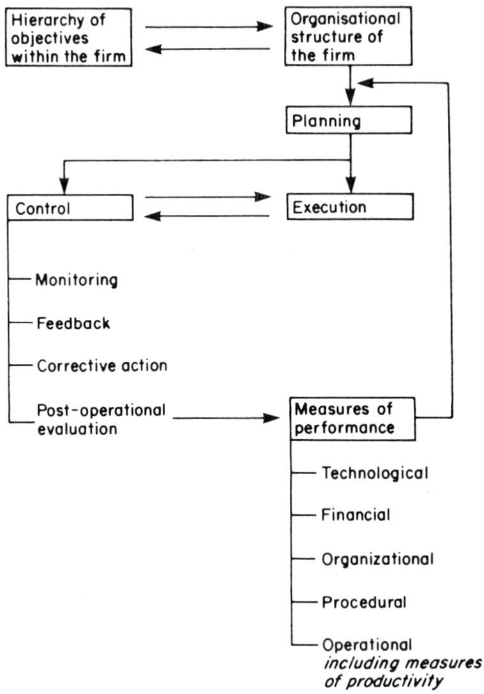

FIG. 1.1. Measures of performance as an ingredient of management control [2].

need for action arises; some relate to operational problems and the effective use of resources (including measures of productivity).

It is clear, therefore, that performance criteria are essential for an evaluation to be carried out. Only when measured against such criteria, is it possible to make a judgement as to whether performance has been good or bad, and whether an improvement has taken place over a period of time. Furthermore, it is possible to contemplate the consequences of various external factors and to consider alternative courses of action with respect to their expected effect on the company.

Figure 1.2 lists some of the main external factors that affect the enterprise: market demand, the economic environment, the force of competition, the cost of resource inputs (e.g. materials and labor), the cost of borrowings, the impact on exchange rate (on the cost of imported materials and the revenue realized from exports). All these factors impinge on management decisions on how to run the plant, and the results are reflected in performance criteria, such as revenue, cost, profit, return on investment, etc.

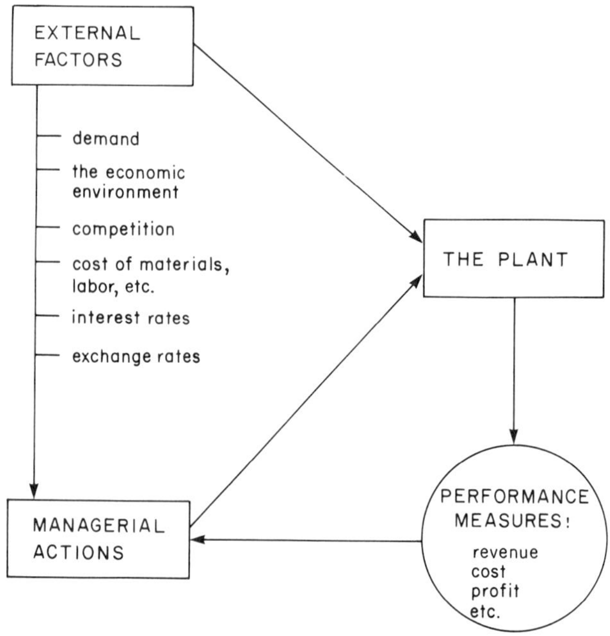

FIG. 1.2. Effect of performance on decision making.

THE COMPLEXITY OF A BUSINESS SYSTEM

An industrial enterprise is a complex system with many variables and interrelationships that affect performance. Some of the factors involved are more predominant in certain industries than in others, and it is therefore difficult to produce a detailed and general purpose model to take account of all the possible variations on a theme that are likely to be encountered in each case. It follows that the development of general models is based on assumptions and simplifications which need to be scrutinized for their validity and applicability in any given situation, so that appropriate modifications and further detail can then be incorporated as necessary.

A general impression of the relationships involved in a manufacturing enterprise can be obtained from the chart in Fig. 1.3. Although the chart seems complicated at first sight, it is, in fact, a simplified representation of the interaction between some of the main factors often identified as having an effect on performance, and it must be stressed that the picture given here is neither comprehensive nor sufficiently detailed to account for the many intricacies found in real-life operations. Nevertheless, the chart is useful in

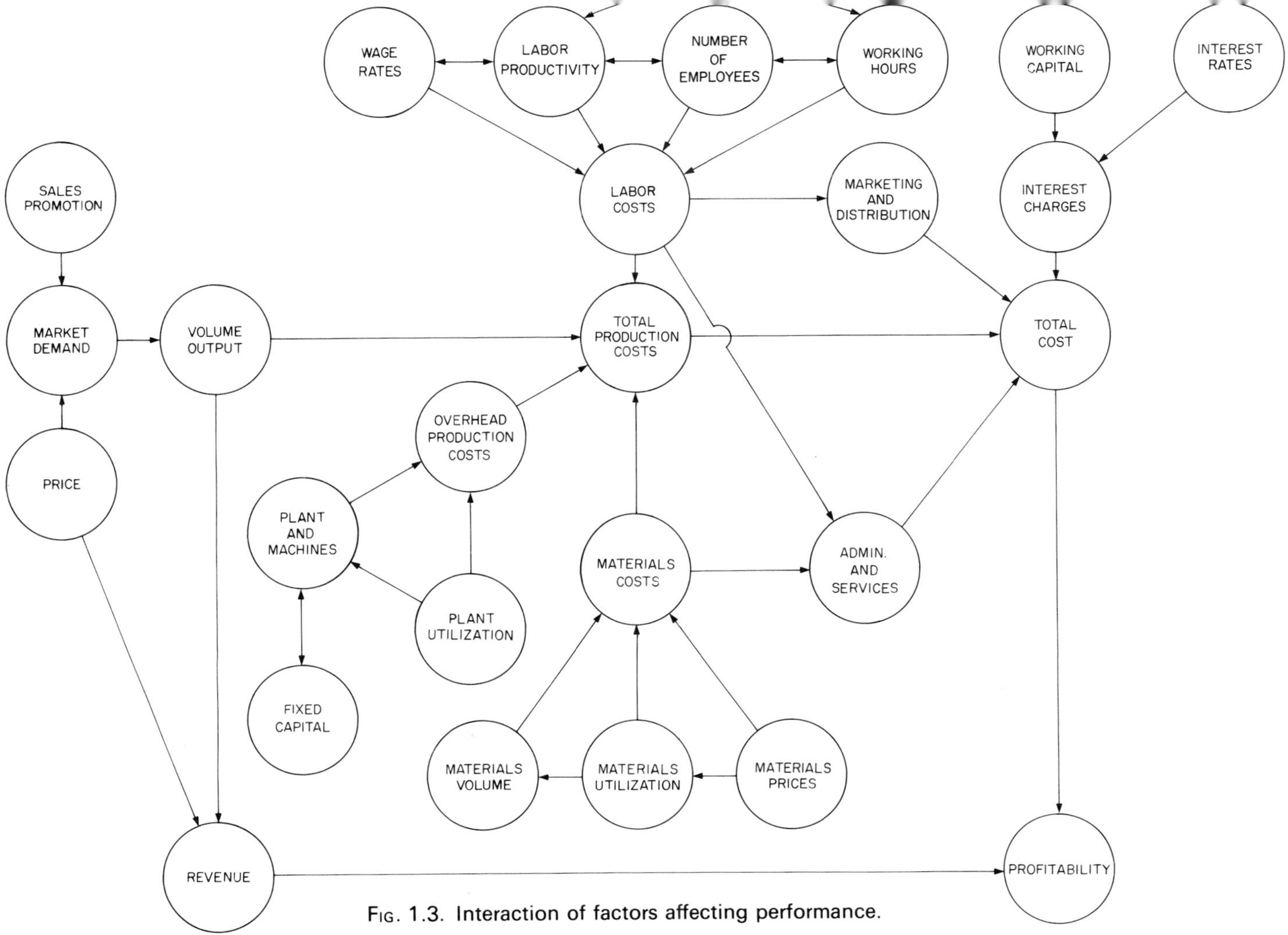

FIG. 1.3. Interaction of factors affecting performance.

highlighting the fact that the system consists of interlocking parts, and that each part affects other parts and, in turn, can be affected by them.

Starting from the left hand side of the chart in Fig. 1.3 we are first reminded of the fact that the output of the plant is generated to meet market demand, and that this demand is affected by market potential, by competition and by price. Further stimulation can be given to the demand level by marketing and sales promotion. The total production costs are affected by the level of output and consist of labor costs, materials and overheads attributable to the production function, and as can be seen from the chart:

- Labor costs are affected by the number of employees, the number of hours worked, the wage rates, and labor productivity
- Material costs are determined by the volume of materials acquired as inputs, the factor prices for materials and the level of utilization (i.e. the productivity of materials)
- Overhead production costs are affected by the availability of suitable plant and machinery (provided by fixed capital), their level of utilization and depreciation charges.

Many of these factors are interlinked: wage rates and wage incentive schemes affect labor productivity, and in turn the latter may influence the level at which the rates are set (see the top of the chart); the number of employees affects the number of overtime hours that employees have to work, and in turn the number of basic working hours affects the number of employees needed to perform the job, and both affect (and are affected by) labor productivity. The chart does not show a link between wage rates and working hours or the number of employees, but circumstances may arise where such a link exists.

Similarly, the factors affecting materials costs (at the bottom of the chart) are linked: high material prices often provide an incentive to reduce scrap and improve materials utilization, which in turn affects the amount of materials needed for production.

Not shown in the chart are the possible interactions between the factors affecting labor costs and those affecting materials costs and/or overheads. For example, the type of machines used and plant utilization may well affect the utilization of materials and labor productivity, and vice versa. Similarly, wage rates and working hours may spur the development of new working methods that may have an impact on plant utilization.

As shown in the chart, labor costs are not only incurred in the production function but in other functions as well, such as marketing, distribution, research and development, management services and general administration (the last three are grouped in the chart under the title "admin. and services"). These functions often have a substantial manpower cost element, but the cost of materials and the deployment of fixed assets may also be important factors, for example the use of vehicles by sales and distribution personnel, use of equipment and materials in research and development, use of stationery, computers and communications facilities. Again, most of these interrelationships are not shown in the chart merely for the sake of simplifying the presentation.

All these costs contribute to the magnitude of the total cost. A further cost element is that of interest charges, which are affected by the level of working capital, the amount of borrowings and the prevailing interest rates. Finally, the total cost and revenue (which is determined by the level of sales and the price of the output) determine the overall performance and the measures of profitability of the enterprise.

PERFORMANCE OF A BUSINESS SYSTEM

The basic question that needs to be asked is: *What is meant by performance?* A simple representation of an enterprise is given in Fig. 1.4(A) as a black box with an input and an output. Since the enterprise converts the input into an output, a direct evaluation of performance would involve an examination of the manner in which the conversion takes place. This conversion consists of many varied and intricate activities, each of which can be examined at length on its own (for example, the operation of each machine in the plant). In addition, there is a need to relate the individual activities to an overall measure

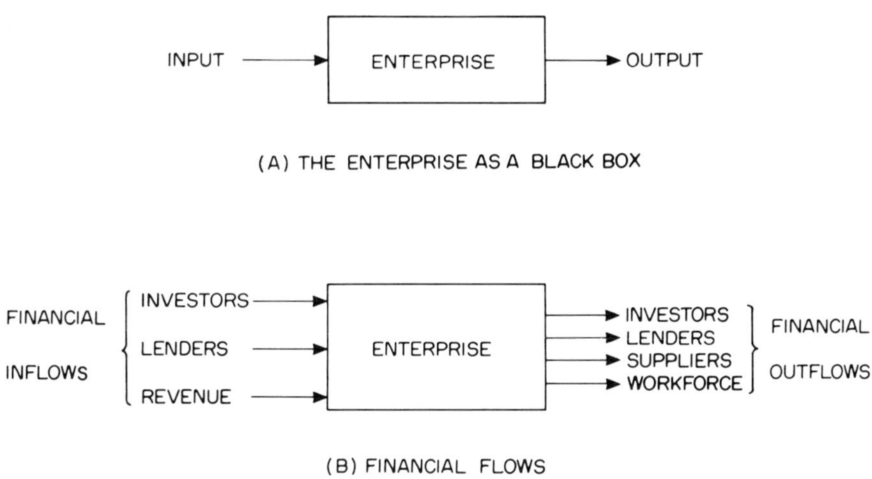

FIG. 1.4. The enterprise.

of performance, in other words to show how each part of the system affects the other parts and how the many diverse activities are integrated into a whole. In a complex system this is a tall order, and this is why analysts are often tempted to regard the whole enterprise as a black box and judge its performance in terms of its total output and relate it to the input. By way of an analogy we are reminded of an engine, which consists of many intricate parts, each of which can become the subject of a separate detailed investigation. But one of the standard measures of performance of an engine as a whole is its efficiency under various running conditions on a test bed, where the efficiency is determined as the ratio of energy output to energy input.

In the case of an enterprise, the input and output may be measured in units of money, though it is important to distinguish between various types of financial flows in terms of their sources and destinations, as shown in Fig. 1.4(B). While this representation is more informative than that in Fig. 1.4(A), it does not explain what the financial inflows are used for within the enterprise, nor does it explicitly state that the conversion process results in a physical output, which is subsequently converted into revenue. In this sense, therefore, revenue may be regarded as the outcome, or *output*, so that instead of concentrating on financial flows, as in Fig. 1.4(B), we can depict the system as receiving a financial input (i.e. cost), which is converted into a series of physical inputs, such as labor, materials and capital expenditure for machinery to provide the necessary production capacity. These physical inputs, or resources, are fed into a manufacturing conversion plant, as shown in Fig. 1.5, and the outcome is a series of physical outputs, sold to customers to yield revenue.

A more detailed financial and materials flow is described in the various layers of Fig. 1.6. The top layer relates to finance. The second layer represents the factor prices, both at the input side (in converting money to physical inputs) and at the output side (in converting products to revenue). The third layer denotes the physical goods, being transformed from raw materials to semi-finished and finally to finished goods, while the next two layers indicate

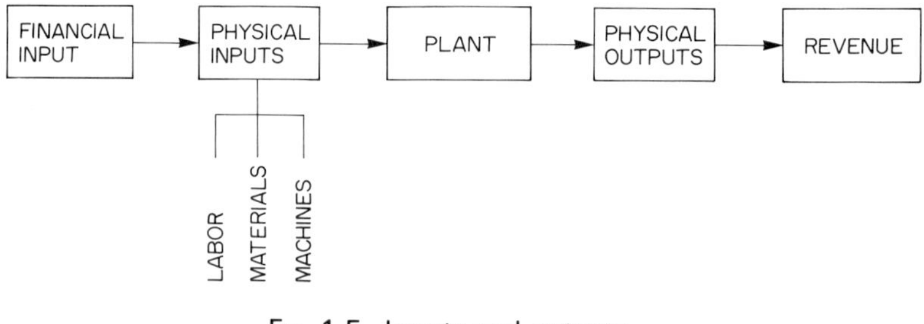

FIG. 1.5. Inputs and outputs.

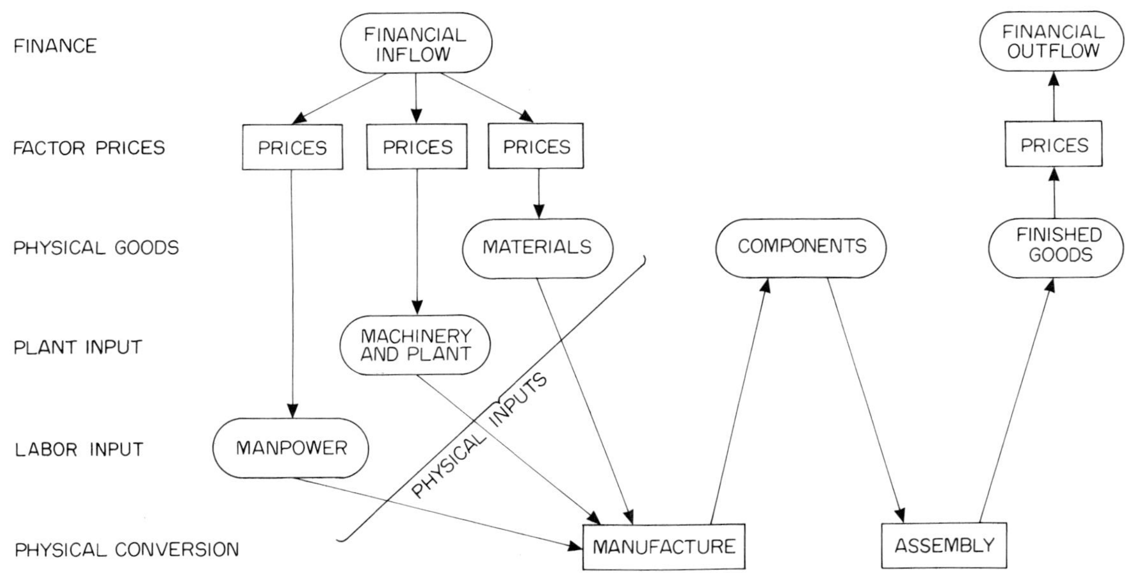

Fig. 1.6. Financial and materials flow [2].

the two inputs of machines and manpower respectively. The bottom layer represents the physical conversion process (shown here as involving production and assembly operations, though each may consist of several stages with intermediate stocks of work-in-progress).

TYPES OF PERFORMANCE ANALYSES

It is possible for every single element and every activity in the flow diagram in Fig. 1.6 to be analysed as part of a performance evaluation. A relatively crude categorization of the various possible analyses is shown in Fig. 1.7, which serves to highlight the following hierarchy:

Overall performance—is a measure in which the financial outcome is compared with the total financial outlay, sometimes referred to as the "productivity of total investment". This global measure can be subdivided into "the productivity of fixed investment" and the "productivity of working capital". Note

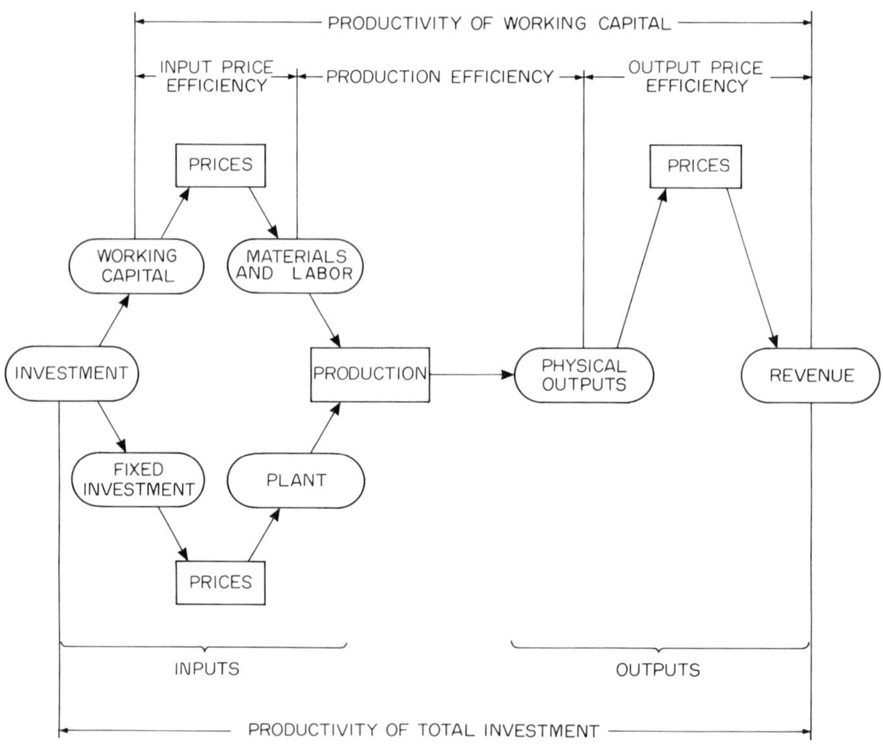

Fig. 1.7. The concept of efficiency [2].

that the term "productivity" in this context is used to denote a ratio of financial output to financial input (in the case of "productivity of total investment" and "productivity of working capital") or potential physical output to financial input (in the case of "productivity of fixed investment"). These terms are cited here for convenience, because they are in popular use. A more rigorous definition of productivity relates physical output to any given input, as discussed further in the next chapter. In Fig. 1.7:

Productivity of fixed investment—expresses the amount of production capacity provided per unit of investment in machinery and plant.

Working capital performance—refers to the financial output in relation to the working capital input and can be disaggregated into three measures through the simple identity:

$$\text{Working capital productivity} = \frac{\text{Financial output}}{\text{Working capital}}$$

$$= \underbrace{\frac{\text{Physical input}}{\text{Working capital}}}_{\substack{\text{ratio relating}\\ \text{to input price}\\ \text{efficiency}}} \times \underbrace{\frac{\text{Physical output}}{\text{Physical input}}}_{\substack{\text{physical}\\ \text{conversion}\\ \text{efficiency}}} \times \underbrace{\frac{\text{Financial output}}{\text{Physical output}}}_{\substack{\text{output price}\\ \text{efficiency}}}$$

This disaggregation yields three types of efficiency ratios (see Fig. 1.7);

Input price efficiency—given by the factor prices for acquiring physical inputs, such as materials and manpower.

Physical conversion efficiency—to measure the use of resources in converting physical inputs into physical outputs.

Output price efficiency—given by the product prices to convert the physical outputs into revenue.

Flow network of inputs and outputs

The array of possible performance analyses is further highlighted by the simplified summary flow chart in Fig. 1.8, in which the network consists of four stages (or nodes), denoting the inputs and outputs, and three arcs, depicting the

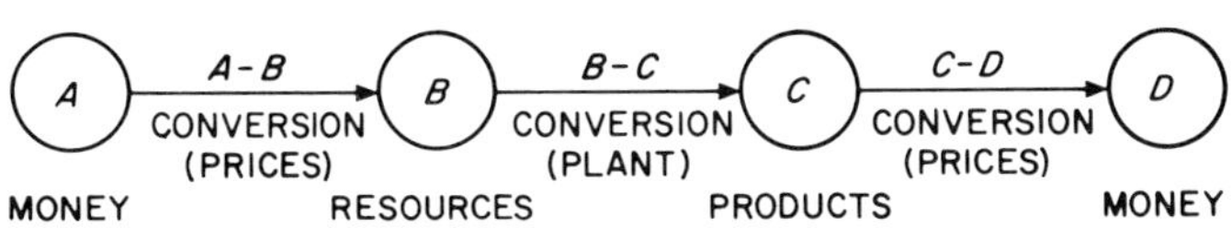

FIG. 1.8. Flow network of inputs and outputs.

conversion processes: Nodes *A* and *D* are the financial input (cost) and financial output (revenue) respectively, while nodes *B* and *C* are the physical input and physical output respectively; the first conversion process *A-B* (from node *A* to node *B*) converts money into other resources, the second process *B-C* uses the resources to produce the physical output, and the third process *C-D* converts the physical output into money.

This representation in Fig. 1.8 gives us a clue as to the types of analysis that can be pursued, namely for each node (concerning its structure and composition), for each arc (regarding the efficiency of the conversion process), and for each combination of an output node in relation to an input node. We thus have a variety of possible analyses, as exemplified by the following (with reference to Fig. 1.8):

Analysis at each stage (or node in the network)

Node A. Here we are concerned with the mixture of financial instruments and sources of finance in order to ensure a reasonable balance between equity capital, bonds and borrowings, bearing in mind the short-term and long-term cost of capital, its flexibility (some loans stipulate what the money can be used for), as well as the degree of assurance of its continued availability (borrowings in the form of bank overdrafts, for example, always carry the risk that the banks may demand payment at short notice).

Node B. It is obviously important to ensure that the facilities employed by the company should be fully utilized. As a first step to that aim it is necessary to have the right facilities in the right balance. For example, a highly automated plant requires a different labor force (in terms of numbers and skill) from a manual–labor–intensive plant. Similarly, the type and quality of materials used may greatly affect the plant and machines required for processing. In Fig. 1.5 three categories of resource inputs are mentioned: materials, labor and plant. Needless to say, each can be disaggregated into several subcategories, such as raw materials (or even various separate types of raw materials) versus bought-in components and assemblies, direct versus indirect labor (or various skills of labor), etc., so that the question of the appropriate balance between the major resource categories may be widened to cover the various subcategories as well.

Node C. At this stage our concern focuses on the product mix, the range of products on offer and their specifications. Again, the purpose of the analysis is to determine the desirable internal balance, i.e. within the node itself.

Node D. The final output is the resultant revenue, which may be regarded as the aggregate cash or cash equivalent of the physical output. Management of this revenue may be helped by several analyses in an attempt to answer such questions as: how to collect the revenue (associated with terms of trade), how to keep it (such as in the form of cash, deposits, foreign currencies, etc., and related problems associated with cash management in an enterprise spread over a number of locations), and how to dispose of it (as shown in the outflows in Fig. 1.4).

Analysis for each conversion process

Arc A-B. This conversion process consists of a series of factor prices, such as those paid for materials, wage rates, and the cost of acquiring plant and machinery.

Arc B-C. Here materials are converted into finished goods and the analyst is naturally interested in the efficient use of manpower and manufacturing facilities, namely in the proper allocation of resources to activities that culminate in the emergence of the required product-mix. Traditionally, this is the area in which industrial engineers can demonstrate their expertise by a variety of cost cutting exercises aimed at reducing waste (of materials), reducing idle time (of labor and machines) and reducing processing time (to increase throughput for given resources).

Arc C-D. This arc is a price mechanism, which translates the physical output to revenue, similar to the function of the arc *A-B* at the input end.

Analysis of ratios

Each of the seven types of analysis described above is ostensibly self-contained. However, decisions at each of the four nodes and three arcs in Fig. 1.8 must take account of what actually happens, or what is expected to happen, elsewhere in the network, and not just at the adjoining nodes or arcs. Thus, the product-mix at node *C* is influenced by the relevant prices in the arc *C-D* and their relationship with the unit costs incurred in the arc *B-C*, and these depend in turn on the resources-mix at node *B* and the factor prices in *A-B*. Similarly, the efficiency of the conversion process *B-C* depends both on the specifications of the product-mix at *C* and on the available resources at *B*.

It is worth noting here that while the performance of the conversion process *B-C* is determined by the way operations are conducted internally, and hence by management decisions and control regarding the allocation and disposition of available resources, the other two conversion processes (*A-B* and *C-D*) are largely dominated by external forces, relating to prevailing wage rates (which may be negotiated nationally or depend on local labor market conditions), to interest rates, to currency exchange rates, to available capacity of subcontractors, to fuel costs, and to supply and demand for raw materials and for the final product respectively. All these market forces culminate in price mechanisms: factor prices at the input end *A-B*, which determine the cost at which the resources can be provided, and product prices at the output end *C-D*, which determine how much revenue can be generated. While management skills are obviously needed to acquire resources and to sell the final product at the best possible prices, the ability of management to influence the outcome with respect to the links *A-B* and *C-D* is more limited. As the final outcome at node *D* is dependent on the combined contributions of all three conversion processes, it is well to bear in mind that the cause of a significant improvement or deterioration in the level of revenue may be entirely due to changes in price levels outside the direct control of the management of the firm.

RATIO OF INPUTS AND OUTPUTS

On the assumption that an industrial activity needs to be purposeful and that it is directed in an attempt to achieve given goals, it follows that its measure of effectiveness is constantly monitored by the gap between actual and planned results, so that suitable corrective action can be taken from time to time. But this gap on its own does not constitute a complete or adequate metric of performance. Since any contemplated corrective action can only be based on a change in the provision and mix of resources, it is necessary for the *utilization of resources* to be examined, namely to incorporate the concept of *efficiency* in the evaluation of performance.

This is why attention is often focused on ratios of outputs to inputs, so that performance is said to improve when more output can be produced for a given level of input, or when the same output can be achieved by a lower level of input (or both). The questions that remain for the analyst to resolve are what to identify as the input and the output respectively and how to measure them.

Various alternatives to the first question come to mind from the presentation in Fig. 1.8. As stated earlier, the input can be identified either as money (node *A*) or as a series of physical resources (node *B*), whereas the output is a product-mix (node *C*) or money (node *D*). These give rise to the following possible ratios between output(s) and input(s):

D/A Ratio between the financial outcome and the financial input. The financial outcome may be interpreted as revenue, or profit (= the difference between revenue and cost), or value added (= the revenue less materials input, discussed further later). The financial input may be measured as total investment in the enterprise or the total costs.

C/B Ratio between the physical outputs and the physical inputs. This ratio sets out to ignore the price conversion arcs *A-B* and *C-D*, so that it can concentrate on the conversion *B-C*. If the input and output were to consist of a single resource and a single physical product respectively, there would be no measurement problem. In reality, however, even in the case of a single product output, there would be various physical inputs, such as labor, materials and plant capacity (see Fig. 1.5), so that instead of a single ratio C/B we would need to have several ratios, such as C/B_1 (to measure the output per unit of resource B_1), C/B_2 (output per unit of resource B_2), etc. Furthermore, the case of a single product output is rarely encountered in practice. More commonly, the output consists of several products, but attempts to proceed with ratios of each of the products to each of the resource inputs are bound to be inconclusive. It is necessary, therefore, to devise an efficiency measure based on total capacity utilization, as discussed later in this chapter, or to design a means of measuring changes in the global output for the multi-product case, and this topic is pursued in Chapters 2 and 3. The ratios of physical output to specified physical inputs are called *productivity ratios* (for example, the ratio of total output per man or man-hour is known as the *productivity of labor*).

D/B Ratio between the financial outcome and each of the physical inputs. This ratio encompasses the price conversion stage *C-D* and overcomes the problem of measuring the overall physical output at node *C*, the measure of output being described in monetary terms at *D*. Again, as there are several inputs, the ratio D/B is often replaced by several ratios D/B_1, D/B_2, etc., to reflect the contribution of specific inputs. For example, many companies single out manpower as an important input and pay particular attention to such ratios as revenue/employee, profit/employee and added value/employee.

A/C Ratio between the financial input and the physical output. Here, the market price for the final product is ignored (i.e. the arc *C-D*) and interest is usually focused on the total cost per unit, for example the cost of producing a ton of steel or a barrel of oil. Such a measure is sought in the multi-product case as well, both when an aggregate measure of output is used and when the cost per unit for each of the products is to be ascertained (on the assumption that the total costs can be distributed among the various product groups).

Clearly, then, an input–output analysis can be guided by several possible ratios belonging to the four groups listed above. Bearing in mind that each group can itself consist of a number of ratios, depending on the degree of disaggregation of the input and output data, we conclude that there is a potential for a large number of measures to be constructed as part of the performance evaluation process.

THE CONCEPT OF EFFICIENCY

An examination of the literature reveals that the terms "effectiveness" and "efficiency" are often used interchangeably. Strictly, there is a difference between them. *Effectiveness* relates to the extent that an activity manages to achieve a prespecified purpose, whereas *efficiency* is a measure of the utilization of available resources (as alluded to in Fig. 1.7). Hence, effectiveness is concerned with aims, while efficiency concentrates on means. It is possible for action to be effective, in that it brings about the desired result, and yet for it to be inefficiently carried out, with resources being idle or squandered. Similarly, it is possible for the converse to happen, namely for an activity to be efficient but ineffective.

How are effectiveness and efficiency to be measured? Following the distinction just stated, we surmise that effectiveness may be measured as the ratio of *actual output to planned output* (either physical or financial), whereas efficiency may be defined as the ratio of the *resources used to the resources available* or *actual output to potential output.*

Thus, if planned output is 120 units and actual output is 90 then the effectiveness measure may be said to be 90/120 or 75%. However, if the resources are capable of producing 100 units, then efficiency may be said to be 90%. In this case, the plan calls for an output that it is beyond the capability of the resources to deliver and the target may therefore be regarded as unrealistic, although in many circumstances the planned output falls well short of the available capacity. In general, it must be borne in mind that planned output cannot be specified without any reference to the demand for the product and to the resource inputs.

When we come to measure efficiency, we need clear definitions of what entities or values are to be compared. Amey discusses "economic efficiency" at some length, since "under *certain conditions* (i.e. under conditions of what the economist calls 'perfect competition') there is a close connection between efficiency and profitability" [1, p. 2]. He suggests that "economic efficiency is best measured by a comparison of actual and budgeted profitability, the budget being set in a particular way, viz in economic terms. That is, instead of judging efficiency by comparing the ex post measures, we are proposing that the comparison should be between ex post and ex ante profits in the same firm" [1, p. 40]. This is the essence of comparing actual performance with budgets, not only for profit but in relation to other measures as well (as further discussed in Chapter 13), and essentially what was referred to above as a measure of effectiveness. For the construction of measures for operational efficiencies, a great deal depends on what is meant by such terms as "available capacity" or "potential output", and this gives rise to several alternative definitions of efficiency, which we shall now proceed to examine.

Time and capacity utilization

To illustrate this point, consider the case of a single machine. A common method of measuring its efficiency is based on the concept of *time utilization*, namely the ratio of the actual operating time of the machine to the total time available. For example, if a machine tool is available for eight hours per working shift and is found to be operational for six hours, its utilization is said to be 75 %. Such a measure is totally divorced from the output side; it does not tell us anything about the purpose for which the machine is used and whether it produces the right kind of product of the specified quality. Indeed, the term "operational" to describe the fact that the machine is busy may raise certain difficulties in definition. If the machine were to be speeded up by 50 % and instead of producing, say, 100 units per hour it could then produce 150 per hour, and if as a consequence it would work only four hours per shift, its total output would remain unaffected but its time utilization would drop from 75 % to 50 %. (Note, incidentally, that in this example a reference is made to the *level of output* in an attempt to make a judgement on the degree of utilization of the machine.) It is, therefore, necessary for the term "operational" to carry with it a set of operating specifications, such as spindle speed, feed, depth of cut, type of coolant to be used, whether set-up time between cycles is included or not, etc. [3, p. 125].

If the machine is employed in the manufacture of a single product, it is a simple matter to convert time availability to the concept of *capacity*. It is possible then to state that the machine, when run under specified conditions defined as "normal" or "good operating practice", is capable of producing, say, 800 units per shift, so that when the machine actually produces 600 units its efficiency (in terms of *capacity utilization*) is said to be 75 %. This ability to convert time directly into capacity implies that there is no difference in this case between *time utilization* and *capacity utilization*.

An interesting point to note is that, as capacity is defined here as the "expected output when good operating practice is followed", and when no idle time or breakdowns occur, it need not always represent the absolute maximum output that can possibly be attained. As in the example mentioned earlier, circumstances may arise when deviations from normal operating specifications allow output to exceed the stated capacity, so that utilization then exceeds 100 %. In a sense, therefore, capacity is a notional convenient yardstick, a standard against which the actual resultant output is measured.

An absolute measure of capacity utilization requires that, in order to compare like with like, actual operating conditions conform to those for which the capacity is determined. However, in most cases managers are less concerned about the absolute measure than about the *change in utilization*, and for that purpose the concept of a notional capacity is perfectly adequate*. Thus, irrespective of prevailing operating conditions, if utilization in one week is 90% (for a given definition of capacity), whereas in the next week it is 108%, the improvement in plant utilization is $[(108-90)/90]\,100 = 20\%$. For the purpose of the following discussion we shall assume, for the sake of convenience, that "capacity" denotes the maximum output that can be obtained from a given production facility, although it is clear from the foregoing remarks that the term "maximum" can be substituted by "notional" (or "standard" or "specified") without any loss of generality. We have thus established that while efficiency is intended as a measure for the utilization of the input, it is necessary first to determine the capacity of resources and hence to rely on figures for actual and potential outputs as a basis for the computation of efficiency.

RATIOS OF PRODUCTION EFFICIENCY

The various alternatives of constructing efficiency ratios were discussed earlier. Leaving aside the overall measures associated with total investment or working capital as inputs (see Fig. 1.7), we shall pursue some of the other ratios alluded to in conjunction with Fig. 1.8 and consider the following approaches:

(1) Given the actual physical output of a plant (consisting of several resources), we ask: what is the combination of physical resources needed to produce this output efficiently? By "efficiently" we mean that in the "ideal", or "best", combination of resources, none can be reduced on its own for the desired output to be obtained. A comparison of the ideal physical input with the actual physical output (i.e. best combination with the actual level of resources used) is called *the technical efficiency* of the operation.
(2) If in (1) the lowest cost to produce the given output is compared with the actual cost, then we have a measure of the *cost efficiency*.
(3) Given the actual physical resources, we ask: what is the maximum volume of product-mix that can be produced with these resources? The answer to this question defines the physical capacity of the plant, and a comparison of the actual physical output with that of the capacity yields the *capacity utilization* as a measure of efficiency.

---

* The subject of yardsticks, through the use of index numbers, is further discussed in Chapter 3, and the analysis of changes is pursued in Chapter 4.

(4) If in (3) a comparison is made between the actual revenue (or profit) with the maximum that can be obtained from the available resources then we have the *revenue efficiency* of the current operation (or profit efficiency, if profit is adopted as the output measure).

Thus, approaches (1) and (2) start with a given output and seek to compare the actual input with the best (or most efficient) input, whereas approaches (3) and (4) start with a given input and proceed to compare the actual with the best output. The four approaches and their definition of efficiency are summarized in Table 1.1.

TABLE 1.1
Four measures for production efficiency.

Method	Input	Output	Units of measurement
(1)	$e(1)$ = technical efficiency $= \dfrac{\text{min. resources}}{\text{actual resources}}$	Given	Physical
(2)	$e(2)$ = cost efficiency $= \dfrac{\text{min. cost}}{\text{actual cost}}$	Given	Financial
(3)	Given	$e(3)$ = capacity utilization $= \dfrac{\text{actual output}}{\text{max. output}}$	Physical
(4)	Given	$e(4)$ = revenue efficiency $= \dfrac{\text{actual revenue}}{\text{max. revenue}}$	Financial

Definitions for the four measures of efficiency

In the single-input-single-output case, let

V = actual physical output
V_{max} = maximum physical output, being the capacity of the production facility
p = unit price for the output
Q = actual physical input
Q_{min} = minimum physical input
m = cost per unit of physical input

The four efficiency measures in Table 1.1, denoted as $e(1)$, $e(2)$, etc., are as follows:

$$e(1) = \frac{Q_{\min}}{Q}$$

$$e(2) = \frac{mQ_{\min}}{mQ} = e(1)$$

$$e(3) = \frac{V}{V_{\max}}$$

$$e(4) = \frac{pV}{pV_{\max}} = e(3)$$

The result $e(2) = e(1)$ follows from the assumption that the factor price m paid per unit of input is the same in the numerator and the denominator, i.e. that this factor price is unaffected by the amount of physical resource used. Similarly, the result $e(4) = e(3)$ is based on the assumption that the factor price p for the output is not affected by the fact that the actual output V differs from the maximum output $V_{\max}$. We further note that if a units of input are needed for each unit of output to be produced (for example, a tons of raw material per ton of product), then by definition the maximum output for a given input Q is $V_{\max} = Q/a$ and the minimum input needed for a given output V is $Q_{\min} = Va$, from which we conclude that

$$\frac{Q_{\min}}{Q} = \frac{V}{V_{\max}}$$

i.e. all the four efficiency measures have the same value. This conclusion does not, however, hold in the multi-input and/or multi-output case, as shown below.

Technical efficiency and cost efficiency

Consider the example of a plant consisting of two inputs used to produce a single product (which may be defined as a given product-mix) at a given level of output. For any given quantity of input 1 there is a corresponding minimum quantity of input 2 needed in order to achieve the specified output. All the combinations of the two inputs determined in this way define what Farrell [5] calls the "efficient production function", or the "production possibilities line" represented by the line EXE' in Fig. 1.9 [2]. As all the points on the line are said to be technically efficient, it follows (by definition) that points below the line are infeasible, while points above the line are clearly inefficient, since for any point X' there is a better point X, which uses the two inputs in the same proportions as at X', but requires smaller amounts of these inputs.

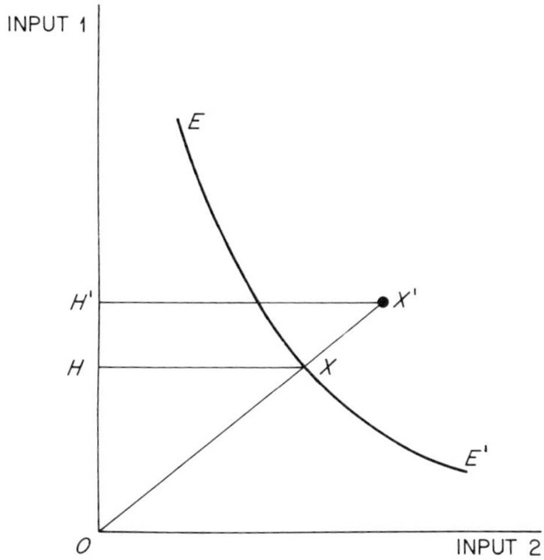

FIG. 1.9. Measurement of technical efficiency for two inputs.

As stated earlier, the concept of *technical efficiency* is based on comparison of efficient use of resources for production (represented here by OX) with the actual use of resources (represented by OX'), namely by the ratio OX/OX'. Notice that this ratio also measures the technical efficiency of each of the inputs: in our case the point X' corresponds to the use of OH' of input 1, whereas point X on the efficient production line calls for the use of OH, so that the technical efficiency of input 1 is $OH/OH' = OX/OX'$. The same considerations apply to input 2, and in general to any input in the multi-input case.

It should be noted that in this case *technical efficiency* is not synonymous with *cost efficiency*. If the line EE' in Fig. 1.10 represents the iso-cost line for the two inputs (i.e. all the combinations of the two inputs on this line incur the same total cost), its intersections with the efficient production function at points E and E' are technically efficient, while all the intervening points on the straight line between E and E' are technically inefficient, even though the cost is unchanged. On the assumption that the cost function is linearly dependent on the amount of resources used, it follows that other iso-cost lines will be parallel to EE', lines above EE' incurring higher costs and those below EE' incurring lower costs. The iso-cost line which is a tangent to the efficient production line at point Y stands for the lowest possible cost, since lines below YY' are in the technically infeasible domain. We may therefore summarize the ideas presented in Fig. 1.10 as follows:

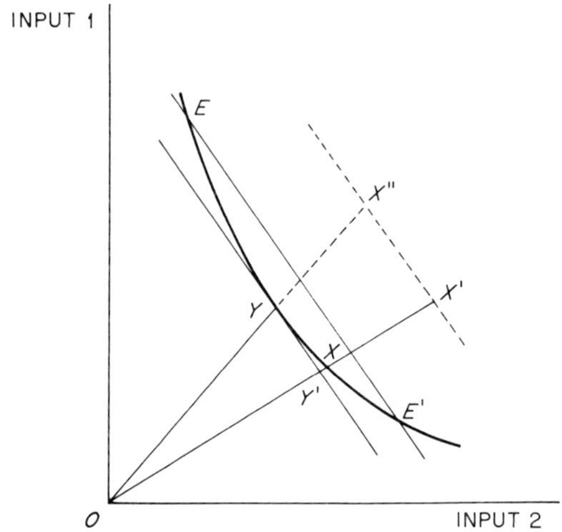

FIG. 1.10. Technical efficiency and cost efficiency for two inputs.

$EYXE'$ is the technically efficient production line, for which any given input is coupled with the minimum amount of the other input needed to produce the specified output

EE' is an iso-cost line, representing equal total cost for various combinations of the two inputs

$X'X''$ is the iso-cost line corresponding to the cost incurred at X'

YY' is the lowest feasible iso-cost line

X' is the combination of the two inputs corresponding to actual performance

X is the most efficient combination of the two inputs used in the same proportion as at X'

$e(1) = \dfrac{OX}{OX'}$ is the *technical efficiency* of operating at X'

Y is the least cost point on the technically efficient production line

Y' is the notional point at which input proportions are the same as at X and X' and at which the cost is the same as at Y (since both Y and Y' are on the same iso-cost line); Y' is not a feasible combination of the two inputs (since it lies below the efficient production line) and is merely used to determine the least cost operation that prevails at Y

$e(2) = \dfrac{OY'}{OX'}$ is the *cost efficiency* of operating at X', comparing the least cost at Y' (or Y) with the actual cost

X'' is the point at which input proportions are the same as the optimum point Y and where the cost is the same as at X'

$\dfrac{OY}{OX''}$ is the ratio between the minimum feasible cost and the actual cost at X'' (or X'); this is an alternative for defining the *cost efficiency* and (as all the iso-cost lines are parallel) it follows that

$$e(2) = \frac{OY}{OX''} = \frac{OY'}{OX'}.$$

The difference between the technical efficiency and the cost efficiency stems from the fact that at X' the two inputs *are not used in the optimal proportions.* From these definitions we conclude that technical efficiency is always greater than cost efficiency for a convex efficient curve such as in Fig. 1.10 (since $OX > OY'$), except when actual production lies on the line OX'', in which case both efficiencies coincide.

Capacity utilization

Figure 1.10 is concerned with the use of resources for a *given output* (or set of outputs), for which the corresponding efficient production line is drawn. An alternative approach, as stated earlier, is to consider the *resources as given* and to compare the actual output with the maximum that can be envisaged.

Consider the case of an input (such as a production facility) used to produce two products, as shown in Fig. 1.11 [3, p. 126]. The full capacity line depicts all the combinations of the maximum output of the product-mix that can be produced on the given production facility, thus representing the trade-off between the two products. Any point U' below the line corresponds to an under-utilized plant, so that the extension of OU' intersects the full-capacity

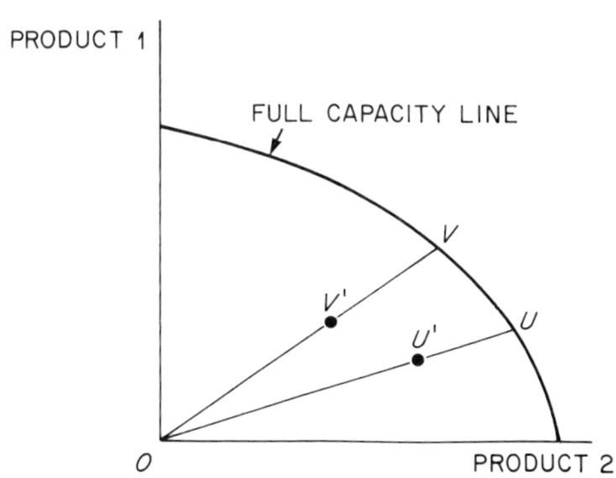

FIG. 1.11. Plant utilization.

line at U, where the proportions of the two products in the product-mix are the same as at U' and where the quantities produced are greater than at U'. Similarly, when production moves from U' to V', the corresponding point to V' on the full capacity line is V. The position of U' relative to U and of V' relative to V can be used as a measure of plant utilization for the points U' and V' respectively.

The case of a two-product plant, which consists of several facilities or resources, can be similarly described, the approach being akin to that employed in linear programming models applied to production problems. If, for example, there are four resources (or processes), then Fig. 1.12 shows the relationship between any given product-mix and these resources. The capacity of each resource (in terms of the maximum amount that it can produce of one product for a given level of the other) can be described as a constraint (see lines I, II, III, IV in Fig. 1.12) and the resultant envelope $PQWST$ is the maximum production capacity line. Points above it are infeasible, since they violate at least one constraint, whereas points within the feasible domain $OPQWST$ involve idle resources.

As in Fig. 1.11, plant utilization at any point may be defined in relation to the full capacity line. If U' in Fig. 1.12 represents an actual product-mix, then the same product proportions apply at U on the full capacity line. A measure of *plant* or *capacity utilization* may then be given by the ratio $e\,(3) = OU'/OU$. If during a subsequent period the plant operates at V', its capacity utilization may be said to have increased if $(OV'/OV) > (OU'/OU)$. It should be noted that capacity utilization is unrelated here to the monetary value or revenue derived

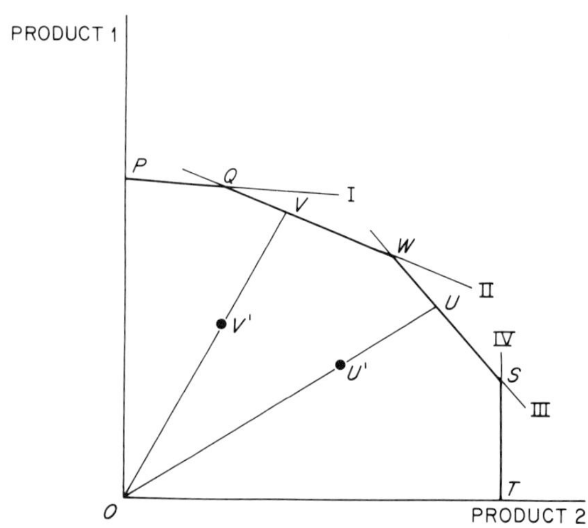

FIG. 1.12. Measurement of capacity utilization for two products.

from the product-mix, so that it is possible for utilization to be higher at V' than at U' and yet for the resultant revenue or profit to be lower.

Revenue efficiency

Like technical efficiency, capacity utilization concentrates on a physical interpretation of the conversion process in the plant and ignores the price mechanism. Similar to the analysis associated with Fig. 1.10, where factor prices for the inputs are accounted for, we can extend the analysis here on the output side to include product prices and to compare the actual revenue (or profit) obtained for the actual output with the potentially best outcome.

In Fig. 1.13 the line FF' denotes the iso-revenue line (or iso-profit line, if profit is taken as the focus for the analysis), namely the same revenue is derived from a product-mix at F and at F', or at any point in between on that line. On the assumption that prices are positive and increase linearly with product quantity, parallel iso-revenue lines closer to the origin O denote lower revenues and the line WW', being the furthest from O, represents the maximum revenue obtained, which is at W (lines further away are in the infeasible domain). Now, $U'U''$ is the iso-revenue line through the actual product-mix U', and U'' is where the product proportions are the same as at the optimal point W. Hence $e(4) = OU''/OW$ is the ratio of the actual revenue to the maximum possible revenue and may, therefore, be regarded as the *revenue efficiency* of the plant operating at U'. Alternatively, we can extend the line OU' to intersect the

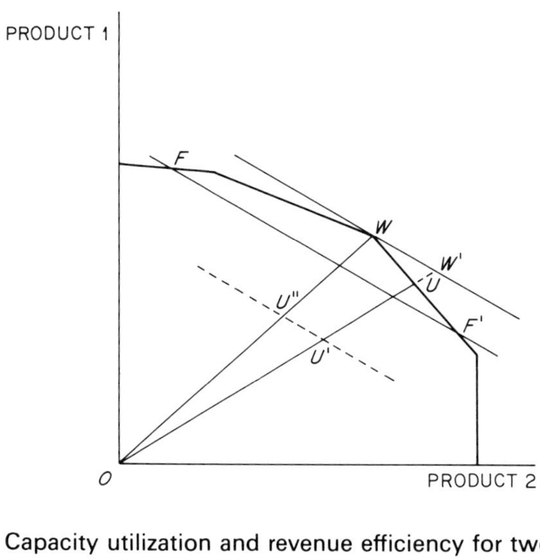

FIG. 1.13. Capacity utilization and revenue efficiency for two products.

maximum iso-revenue line at W' and consider the ratio OU'/OW' which compares the actual revenue at U' with the hypothetical revenue at W', had it been possible to maintain the product proportions at U' and achieve the potential revenue at W. In contrast (as noted above) we find that the capacity utilization at point U' is $e(3) = OU'/OU$.

It may be useful to summarize the salient points brought up in the discussion of Fig. 1.13:

$FWUF'$	is the full capacity line of the plant
FF'	is an iso-revenue line for various product-mixes
WW'	is the maximum iso-revenue line
W	is the maximum obtainable revenue
U'	is the actual product-mix
U	corresponds to full capacity and has a product-mix in the same proportions as at U'
U''	has the same revenue as at U' and a product-mix in the same proportions as the optimal one at W
W'	is a notional point in the infeasible domain (i.e. it is above the full capacity line) on the maximum revenue line (with the same revenue as at W) and product proportions as at U'
$e(3) = \dfrac{OU'}{OU}$	is the capacity utilization for the actual operating point U'
$e(4) = \dfrac{OU''}{OW} = \dfrac{OU'}{OW'}$	is the revenue efficiency at point U'

It was concluded earlier from Fig. 1.10 that technical efficiency is greater than cost efficiency, and similarly from Fig. 1.13 it is clear that the capacity utilization is greater than revenue efficiency (since $OW' > OU$). The reasons are the same in both cases, namely they relate to whether the comparison is based on physical or financial measures. *Technical efficiency and capacity utilization compare the actual physical situation (of input proportions or product mix) with the best physical scenario, whereas cost efficiency and revenue efficiency compare actual financial operations with optimal conditions (which yield minimum cost for a given output level or maximum revenue for a given array of resources).*

Another point to realize is that the full capacity line in Figs 1.12 and 1.13 represents the maximum amount of output (in terms of the two-product mix) that can be obtained from the plant as a whole, and this is determined by the bottlenecks attributed to one or two of the resources. Any point between P and Q in Fig. 1.12 denotes full use of resource I, whereas all the other resources are under-utilized. At the point Q both I and II are fully employed, while any point between Q and W denotes full use of II only, and so on. This means that full

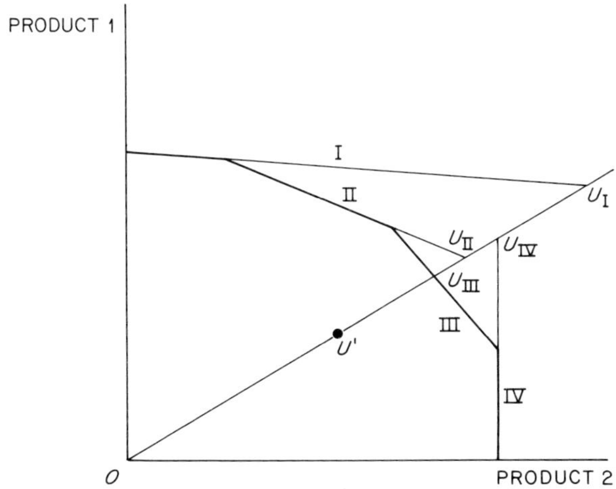

FIG. 1.14. Utilization of individual resources.

capacity at point U (which we employ in the definition of capacity utilization OU'/OU) refers only to resource III being fully used, which acts here as the bottleneck. To find out how the other resources fare, we turn to Fig. 1.14, where OU' is extended to intersect the four resource lines at $U_{\rm I}$, $U_{\rm II}$, $U_{\rm III}$ (which is identical to U in Figs 1.12 and 1.13) and $U_{\rm IV}$. Now, although $U_{\rm I}$ is above the full capacity line and hence lies in the infeasible domain (because it violates the constraints II, III and IV), it denotes full use of I for product proportions which are the same as at U'. Hence, the utilization of resource I at U' may be measured by the ratio $OU'/OU_{\rm I}$. Similarly,

$$\frac{OU'}{OU_{\rm II}} \text{ measures the utilization of II}$$

$$\frac{OU'}{OU_{\rm III}} \text{ measures the utilization of III}$$

and $OU'/OU_{\rm IV}$ is the ratio for IV. Of these four ratios the third (which refers to the bottleneck III) is identical to the ratio of the overall capacity utilization, and because

$$OU = OU_{\rm III} < OU_{\rm I}, OU_{\rm II}, OU_{\rm IV}$$

this ratio is also the highest of the four individual ratios.

Technical efficiency in a multi-input-multi-output plant

We may now turn to the question of measuring technical efficiency in the multi-input-multi-output product case. We recall that in Fig. 1.9, which relates to multi-input-single-output, the definition of technical efficiency relies on a comparison of use of resources on the efficient production possibilities line with the actual use of resources, where the input proportion in both cases is the same. In Fig. 1.12 there are several inputs and two products and we may proceed to use the same definition of technical efficiency by posing the following question: for a given output (specified by the product-mix at point U') what would be the minimum amount of resources needed and what is the actual amount used?

The minimum amount needed is equivalent to every resource being fully utilized in producing the product-mix, and that is obtained when all the constraints pass through the point U', as shown in Fig. 1.15. The technical efficiency of resource I is now derived as OU'/OU_{I}, being the ratio of the minimum required to the actual amount available, and this ratio is precisely the one obtained above for the capacity utilization of resource I. Similarly, the technical efficiency for resource II is OU'/OU_{II} which is the same as the utilization of II, and so on.

If instead of determining the technical efficiency of individual resources or processes we have to find it for the plant as a whole, then we first need to define the overall minimum capacity required to produce the product-mix at U'. If we assume that we seek the overall minimum capacity which has the four resources in the same proportions as in current use, then the required capacity is represented in Fig. 1.16 by the lower polygon, which has two properties: it is geometrically similar to the original full-capacity line (to preserve the original

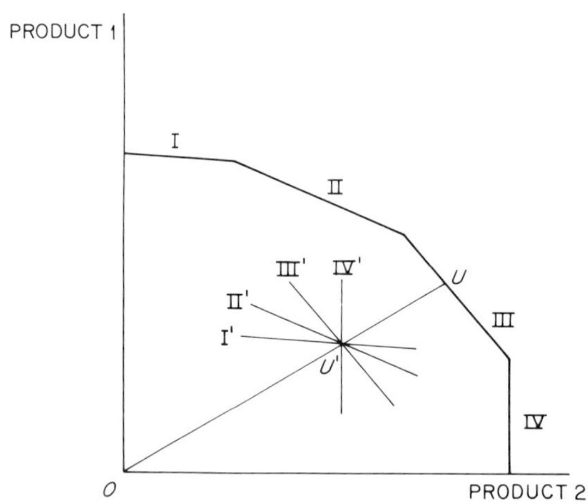

Fig. 1.15. Full utilization of the individual resources when all the constraints pass through U'.

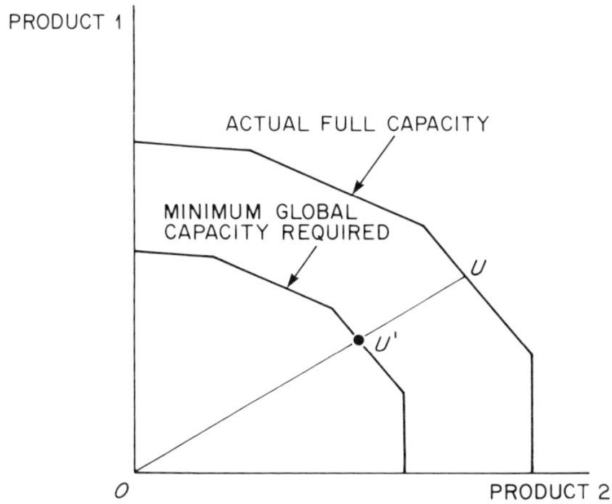

FIG. 1.16. Minimum overall capacity requirement for operating at U'.

input proportions) and it passes through U' (a polygon above U' has slack for all the resources whereas a polygon below U' is infeasible). The overall minimum capacity requirement is given by its bottleneck and is represented by OU' and the actual capacity is similarly represented by OU. Hence, the overall technical efficiency is OU'/OU which is the same as the capacity utilization.

CONCLUSION

Many interested parties are concerned about performance measures of an enterprise: investors, lenders, customers, employees, suppliers and management. Performance analysis is, therefore, crucial in evaluating past operations, as well as in the planning process.

Various types of performance analyses can be undertaken. Their common denominator is a comparison between some measure of output to some measure of input, where the input is measured either in financial or physical terms, and so is the output. This broad classification yields four output to input ratios, in addition to examination of the effectiveness of factor prices and of the conversion processes in the plant (see Fig. 1.8). These ratios proliferate when consideration is given to measures of output for each individual input, or inputs required for each output in the product range (an example of the latter is the number of man-hours involved in producing a unit of output, or the total cost per unit).

Four measures of production efficiency are defined in Table 1.1 for given outputs and given inputs on the one hand and for physical and financial comparisons on the other, as summarized below:

	Given output	Given input
Physical	$e\,(1)$ = technical efficiency	$e\,(3)$ = capacity utilization
Financial	$e\,(2)$ = cost efficiency	$e\,(4)$ = revenue (or profit) efficiency

On the assumption that factor prices in the optimal situation are the same as in the actual case and that the amount of physical input per unit of output is also the same in both cases, we have established the following results:

(1) For a single-input-single-output plant the four efficiency measures are the same: $e\,(1) = e\,(2) = e\,(3) = e\,(4)$.

(2) For the multi-input and/or the multi-output plant it is necessary to consider the resources-mix and the product-mix in the optimal case when it is compared to the actual case. For measures of physical efficiencies (i.e. technical efficiency and capacity utilization) we assume that each of the mixes remains the same in both cases, and this is the best way to compare like with like. However, in measures of financial efficiency (such as cost efficiency and revenue or profit efficiency) the optimal input-mix and/or product-mix are an intrinsic part of the optimal operating conditions (defined in financial terms).

(3) For the multi-input-multi-output plant we generally find that

$$e\,(1) > e\,(2)$$
$$e\,(3) > e\,(4)$$

and for given resource mix and product mix

$$e\,(1) = e\,(3)$$

In the special case where the actual and the optimal input proportions are identical, then $e\,(1) = e\,(2)$; and in the case where the actual product-mix is optimal, then $e\,(3) = e\,(4)$.

REFERENCES

[1] Amey, R. L. (1969). "The Efficiency of Business Enterprises". George Allen and Unwin.

[2] Eilon, S. and Teague, J. (1973). On measures of productivity. *Omega* **1**, 505–11.

[3] Eilon, S., Gold, B. and Soesan, J. (1976). "Applied Productivity Analysis for Industry". Pergamon Press.

[4] EILON, S. (1979). "Management Control". Pergamon Press.
[5] FARRELL, M. J. (1957). The measurement of productive efficiency. *J. R. Statist. Soc. Series A* **120**, 253–90.

EXERCISES

1. Discuss the need for performance analysis and how it can be of assistance to management.
2. Define the four ratios of production efficiency and discuss how they relate to each other and to a measure of effectiveness.
3. What is the role of industrial engineering in improving plant performance?
4. In a quarrying operation it was found that a trade-off can be established between labor and machinery: the more machinery is provided, the less labor input is needed to operate the quarry for a given level of weekly output. This trade-off, the actual operation and cost are shown in the following table, where labor input is measured in man-days per week and machinery in terms of its installation costs.

Alternative	Input 1: labor	Input 2: units of investment in machines
A	500	2
B	400	3
C	300	4
D	250	5
E	100	11
Actual operation	450	4
Cost/unit of input	1	30

As suggested by the data, a saving of 100 units of labor input is envisaged by moving from alternative A to B and a similar saving by moving to C, while a substantial reduction of labor input is possible at E. The ratio of operating costs (per unit of input) is estimated as 30:1 (for the machines input this includes running costs, financing costs and depreciation).

Find the technical efficiency and cost efficiency and comment on the results. If the operation is to be moved from the point stated to a new point involving 400 units for input 1 and 5 units for input 2, will the two efficiencies improve?

5. Two products are produced in a plant, which consists of six processing facilities, labeled as I, II, etc. Facility I is not required for the manufacture of

product 1 and facility VI is not required for product 2; all the other facilities are required for both. The capacities of these facilities are shown below:

Facility	Units of product 1	Units of product 2
I	100	—
II	110	270
III	140	130
IV	280	110
V	120	720
VI	—	120

The current operation involves a product mix of 80 and 20 units for product 1 and 2 respectively, and it is scheduled to be changed to a new mix of 50 and 60 respectively. The revenue per unit is 10, irrespective of which product is sold.

Find the capacity utilization

(i) of the plant as a whole
(ii) of each of the facilities

and the revenue efficiency of the current operation and of the scheduled new product-mix.

2 Profitability and Productivity Measures

Themes in this chapter

Ratios in practice
Ratios between:
outputs and inputs
inputs (resource mix)
outputs
Revenue proportions
Profit and return on investment
Productivity ratios
Network of productivity relationships
The cost pyramid
Ways of measuring labor productivity

RATIOS IN PRACTICE

The theoretical background set out in the previous chapter suggests that there are three principal ways in which actual or potential performance can be measured, as indicated in this simple matrix:

<table>
<tr><td></td><td></td><td colspan="2">OUTPUT</td></tr>
<tr><td></td><td></td><td>Actual</td><td>Maximum</td></tr>
<tr><td rowspan="2">INPUT</td><td>Actual</td><td>1</td><td>2</td></tr>
<tr><td>Maximum</td><td>3</td><td></td></tr>
</table>

Cell 1 in the matrix refers to a comparison of *actual output* with *actual input*;
Cell 2 takes the *actual input* as given and seeks to determine the *maximum output* that can be obtained;
Cell 3 assumes the *actual output* as given and attempts to find the *minimum input* needed to produce the output.

Each of these cells does not provide an adequate measure of performance on its own. When the ratio of actual output to actual input in cell 1 is determined, it is not at all clear whether the result is good or bad, and in order to make a judgement it is necessary to compare it with a similar ratio relating to the operation of the plant during a previous time period:

$$\frac{\text{Cell 1 now}}{\text{Cell 1 previously}}$$

In this way it is possible to ascertain whether an improvement over time has taken place. Another way is to compute the ratio for cell 1 for a given plant with that relating to another similar plant elsewhere, or to the industry as a whole (this being the essence of inter-firm comparison), so that the performance of the plant in question can be inferred from its position in a league table.

Alternatively, instead of comparing the ratio in cell 1 with itself in this way, it can be compared with ratios implied in the other cells:

$$\begin{aligned}\frac{\text{Cell 1}}{\text{Cell 2}} &= \frac{\text{(actual output)/(actual input)}}{\text{(maximum output)/(actual input)}}\\ &= \frac{\text{actual output}}{\text{maximum output}} = \frac{\text{actual output}}{\text{capacity}}\\ &= \text{capacity utilization (or revenue efficiency)}\\ \frac{\text{Cell 1}}{\text{Cell 3}} &= \frac{\text{(actual output)/(actual input)}}{\text{(actual output)/(minimum input)}}\\ &= \frac{\text{minimum input}}{\text{actual input}}\\ &= \text{technical efficiency (or cost utilization)}\end{aligned}$$

The interesting facet of these results is that, although the starting point of performance analysis is the notion of a ratio between output and input, we find that it culminates in comparing output with output (actual and capacity respectively), or input with input (minimum and actual respectively). Thus, the ratio of cell 1 to cell 2 in the above matrix becomes the familiar measure of capacity utilization (or revenue efficiency, if capacity and actual output are measured in financial terms) and the ratio of cell 1 to cell 3 becomes technical efficiency (or cost utilization).

In both cases an attempt is made to compare actual operations with the best results that "can be conceived under the circumstances" (for given inputs or for given outputs). The optimal solution is seen as an objective standard to measure against, so that judgements about actual achievements can be made. The goal is to attain performance of 100%, namely to operate at the optimal point, and the degree to which actual performance falls short of that is a measure of inefficient use of resources.

Practical difficulties

This method suffers from several practical shortcomings. To start with, information is rarely adequate to draw a production-possibilities curve or a full-capacity line, such as in Figs 1.9–1.12, particularly in the multi-input-multi-output case, which is the one most commonly found in practice. More often than not, a given plant is not likely to have operated over a wide enough range of conditions to provide sufficient historical data for this purpose. The alternative approach of assembling data from several plants (as adopted in some empirical research) incurs the risk that the data would not be homogeneous, since no two plants are alike and working practices may vary greatly [1, p. 13].

This means that the theoretical construction of production-possibilities and full-capacity lines is largely extrapolated from experience based on current operations. When optimal solutions are sought, it is commonly assumed that factor prices and the amount of resource needed for each unit of output remain unchanged, whereas in reality they may well be affected by a change in operating conditions. Thus, a theoretical optimal solution (e.g. in relation to the optimal product-mix) may be unachievable in practice, since the basic assumptions on which the optimal solution is computed no longer hold, and the objective standard may therefore be no more than a notional yardstick. Furthermore, as operating conditions in a given plant change, and as the data and the underlying assumptions are updated, a new optimal solution may emerge, so that the "objective standard" can easily become a moving target. While in principle measurements against such a notional yardstick are perfectly valid, the inherent difficulties of maintaining a reliable data base are such that this method is rarely adopted in practice as part of performance evaluation.

As a substitute to the first method of comparing actual results with optimal performance, management seeks to define budgets or targets for its forward projections, and performance is then compared with these targets. There is no suggestion that such budgets or targets are in any sense optimal. They are merely expectations of results that higher echelons in the managerial hierarchy impose on various operating units; alternatively, they represent indications from the managers of those units of what they hope to achieve. Thus, irrespective of whether the corporate planning process is primarily top-down or bottom-up, these targets become official expectations of good operating practices, based on the knowledge and experience of the management, and they are determined without having to employ elaborate data collection and sophisticated analytical tools to seek optimal solutions. A format for this corporate planning exercise is given in Chapter 13.

Reliance on actual results

Because of the commonly found practical difficulties in ascertaining optimal levels of inputs or outputs, we have to revert to the other method for the measurement of performance, cited above, which relies on ratios of actual outputs to actual inputs, shown in cell 1 in the above matrix. In trying to make a judgement as to whether such a ratio is satisfactory or not, a comparison is often made between cell 1 and itself, namely between the actual current ratio and

- the same ratio in a previous time period (e.g. the ratio this month compared with the ratio last month, or this month last year)
- the same ratio for a comparable plant or company
- the same ratio for the industry as a whole

Such comparisons, carried out over a period of time, may reveal trends that management should be aware of, so that adjustments, or changes in direction, can be undertaken when appropriate.

As stated earlier, the two methods are not mutually exclusive; in fact, they complement each other and provide a multi-faceted view of how well the enterprise is doing. Different functions and levels in the managerial hierarchy are naturally interested in different performance measures. For example, senior managers in the production function are interested in production volumes, in the breakdown of production costs, in productivity of labor and in the utilization of production facilities. Junior production managers are concerned mainly with the operations of their own departments and with costs relating to products or components under their own control. Senior sales managers, on the other hand, concentrate on sales volumes, prices, revenue and its breakdown (according to product groups, territories, or types of customers); they are further interested in the cost of the sales effort, in the ratio of revenue to marketing costs, in the effect of advertising, and so on, while junior sales managers are anxious to meet their own targets and keep a watchful eye on the expenses of their sales force. Thus, the diversity of measures stems from the different responsibilities of the various functions within the organizational structure and hence from the different needs of members of the hierarchy within each function.

Use of ratios

The number of ratios that has been suggested by accountants and business analysts is legion. Westwick defines several hundreds of such ratios [12], of which the small selection in Table 2.1 is a good illustration of the criteria suggested for use by the chief executive alone. There are many other ratios in the same vein for marketing management, production, purchasing, stock-holding, finance and personnel. Furthermore, there are ratios of particular interest to certain industries, and some are specifically designed for non-manufacturing organizations, such as merchanting, retailing, hotels, catering, accountancy and other professional firms [12].

The variety of ratios to choose from is truly enormous, and it is not surprising that different firms follow different practices, depending on tradition and the preferences of their senior managers. The list of ratios that follows, is, therefore, neither universal nor comprehensive, but it does represent the type of ratios most commonly found in practice as part of the overall corporate planning and control function of a manufacturing enterprise (see also Chapter 13). Needless to say, individual executives and particular circumstances generate their special needs, so that control mechanisms and performance ratios are constantly adjusted and refined to take account of these needs.

TABLE 2.1
A selection of ratios for the chief executive (from [12]).

Area of concern	Ratios
Sales	$\dfrac{\text{Sales this year}}{\text{Sales last year}}$
	$\dfrac{\text{Sales this month}}{\text{Sales last month}}$
	$\dfrac{\text{Sales this month}}{\text{Sales this month last year}}$
	$\dfrac{\text{Market share now}}{\text{Market share last year, last month, etc.}}$
	$\dfrac{\text{Value of outstanding orders}}{\text{Average sales value per day}}$
Source of capital	$\dfrac{\text{Net profit after tax}}{\text{Equity capital}}$
	$\dfrac{\text{Price per share}}{\text{Earnings per share (after tax)}}$ (called the P/E ratio)
	$\dfrac{\text{Interest paid}}{\text{Borrowed capital}}$
	$\dfrac{\text{Total profit}}{\text{Interest paid}}$ (= interest cover)

TABLE 2.1 (*Contd.*)

Area of concern	Ratios
	$\frac{\text{Borrowed capital}}{\text{Equity capital}}$ (= gearing ratio)
Suppliers	$\frac{\text{Suppliers' price}}{\text{Suppliers' price at a base period}}$
	$\frac{\text{Outstanding orders with suppliers}}{\text{Average daily purchases}}$
	$\frac{\text{Creditors}}{\text{Average daily purchases}}$
Liquidity	$\frac{\text{Cash} + \text{debtors} + \text{marketable securities}}{\text{Current liabilities}}$
Quality of purchases	$\frac{\text{Value of returned goods}}{\text{Purchases}}$
Employees	$\frac{\text{Number of leavers}}{\text{Average number of employees}}$
	$\frac{\text{Average age of senior staff}}{\text{ditto five years ago}}$
Use of assets	$\frac{\text{Operating profits}}{\text{Operating assets}}$
Overall performance	$\frac{\text{Profit}}{\text{Capital employed}}$
	$\frac{\text{Profit}}{\text{Sales}}$
	$\frac{\text{Sales}}{\text{Capital employed}}$
	$\frac{\text{Sales}}{\text{Fixed assets}}$
	$\frac{\text{Sales}}{\text{Stocks}}$
	Sales per employee Profit per employee Added value per employee Wages and salaries per employee

When an analysis of the operations calls for an investigation in great detail, many of the measures have to be disaggregated, either by dividing the enterprise into several operating entities and considering each on its own, or by dividing the inputs and outputs into their constituent elements. While it is not possible, therefore, to be dogmatic in stating which performance measures should always predominate, and what level of aggregation or disaggregation is always best, the approach and methodology advocated in this book can easily be adapted to help in the required analysis.

COMMONLY USED RATIOS

The following notation (based on [2]) is adopted:

Inputs

$I =$ total investment, or capital employed $= I_F + I_W$ (see Fig. 2.1)
$I_F =$ investment in fixed assets, such as buildings, machinery and plant
$I_W =$ working capital, being the difference between current assets and current liabilities, and usually comprising cash, stocks (of materials, work-in-progress and finished goods) and debtors less creditors
$N =$ average number of employees during the period in question (a month, a quarter, a year). N may be subdivided into several categories, for example,

$$N = N_1 + N_2$$

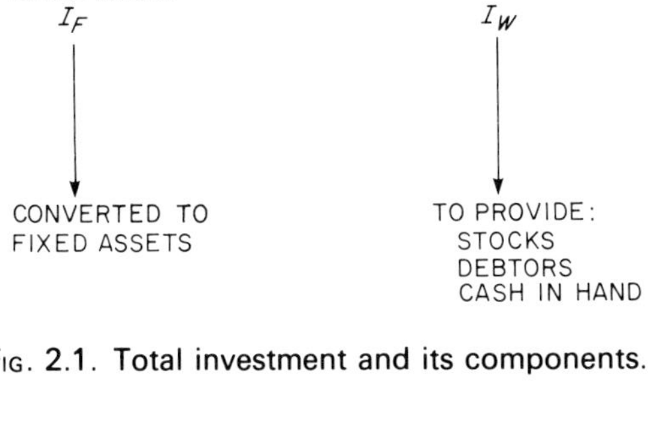

FIG. 2.1. Total investment and its components.

where N_1 = number of indirect workers
N_2 = number of direct workers

Each of these categories may be disaggregated further, for example,

$$N_2 = N_{21} + N_{22}$$

where N_{21} = number of skilled workers, and N_{22} = number of unskilled workers. Another common measure of labor input is in terms of the total number of hours NT, where T = average number of hours per employee, and this figure may be disaggregated into

$$T = T_b + T_o + T_h$$

where T_b = number of hours worked at the basic wage rates
T_o = number of overtime hours
T_h = number of holiday hours entitlement

W = total wages and salaries, given by

$$W = wN$$

where w = average wage rate per employee. W may be divided into

$$W = W_1 + W_2$$

where W_1 and W_2 = the wages and salaries of the N_1 and N_2 employees respectively with different average wage rates w_1 and w_2, so that

$$W = w_1 N_1 + w_2 N_2$$

Q = physical volume of materials

M = cost of materials input. Both Q and M may be divided into several categories, for example,

$$Q = Q_1 + Q_2$$

where Q_1 = physical volume of raw materials, costing m_1 per unit
Q_2 = bought-in components and assemblies, at a cost of m_2 per unit

Thus

$$M = M_1 + M_2 = m_1 Q_1 + m_2 Q_2$$

where M_1 = cost of raw materials
M_2 = cost of bought-in components and assemblies

E = other expenses, covering all items of expenditure not included under W and M. Disaggregation of E may take the form

$$E = E_1 + E_2$$

where E_1 = depreciation charges for fixed assets
E_2 = other costs, and this category may be subdivided into

$$E_2 = J + E_3$$

where J = interest charges for borrowings
E_3 = the residual expenses, not included elsewhere in the above cost schedule

The total cost of the inputs listed here is then

$$C = M + W + E \tag{2.1}$$

and a simple breakdown of these costs is shown in the lower part of Fig. 2.2.

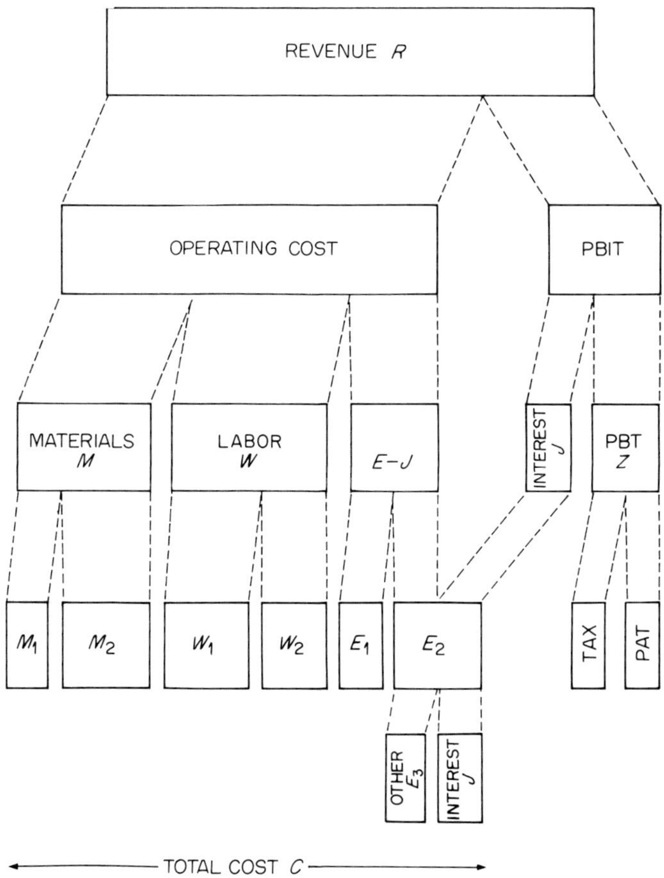

FIG. 2.2. Revenue and its components.

Outputs

V = physical output volume. A distinction sometimes needs to be made between the production volume and the sales volume, since the production output goes into finished goods inventory, from which sales are drawn. It is, therefore, possible for sales volume and production volume to differ, the difference between them being reflected in fluctuations in the stock level. Throughout the analysis in this book we shall assume for the sake of simplicity that the two volumes are the same, but when in practice the difference between the two is found to be significant, and when this difference has an effect on performance measures (for example, on costs associated with working capital, as discussed in Chapter 7), then various models developed in this book would need to be modified accordingly.

$R =$ the total sales revenue $= pV$

where $p =$ the average price per unit of output (or the "unit price" for short)

$Z =$ the profit, obtained from

$$Z = R - C \tag{2.2}$$

As shown in Fig. 2.2, profit may be defined at different levels:

PBIT = Profit Before Interest and Tax, being the gross profit (namely, the difference between gross revenue and operational costs)

PBT = Profit Before Tax, which is what remains of PBIT after payment of interest charges on borrowings

PAT = Profit After Tax, being the net profit after payment of interest and tax

Each of these profit definitions has its own significance to the people concerned. PBIT is the figure of concern to operating managers, both in the production and sales functions, since they regard interest charges and taxes as being completely outside their spheres of responsibility or influence. PAT is the figure that investors and financial analysts generally focus on, as this is the amount that is available for dividend payments, for renewal of fixed assets and for expansion. PBT is an intermediate figure, which recognizes that interest charges must be included as part of the total cost of the activities of the plant, but regards tax as an external factor, outside the immediate control of those responsible for total operations. Throughout this book the profit Z is defined as PBT, although it is not difficult to modify the various models to cater for other definitions of profit. As suggested in Fig. 2.2 the term "operating costs" in the accounts is often interpreted as a figure that excludes the interest charges, which are shown as part of PBIT. According to this definition,

$$\text{Operating cost} = C - J$$

so that while J is excluded from the calculation of the operating cost, it is included in the total cost. Figure 2.2 illustrates the well-known maxim among accountants that there are many ways of dividing a cake, and it is therefore important to be precise about the definitions used in any one instance and to be consistent in the analysis undertaken thereafter.

$$A = \text{added value} = R - M = W + E + Z \tag{2.3}$$

Added value has been suggested as a crucial measure of the outcome of an industrial activity. Once capital has been sunk in fixed facilities, the plant may be regarded as taking materials with a value M and converting them to an output with a value R. The conversion process is, therefore, responsible for *adding value*, measured by the difference between the output and the material input, namely by $R - M$. As the last equation suggests (see also Fig. 2.3), the added value is used to meet all the costs, other than costs of materials and services bought from the outside, as well as to provide for a residual profit. Hence, the higher the added value, the greater is the scope to increase W, either by increasing the level of remuneration to employees, or by increasing the number of employees (or both). This is the reason for some countries paying special attention in recent years to developing industries with high added value, in order to improve the employment prospects of the local population. Some analysts go as far as to prefer the measure of added value to that of profit. Others are somewhat dubious and point out that, while profit takes account of all the input costs ($Z = R - C$), added value focuses just on the cost of materials ($A = R - M$), which is only one of several inputs to the system.

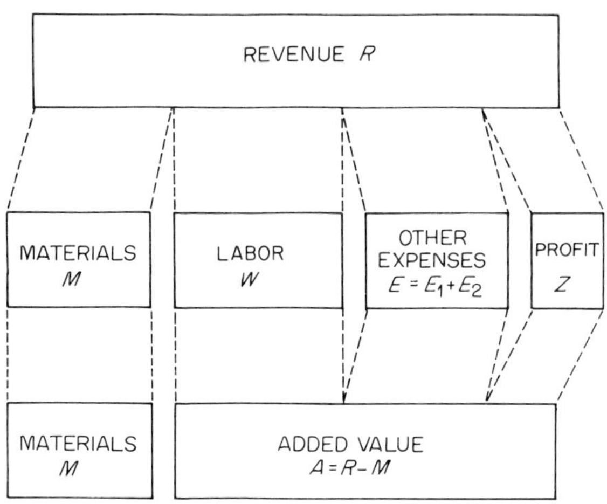

FIG. 2.3. Added value.

A word of caution is appropriate here regarding the definition of M in Equations (2.1) and (2.3). In the former M is defined as the cost of purchased materials (raw materials, components, etc.), whereas in the latter M often includes certain services provided by external contractors as well. For example, if a company is considering the closure of its transport department and awarding a contract to a haulage firm to undertake the delivery of its goods, then such a move will have little effect on M in Equation (2.1), since the cost of a transport department is not included as part of materials cost, except for fuel and spare parts. The closure of the department will, therefore, reduce M in Equation (2.1) only by the amount spent on fuel and spare parts. However, in computing the added value, the effect of replacing the transport department with a service bought from a contractor increases M by the corresponding value of the contract and causes the added value to decline. This example illustrates again the importance of definitions and the pitfalls that may be encountered in not comparing like with like. The example also demonstrates how added value can fall while the net profit may rise (if the transport cost component in this case is reduced by use of the haulage firm).

Ratios between outputs and inputs

The most common ratios of outputs to inputs are listed in Table 2.2. The four methods shown for measuring output are physical volume, revenue, profit and added value. The inputs are listed as total investment, fixed investment, number of employees (or the total hours of the workforce), total wages and salaries, materials costs and total costs.

The twenty ratios fall into four groups (shown in the columns in Table 2.2) with reference to each output.

(a) The six ratios in the first column, specifying the *physical output* in the numerator, are used as surrogate production efficiency measures, often referred to as *productivity ratios* (discussed in greater detail later). Their

TABLE 2.2
Ratios of outputs to inputs.

Input	Output V	R	Z	A
I	V/I	R/I	$Z/I = r$	A/I
I_F	V/I_F	R/I_F	Z/I_F	A/I_F
N (or NT)	V/N	R/N	Z/N	A/N
W	V/W	R/W	Z/W	A/W
M	V/M	R/M		
C	V/C	R/C		

reciprocals, aiming to measure the amount of a given resource input (in physical or cost terms) per unit of output, are also widely used. The sixth ratio in this column is more recognizable in its inverse form C/V, which is the total cost per unit.

(b) The second column relates *revenue* to various inputs, and of the six ratios R/I (revenue to total investment) and R/N (revenue per employee) are often quoted, followed sometimes by R/I_F and R/W. The inverses of the last three ratios are fairly common: W/R and M/R are the proportions of the revenue taken up by wages and materials, and C/R is the proportion of the revenue devoted to total costs.

(c) The third column focuses on *profit*, and the first ratio in this column is the most widely quoted in financial evaluation and is referred to as the *return on capital*, namely the ratio of profit to total investment (denoted here by r). The next three ratios are also popular: Z/I_F (profit to fixed investment), Z/N (profit per employee) and Z/W (profit to wages).

(d) The last column deals with measures of *added value*, and of these the three ratios A/N (added value per employee), A/W (added value to the total wage cost) and A/I (added value to total investment) are the most commonly used. Many analysts prefer the inverse of A/W, namely W/A, to measure the proportion of the added value spent on wages and salaries.

Instead of the ratios of output to input in the first column in Table 2.2, we find that the ratio of a given input to output is often cited as a significant measure. For example, N/V or NT/V is used to make comparisons between plants (such as the number of man-hours required to produce a ton of steel or an automobile in one country compared with another), and Q/V (the quantity of materials needed to produce a unit of output, which is of importance in the mining industry). Of particular interest are cost measures:

$$\frac{M}{V} = c_M = \text{cost of materials per unit of output (called "unit material cost")}$$

$$\frac{W}{V} = c_W = \text{cost of labor per unit of output (called "unit labor cost")}$$

$$\frac{E}{V} = c_E = \text{other costs per unit of output (called "unit other costs")}$$

These ratios represent a breakdown of total unit cost, since

$$c = \frac{C}{V} = c_M + c_W + c_E \tag{2.4}$$

Another useful group of ratios of inputs to output relates to the breakdown of revenue into its component parts:

$\dfrac{M}{R}$ = proportion of revenue consumed in materials

$\dfrac{W}{R}$ = proportion of revenue consumed in wages and salaries

$\dfrac{E}{R}$ = proportion of revenue taken up by other costs

$\dfrac{Z}{R}$ = proportion of revenue left as profit; this is the *profit margin* α

These are called *revenue proportions*, so that

$$\frac{M}{R}+\frac{W}{R}+\frac{E}{R}+\frac{Z}{R}=1 \tag{2.5}$$

and when the proportions are expressed in percentage terms, then

$$\frac{M}{R}+\frac{W}{R}+\frac{E}{R}+\frac{Z}{R}=100$$

Surrogate measures

Instead of relating directly outputs to inputs, management often resorts to surrogate measures, which are more amenable for control purposes, the assumption being that if the surrogate measures are satisfactory, then the overall objectives of profit and rate of return naturally follow.

Some common surrogate measures are:

(i) *Ratios between inputs*, such as

$I_F/I =$ i = ratio of fixed investment to total investment
$I/N =$ investment per employee
$I_F/N =$ fixed investment per employee; this is a useful measure of the degree to which a plant is capital intensive
$N_1/N =$ ratio of the number of a particular category of employees to the total (e.g. direct labor to the total)
$M_1/M =$ ratio of the cost of a given material resource to the total cost of materials (e.g. the cost of energy to the total, or the cost of bought-in components to the total)
$M/C =$ cost of material to total cost
$W/C =$ cost of employees to total cost
$E/C =$ capital charges and other expenses to total cost

The last three ratios are called the *cost proportions* and it follows that

$$\frac{M}{C}+\frac{W}{C}+\frac{E}{C}=1 \tag{2.6}$$

and when the cost proportions are measured in percentage terms, then

$$\frac{M}{C}+\frac{W}{C}+\frac{E}{C}=100$$

(ii) *Ratios between outputs*, such as

$$p = \text{average unit price} = \frac{R}{V}$$

$$z = \text{unit profit} = \frac{Z}{V} = p - c$$

$$\alpha = \text{profit margin} = \frac{z}{p} = \frac{p-c}{p} = \frac{Z}{R}$$

A/R = added value to revenue
Z/A = proportion of profit within added value

RETURN ON CAPITAL

As stated earlier, the return on capital (or total investment or capital employed) is regarded by many as the prime measure of overall performance. This measure is listed in Table 2.2 and defined as

$$r = \frac{Z}{I} = \frac{R-C}{I}$$

But, as we have seen, in practice numerous other measures are used to evaluate particular functions or disaggregated segments of the total operation. Clearly, no one measure is adequate on its own, since all the activities in the enterprise make a contribution to the overall return on capital, and furthermore—as Gold repeatedly emphasizes [4, 5, 7]—a change in the value of any given ratio may be the outcome of changes in other ratios. For example, an increase in the output per man-hour need not necessarily result from greater effort by the workforce, but emanate from an increase in the ratio of fixed to total investment in the plant. It is, thus, important to consider the *network of relationships* between the various measures employed (as shown in Fig. 1.3), and to take account of the way they affect each other, in trying to elicit the causes of changes in performance.

Gold starts with the overall return on capital and proceeds to disaggregate it as follows [4, 5, 1 p. 22–4]

$$r = \frac{\text{Profit}}{\text{Total investment}} = \frac{\text{Profit}}{\text{Output}} \times \frac{\text{Output}}{\text{Total investment}}$$

$$= \text{Unit profit} \times \frac{\text{Output}}{\text{Total investment}} \tag{2.7}$$

Also:

$$\text{Unit profit} = \text{Unit price} - \text{Unit cost}$$

and

$$\frac{\text{Output}}{\text{Total investment}} = \frac{\text{Output}}{\text{Capacity}} \times \frac{\text{Capacity}}{\text{Fixed investment}} \times \frac{\text{Fixed investment}}{\text{Total investment}}$$

where

$$\left.\begin{aligned} \frac{\text{Output}}{\text{Capacity}} &= e \\ \frac{\text{Capacity}}{\text{Fixed investment}} &= \theta \\ \frac{\text{Fixed investment}}{\text{Total investment}} &= i \end{aligned}\right\} \tag{2.8}$$

hence e is the level of capacity utilization (referred to as e (3) in the previous chapter)

θ is the provision of capacity by fixed investment

i is the proportion of total investment allocated to fixed assets

so that

$$\frac{\text{Output}}{\text{Total Investment}} = \frac{V}{I} = e\,\theta\,i$$

The general scheme of this breakdown is shown in Fig. 2.4. Equation (2.7) may be restated as follows [3]:

$$\begin{aligned} r = \frac{Z}{I} &= \frac{Z}{V}\frac{V}{I} = z\frac{V}{I} \\ &= z\,e\,\theta\,i = z\,e\,k \end{aligned} \tag{2.9}$$

where z = unit profit $= p - c$

$$e = \frac{V}{P}$$

P = capacity

k = ratio of capacity to total investment $= \dfrac{P}{I} = \dfrac{P}{I_F}\dfrac{I_F}{I} = \theta\,i$

Gold proceeds to remind us that [1, p. 24]:

"1. top management seeks not to maximize performance within each subsector of operations, but rather to effect the best combination of such

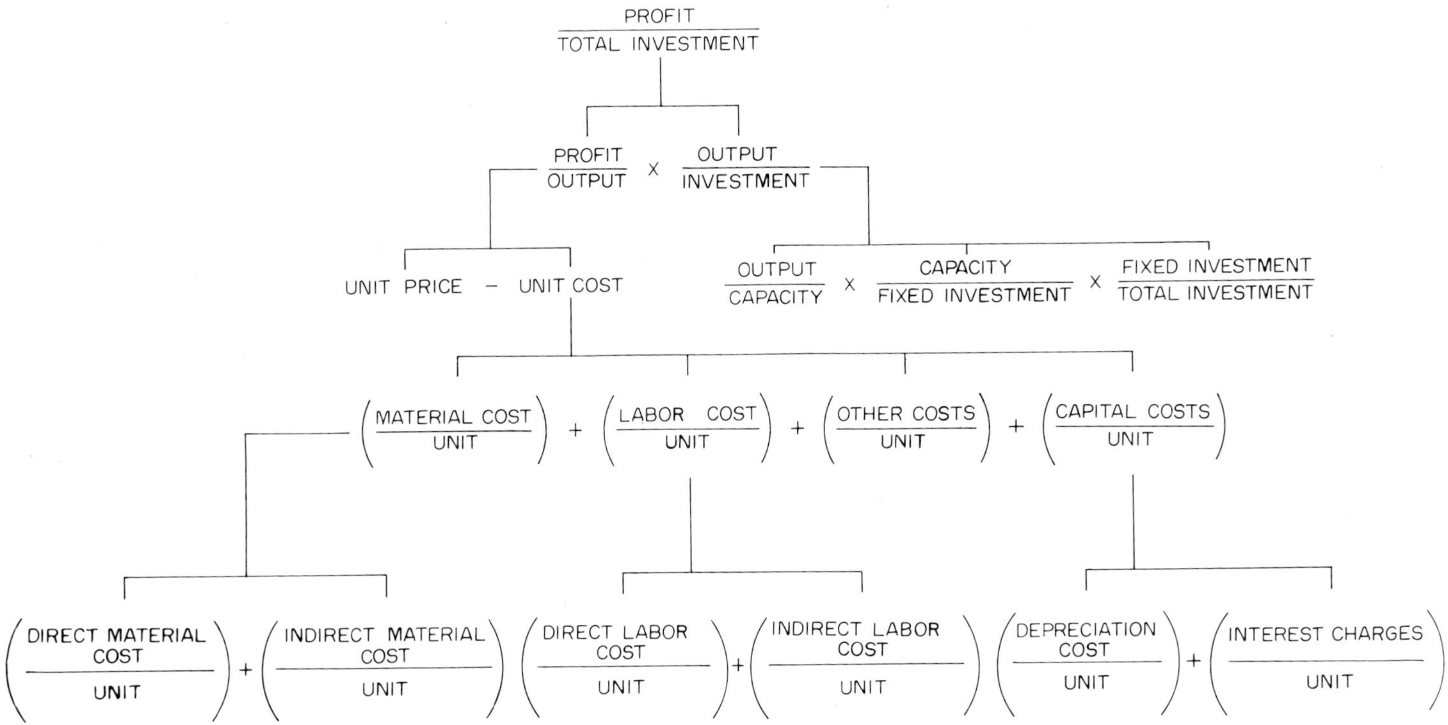

FIG. 2.4. A pyramid of ratios.

component performances—those which most effectively reinforce one another in promoting the objectives of the firm;

"2. hence, management must consider alternative patterns of price, output, investment and productivity adjustments instead of concentrating on productivity improvement efforts alone; and

"3. actual or prospective adjustments in physical input-physical output relationships must, therefore, be appraised within such a larger analytical framework to yield evaluations relevant to managerial decision-making."

Thus, the simple expansion of the expressions in Equations (2.7), (2.8) and (2.9) allows for managerial ratios to be identified in a way that makes sense to particular functions within the enterprise. In Chapter 4 a methodology is developed for the analysis of changes that take place in performance over a period of time, and examples are cited from industrial case studies to highlight the effects of changes in the individual ratios on the overall performance r.

PRODUCTIVITY RATIOS

Productivity has long been a catchword to describe the key factor that determines the performance of industrial enterprises, or whole industrial sectors, in terms of their ability to compete in the market place [11]. It is not uncommon to find long expositions on the subject of productivity, where the author starts with the complaint that productivity is one of the most overworked words in our vocabulary, and that it is an elusive property that defies satisfactory methods of measurement. Nevertheless, the author then proceeds to extol the virtues of productivity and to tell us how important it is to raise it, without attempting to define it or suggest how it should be measured. We are simply told that firms with low productivity incur high production costs, which cause prices to increase and result in loss of sales and in questionable viability.

In practice, productivity is often interpreted as "production per man-hour"* and the general approach to it is summarized as follows†: "Productivity in economics is the ratio of what is produced to what is required to produce it. Usually this ratio is in the form of an average, expressing the total output of some category of goods divided by the total input of, say, labour or raw materials. In principle, any input can be used in the denominator of the productivity ratio. Thus, one can speak of factors of production . . . Labour is by far the commonest of the factors used".

---

* The Concise Oxford Dictionary.

† Encyclopaedia Britannica, 1974.

Apart from the widely held conviction that labor productivity is the major determinant of competitiveness and profitability, it has become a central issue in pay negotiations, sometimes with explicit encouragement from government agencies to link higher wages with improved productivity performance. This raises several fundamental questions [1, p. 3–4]:

"1. Assuming that improvement in overall performance can be measured, how closely can we determine the relative contribution of labour and of other factors to such improvement?

"2. Having determined the relative contributions of the major input factors, what adjustments in their respective rewards would be appropriate?

"3. When improvements in labour's contributions are not accompanied by improvements in the performance of the enterprise as a whole, should labour's compensation be increased nonetheless?

"4. When improvements in the performance of the enterprise are not due to labour contributions, should labour nevertheless have a share in the benefits to the enterprise, and if so how much?

"5. Should a higher productivity-linked compensation become operative only after increases in productivity materialize, or should it be paid on the basis of future expected performance?"

These questions received prominent consideration in the UK in the 1960s and subsequently, when productivity agreements were designed to institute "changes in the working practices for the purpose of improving the utilization of both capital and labour, in return for improvements in pay" [8, 10]. The productivity criterion defined by the British Government allowed for pay increases to take place above a declared norm, which firmly stipulated an otherwise upper limit for pay increases, "where the employees concerned, for example by accepting more exacting work or a major change in working practices, make a direct contribution towards increasing productivity in the particular firm or industry". These general guidelines were then expanded by the National Board for Prices and Incomes in the UK as follows [10]:

"we would, for the guidance of managements, unions, and the Ministry of Labour, rewrite the seven guidelines as follows:

(i) It should be shown that workers are making a direct contribution towards increasing productivity by accepting more exacting work or a major change in working practices.

(ii) Forecasts of increased productivity should be derived by the application of proper work-standards.

(iii) An accurate calculation of the gains and the costs should normally show that the total cost per unit of output, taking into account the effect on capital, will be reduced.

(iv) The scheme should contain effective controls to ensure that the projected increase in productivity is achieved, and that the payment is made only as productivity increases or as changes in working practice takes place.
(v) The undertaking should be ready to show clear benefits to the consumer through a contribution to stable prices.
(vi) An agreement covering part of an undertaking should bear the cost of consequential increases elsewhere in the same undertaking, if any have to be granted.
(vii) In all cases negotiators should beware of setting extravagant levels of pay which would provoke resentment outside."

The intention behind these statements was to make a distinction between productivity of labor and productivity of capital, though no attempt was made to specify precisely how they should be measured. The general preoccupation with the productivity of labor leads Gold to postulate [1, p. 16] that "among the most widely prevailing elements of the mythology relating to productivity, the following four may be most important:

"1. that productivity measures reflect changes in the 'efficiency' of production;
"2. that changes in productivity are reasonably well measured by output per man-hour;
"3. that increases in output per man-hour are invariably desirable because they yield decreases in unit cost; and
"4. that increases in output per man-hour warrants parallel increases in wages per man-hour".

He adds: "And yet, not one of these contentions can be sustained either by theoretical or on empirical grounds". He argues strongly that output per man-hour does not measure productive efficiency as a whole, and that an increase in output per man-hour is neither evidence of the direct contribution of labor nor does it necessarily result in a reduction in the total unit cost.

MISUSE OF PERFORMANCE RATIOS

Many writers identify productivity as a ratio of output to input and suggest that productivity indices can be constructed for any aggregated or disaggregated activity in an enterprise. If, as proposed by Mali [9, p. 6], productivity is defined as "the measure of how well resources are brought together in organizations and utilized for accomplishing a set of results", then all measures of performance (including the efficiency ratios and others discussed in the previous chapter) fall under this general umbrella. A more constrictive definition (but potentially more useful for analytical purposes) is to relate physical output to given factor inputs, as discussed later. Be that as it may, the general definition of an output to input ratio as a measure of performance for any conceivable activity in an organization can lead to numerous arbitrarily chosen criteria, which can easily become rather parochial, even whimsical, and hence of questionable relevance to effective performance analysis. Consider, for example, the following ratios (found among a recommended list of measures [9, pp. 88–9]):

1. Inventory/Advertising costs is one of the proposed cost ratios, though it is none too clear what relationship there might be between the numerator and the denominator. The ratio can presumably be increased by increasing inventory levels and reducing advertizing expenditure, but why should this be interpreted as an improvement in performance?
2. Recruits selected/Costs is a proposed measure in the governmental sector, but recruiting costs per employee are usually minute, compared with the subsequent costs of employment, so that it may in fact be beneficial to increase recruitment costs to improve the quality of intake, rather than embark on a cost cutting exercise. These comments apply equally to recruitment in the private sector.
3. Students graduating/Annual costs is suggested as a measure in the field of education, though again no reference is made to quality, which is obviously of prime importance, even if it is difficult to measure objectively. Clearly, it is possible to reduce the cost of education per student by providing inferior facilities, though whether this would be desirable is another matter.
4. Graduates/Curriculums is another ratio in the field of education, proposed as a work standard, though its likely purpose is somewhat obscure.
5. Research reports/Allocated budget is a cost ratio cited in the field of "health and human services" (and is sometimes suggested for a research and development activity). The naïve notion that the number of research reports can in any way be a sensible measure of output is perhaps symptomatic of the degree of crudity with which some ratios can be constructed. A change in the value of such a ratio can hardly be indicative of a change in the utility, efficiency or even the level of activity of the department concerned.
6. Admission cost per patient per week in a hospital is quoted in a case study where additional office machinery helped to reduce staff and thereby total admission costs [9, pp. 57–8]. However, a great deal depends on what is meant by 'admissions'. It is possible, of course, to reduce processing at the admission stage by deferring some of the work for completion to a later period, with the result that the cost per admission is reduced, but the total work load remains unchanged.

Many other examples can be cited, but these suffice to support two important propositions:

(a) Although, in theory, an infinite number of ratios can be devised, only 'sensible' ratios need be considered for practical purposes, i.e. *ratios that are relevant in the managerial control context.*
(b) It is usually not difficult to conceive of ways of *improving a particular ratio at the expense of another* (or several others), and it is therefore not immediately obvious that the overall outcome would be desirable.

NETWORK OF PRODUCTIVITY RELATIONSHIPS

Reference to productivity ratios was made in Table 2.2. *The productivity of a factor input is defined as the ratio of physical output to that input.* Let

x = Quantity of a factor input
ξ = Factor price for the input, i.e. the price paid per unit of input
$X = x\,\xi$ = Cost of the factor input

then the productivity of this factor input, when defined in relation to the physical amount of the latter, is given by

$$\pi_x = \frac{\text{Physical output}}{\text{Quantity of the factor input}} = \frac{V}{x}$$

Hence

$$\pi_x = \frac{V\xi}{x\,\xi} = \frac{V\xi}{X}$$

but

$$\frac{X}{V} = \text{cost of } X \text{ per unit of output} = c_X$$

leading to the general conclusion that

$$\pi_x = \frac{V}{x} = \frac{\xi}{c_X} \tag{2.10}$$

If the three major inputs are identified as those of *labor*, *materials* and *capital*, then the three productivity ratios are:

$$\text{Productivity of labor} = \pi_N = \frac{\text{Output}}{\text{Number of employees}} = \frac{V}{N} \tag{2.11}$$

Alternatively, productivity of labor may be related to the total number of man-hours $H = NT$, hence

$$\pi_H = \frac{V}{NT} = \frac{V}{H}$$

Similarly,

$$\text{Productivity of materials} = \pi_Q = \frac{\text{Output}}{\text{Materials quantity}} = \frac{V}{Q} \tag{2.12}$$

As for the productivity of capital, we recall that total investment consists of fixed investment and working capital (Fig. 2.1). The purpose of the former is to provide capacity (see Fig. 2.5), and its ability to do so is defined by Gold as the productivity of fixed investment. To be consistent with the format in equation (2.10), we shall adopt here the following definitions:

$$\pi_I = \frac{V}{I} = \text{productivity of total investment} \tag{2.13}$$

$$\pi_{I_W} = \frac{V}{I_W} = \text{productivity of working capital} \tag{2.14}$$

$$\pi_{I_F} = \frac{V}{I_F} = \text{productivity of fixed investment} \tag{2.15}$$

but

$$\theta = \frac{P}{I_F} = \text{efficiency of fixed investment}$$

so that

$$\pi_{I_F} = \frac{V}{P}\frac{P}{I_F} = e\theta \tag{2.16}$$

From equation (2.16) we find that the output is determined by

$$V = I_F e\,\theta \tag{2.17}$$

namely by: the amount of fixed investment I_F, the efficiency θ with which this

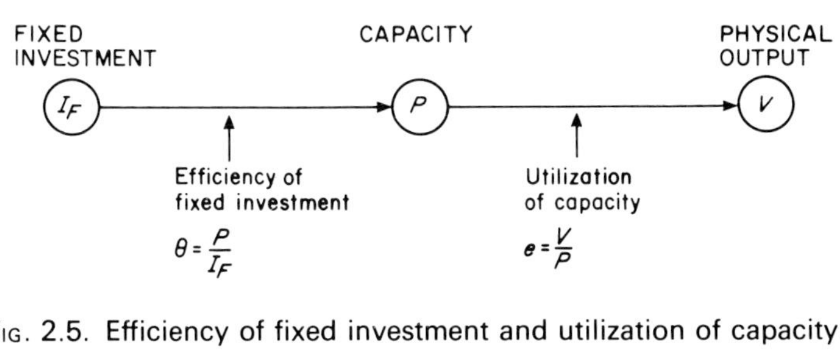

FIG. 2.5. Efficiency of fixed investment and utilization of capacity.

investment is converted into capacity, and the capacity utilization e. Note that if the capacity is fully utilized, then $e = 1$ and $e\,I_F = I_F$, so that the fixed investment is available in its entirety and $e = 1$ implying that $V = P$. The degree to which e deviates from 1 affects the availability of I_F, hence $e\,I_F$ may be regarded as the *effective fixed investment* (or, as Gold calls it, *utilized fixed investment*). Equation (2.17) expresses output as the outcome of the effective fixed investment ($e\,I_F$) and the efficiency of fixed investment (θ).

Gold's productivity network

Gold proposes a network productivity relationship [6, 1 p. 25] as shown in Fig. 2.6, where two nodes represent the productivities of labor and materials respectively, while the third is the efficiency of fixed investment. Three other productivity relationships are represented by the three arcs in the network. The first arc is

$$\frac{\text{Materials volume}}{\text{Man-hours}} = \frac{Q}{N} \text{ is the ratio of materials to labor inputs}$$

The other two arcs have in their denominator:
Fixed investment × Output/Capacity $= eI_F$, namely the effective fixed investment. Hence the two arcs are

$$\frac{\text{Man-hours}}{\text{Fixed investment}\left(\dfrac{\text{Output}}{\text{Capacity}}\right)} = \frac{N}{e\,I_F} \quad \text{is the ratio of the labor input to the effective fixed investment output}$$

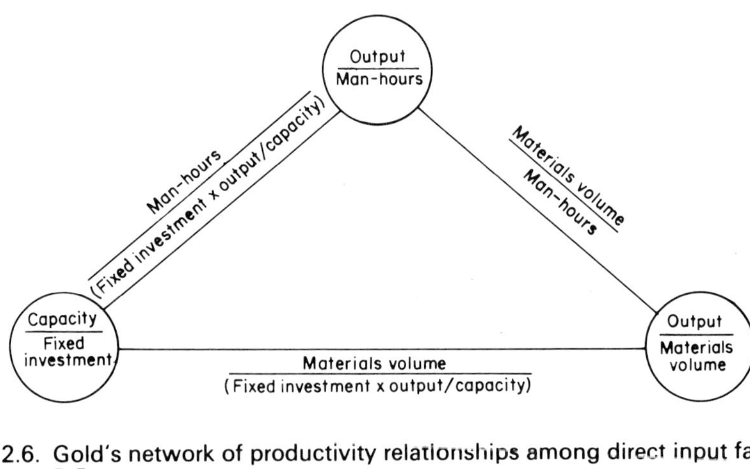

FIG. 2.6. Gold's network of productivity relationships among direct input factors [6].

$$\frac{\text{Material volume}}{\text{Fixed investment}\left(\dfrac{\text{Output}}{\text{Capacity}}\right)} = \frac{M}{e\,I_F}$$ is the ratio of the materials volume input to the effective fixed investment

Gold proceeds to integrate his productivity network into a complete cost structure [6, 1 p. 28], the peak of his pyramid being the total unit cost, as shown in Fig. 2.7. To the total unit cost are added four other managerial control ratios, namely:

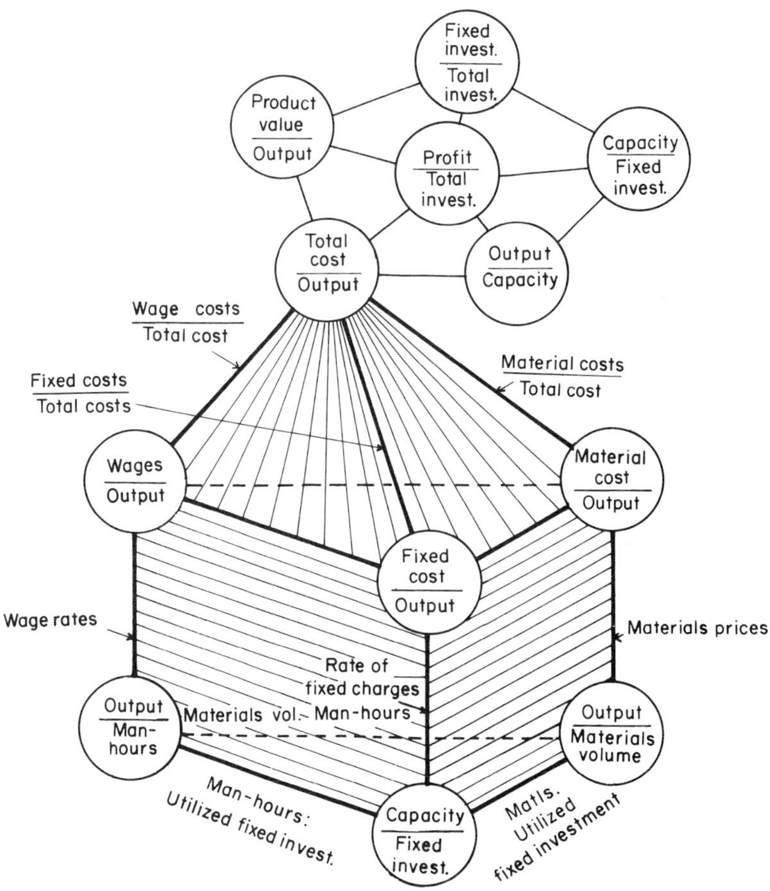

FIG. 2.7. Gold's model: productivity network, cost structure and managerial control ratios [6].

$$\frac{\text{Production value}}{\text{Output}} \ (= p)$$

$$\frac{\text{Fixed investment}}{\text{Total investment}} \ (= i)$$

$$\frac{\text{Capacity}}{\text{Fixed investment}} \ (= \theta)$$

$$\frac{\text{Output}}{\text{Capacity}} \ (= e)$$

These control ratios, shown at the top of Fig. 2.7, and the interactions between them, affect the final outcome for return on total investment.

A modified version of Gold's productivity network is given in Fig. 2.8, where the three nodes represent the three productivity ratios π_N, π_Q and π_{I_F}, relating to the three factor inputs of labor, materials and fixed investment. Here the numerator of each ratio is the physical output V, as suggested in Equation (2.10), and the denominator is one of the factor inputs. The three arcs between the nodes also depict ratios, which refer to the mix of inputs: the arc between the nodes π_N and π_Q denotes N/Q, the one between π_Q and π_{I_F} denotes Q/I_F and the one between π_{I_F} and π_N stands for I_F/N.

Thus if we move in a clockwise direction, then each node multiplied by the subsequent arc yields the next node:

$$\pi_N \frac{N}{Q} = \pi_Q \tag{2.18}$$

$$\pi_Q \frac{Q}{I_F} = \pi_{I_F} \tag{2.19}$$

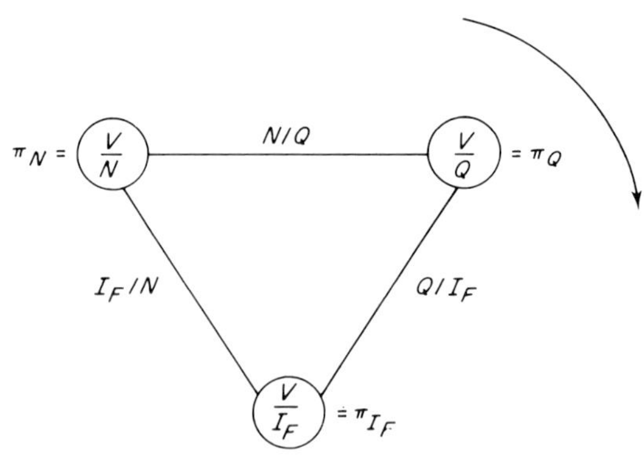

FIG. 2.8. A modified network of productivity ratios.

$$\pi_{I_F} \frac{I_F}{N} = \pi_N \tag{2.20}$$

and the product of the three arcs equals 1.

It is evident, therefore, that a change in the input ratio of any one arc will affect at least one of the productivity ratios adjacent to it and may well affect the third as well. For example, if the level of fixed investment is increased with a view to reducing the labor content in the final product, and if this does succeed in reducing N, then for a given level of output V we find that:

I_F/N	increases, because I_F increases and N declines
N/Q	decreases, provided the decrease in N is greater than the decrease in Q (with an increase in the level of mechanization of the plant due to increased I_F it is conceivable that the quantity of material Q would be affected)
Q/I_F	decreases, provided the increase in Q (if an increase occurs) is smaller than the increase in I_F
π_N	increases, because N decreases
π_Q	increases or decreases if Q decreases or increases respectively
π_{I_F}	decreases, because I_F increases

Thus, while one productivity ratio increases, at least one other declines, and it is not immediately obvious whether this combination of changes is beneficial to the overall performance of the plant.

A modified productivity network

A further modification of Fig. 2.8 is given in Fig. 2.9, where the working capital I_W is shown as another input. This network consists of four productivity ratios

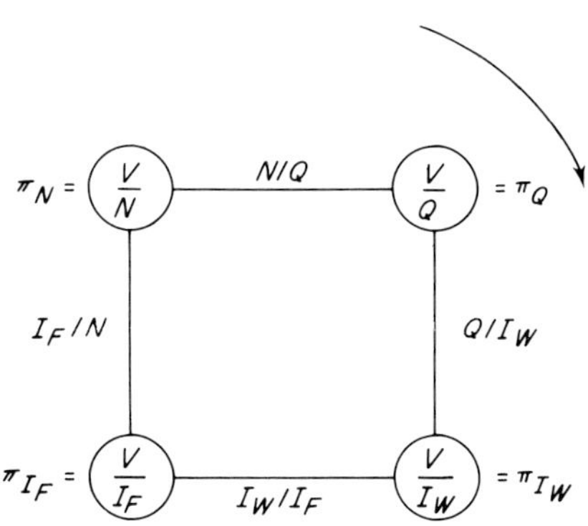

FIG. 2.9. A productivity network with four inputs.

π_N = productivity of labor
π_Q = productivity of materials
π_{I_F} = productivity of fixed investment
π_{I_W} = productivity of working capital

and the four arcs represent the ratios between the inputs:

$$I_F/N \qquad N/Q \qquad Q/I_W \qquad I_W/I_F$$

so that the product of the four arcs equals 1. The last ratio I_W/I_F relates to the split of total investment into I_F and I_W. Since $I_F/I = i$ and $I_W/I = 1-i$ it follows that

$$\frac{I_W}{I_F} = \frac{1-i}{i}$$

Going clockwise in Fig. 2.9, each node multiplied by the next arc yields the next node (as in Fig. 2.8), i.e.

$$\pi_N = \pi_{I_F}\frac{I_F}{N}$$

$$\pi_Q = \pi_N\frac{N}{Q}$$

$$\pi_{I_W} = \pi_Q\frac{Q}{I_W}$$

$$\pi_{I_F} = \pi_{I_W}\frac{I_W}{I_F} = \pi_{I_W}\frac{1-i}{i}$$

Factor unit costs

The reciprocals of the productivity ratios in Fig. 2.9 give the amount of each factor input required to produce one unit of physical output, and this is shown in the next layer up the pyramid in Fig. 2.10. The arcs here are the reciprocals of those shown in the bottom layer. In the third layer the fixed investment and the working capital are added to give the total investment, so that the three nodes in the triangular network are N/V, Q/V and I/V (these being the reciprocals of the three productivity ratios π_N, π_Q and π_I). We now add three factor prices as vertical sides to take us to the next layer of the pyramid. The factor prices are: w (cost of employment per man), m (unit cost of material volume) and j_I (cost incurred in providing the total investment, the cost being per unit of I). If each node in the third layer is multiplied by the corresponding factor price, we get

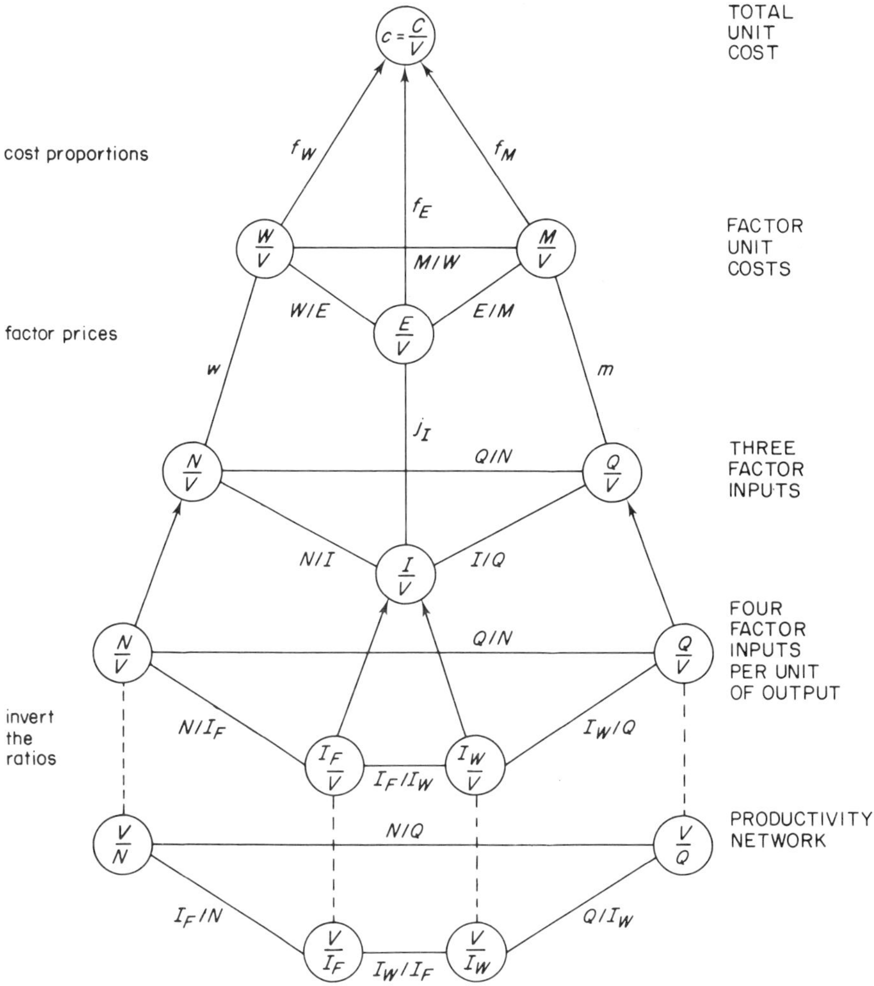

FIG. 2.10. The unit cost pyramid.

the unit cost of labor, materials and other expenses respectively, shown in the network comprising the fourth layer:

$$c_W = \frac{W}{V} = \frac{N}{V} w \tag{2.21}$$

$$c_M = \frac{M}{V} = \frac{Q}{V} m \tag{2.22}$$

$$c_E = \frac{E}{V} = \frac{I}{V} j_I \tag{2.23}$$

These factor costs add up to the total unit cost, as given in Equation (2.4) and shown at the apex of the pyramid. It should be noted that every horizontal network at the four layers in the pyramid has arcs between the nodes to depict ratios between input factors, either in physical or cost terms. This scheme preserves the principle that, going clockwise in each network, a node multiplied by its subsequent arc yields the next node.

As in Gold's model (Fig. 2.7), the links between the apex and the three nodes in the fourth layer represent the cost proportions, being the ratios between each of the nodes and the apex:

$$\frac{W}{C} = \frac{C_W}{c} = f_W = \text{cost proportion of labor} \tag{2.24}$$

$$\frac{M}{C} = \frac{C_M}{c} = f_M = \text{cost proportion of materials} \tag{2.25}$$

$$\frac{E}{C} = \frac{C_E}{c} = f_E = \text{cost proportion of other expenses} \tag{2.26}$$

so that, as in Equation (2.6),

$$f_W + f_M + f_E = 1 \tag{2.27}$$

The construction of the pyramid in Fig. 2.10 suggests a simple relationship between the factor unit costs in the fourth layer and the productivity ratios lower down. If in Equation (2.21) we substitute

$$\frac{N}{V} = \frac{1}{\pi_N}$$

we get

$$c_W = \frac{w}{\pi_N} = f_W c \tag{2.28}$$

similarly,

$$c_M = \frac{m}{\pi_Q} = f_M c \tag{2.29}$$

$$c_E = \frac{j_I}{\pi_I} = f_E c \tag{2.30}$$

and

$$\frac{w}{\pi_N} + \frac{m}{\pi_Q} + \frac{j_I}{\pi_I} = c \tag{2.31}$$

Comments

These relationships lead to the following conclusions:

(1) *The factor unit costs are inversely proportional to the corresponding productivity ratios.* For example, if—all things being equal (which incidentally, they never are)—productivity of labor declines by 20%, then the labor cost per unit of output (i.e. c_W) in Equation (2.28) increases not by 20% but by 25%. Similarly, if the productivity of labor rises by 20%, the unit labor costs declines by only 16.7%. This means that a relative deterioration in productivity causes a greater increase in factor unit costs, while an improvement in productivity results in a lower proportionate reduction in factor unit costs. This relationship conjures up a picture of a climber trying to go up a very slippery slope. He needs to summon up ever increasing efforts to make even modest progress, but the slightest relaxation on his part may lead him to slip a long way down.

(2) *If increases in factor prices are matched by increases in productivity, then unit factor costs are unaffected.* For example, if material prices m increase by 10% but productivity of materials π_Q also increases by 10%, then c_M in Equation (2.29) is unchanged. However, if m and π_Q change by different amounts, then m/π_Q assumes a new value, leading to a change in at least one other term in Equation (2.31), if the unit cost c is to remain unaffected. Similarly, if the rates w, m and j_I are unchanged and if c is to remain unaffected, it follows from Equation (2.31) that any change in one of the three productivity ratios π_N, π_M and π_{I_F} must lead to a change in at least one of the others. This result reiterates our former conclusion that *any single change in the system starts a chain reaction, causing changes in performance ratios that seemingly have nothing to do with the original cause.*

(3) *There is no single "overall productivity ratio".* The nearest that we can get to such a concept is by defining an "overall input". Two alternative "overall inputs" may be considered:

(a) The total cost, which represents the sum of all the inputs in financial terms, leading to the definition

$$\pi_C = \frac{V}{C} = \frac{1}{c} \tag{2.32}$$

namely, the productivity of total cost is simply the inverse of total unit cost. In this sense, overall cost productivity may be said to rise if, and only if, total unit cost falls. As in the case of the factor unit costs, the "slippery slope syndrome" applies here: An increase in overall productivity of x% reduces total unit cost by less than x%, and a decrease in overall productivity of x% increase total unit cost by more than x%.

(b) The total investment, which signifies the total amount of money tied up in the business, so that

$$\pi_I = \frac{V}{I}$$

and since $I_F = iI$, then

$$\pi_I = i\pi_{I_F}$$

The overall ratio π_I thus depends on the productivity of fixed investment (π_{I_F}) and on the internal allocation of investment (i).

(4) If the factor prices in Equation (2.31) do not change, then *an increase in all three productivity ratios will lead to a reduction in total unit cost and a decline in the three ratios will cause the unit cost to rise.* A further discussion of changes in productivity ratios and their effect on costs is given in Chapter 6.

LABOR PRODUCTIVITY

As stated earlier, of all the productivity ratios, labor productivity has assumed the greatest prominence in the analysis of industrial performance. We defined labor productivity as

$$\pi_N = \frac{V}{N}$$

namely, as the ratio of physical output per employee. The question of how to measure physical output in a multi-product plant is discussed at some length in the next chapter. There are, of course, other possible candidates for the numerator of a labor performance ratio, such as: revenue, added value and profit. Similarly, there are alternative ways for measuring the labor input in the denominator, such as the total number of man-hours H (to account for varying amounts of basic working hours and overtime in different circumstances, the total number of man-hours being given by $H = NT$) and the total wage bill. The various combinations of the most common ratios are summarized in Table 2.3.

There is a formidable array of 12 ratios in Table 2.3 to choose from. The most commonly used ratios refer to the physical volume as their measure of output, so that the notation π_N implies $\pi_N(\bar{V})$ namely V/N and similarly π_H implies V in the numerator. Where another output measure is considered, it is necessary to specify it in the notation to avoid any possible misunderstanding, so that $\pi_N(R)$, for example, defines R as the output, namely $\pi_N(R) = R/N$.

In discussing the relative merits of the various ratios, it should first be noted that there is no conceptual difference between N and H as a measure of labor

TABLE 2.3
Various labor performance ratios.

Labor input	Output Physical volume V	Revenue R	Profit Z	Added value A
Employees N	$\pi_N(V) = \pi_N = \dfrac{V}{N}$	$\pi_N(R) = \dfrac{R}{N}$	$\pi_N(Z) = \dfrac{Z}{N}$	$\pi_N(A) = \dfrac{A}{N}$
Manhours $H = NT$	$\pi_H(V) = \pi_H = \dfrac{V}{H}$	$\pi_H(R) = \dfrac{R}{H}$	$\pi_H(Z) = \dfrac{Z}{H}$	$\pi_H(A) = \dfrac{A}{H}$
Total wages W	$\pi_W(V) = \pi_W = \dfrac{V}{W}$	$\pi_W(R) = \dfrac{R}{W}$	$\pi_W(Z) = \dfrac{Z}{W}$	$\pi_W(A) = \dfrac{A}{W}$

input. When the average number of hours T remains stable over a period of time, or is roughly the same in plants or industries for which comparisons are made, then there is no special merit in selecting H as the input in Table 2.3 and the ratio of output per employee may well be adequate. However, when the ordinary length of the working week is a matter for negotiations between management and labor, and when the amount of overtime hours worked is subject to fluctuations, then H rather than N is a truer reflection of the labor input. To illustrate this point, a remark made by an airline president may be cited to the effect that the productivity of the airline could be raised only if members of the crew were willing to work longer hours. What he probably meant was that, if an increase in T allowed H to decrease (by reduction of the size of the work force) for the same output (measured by miles or passenger miles), or for H to increase by a lesser amount than V, then productivity would go up.

The argument for using ratios in the bottom row of Table 2.3 is that they take account of the cost of the labor input, including the differential rates between normal working hours and overtime. However, a definition based on cost has the disadvantage that a change in the value of a ratio does not reveal the cause of the change, and at times may be misleading. For example, if π_W suddenly declines, it is possible that the reason lies in a recent wage settlement rather than in any intrinsic fall in employee performance. Note, also, that

$$\pi_W = \frac{V}{W} = \frac{1}{c_W}$$

i.e. it is the inverse of the labor unit cost. Similarly, R/W and A/W are reciprocals of widely used measures, as noted earlier in Table 2.2.

CONCLUSION

Comparisons of performance based on computations of optimal solutions (for the maximum possible output for given inputs, or for minimum inputs for given outputs) are generally difficult to implement in practice. The reason lies partly in problems of data collection and partly in the validity of the assumptions used in optimization models. Instead, evaluation of performance in practice relies on comparisons between actual results with past performance and/or notional planning objectives determined by managerial judgement and experience. Most of the performance measures are represented as ratios.

There are numerous ratios to choose from. Some are relevant to particular functions and echelons in the organizational hierarchy; some are designed to meet the specific needs of a given firm or industry. The most prominent general ratios between outputs and inputs are summarized in Table 2.2, the one commonly cited for evaluation of overall financial performance being the return on capital $r = Z/I$.

In addition, there are many surrogate measures for the proportions of inputs and outputs respectively, some of those often used being: *cost proportions,revenue proportions, added value proportions, profit margin* and *productivity ratios.* The reliance on surrogate measures is based on the proposition that improvements in their values must be beneficial to the overall objectives of the enterprise in terms of profit, cost and return on capital. It must be realized, however, that this proposition does not always hold.

Productivity is defined as the ratio of physical output to a given factor input.

TABLE 2.4
Productivity ratios for physical volume.

	$\pi = \dfrac{V}{\text{Factor input}}$
Labor	$\pi_N = \dfrac{V}{N}$ or $\pi_H = \dfrac{V}{H}$
Materials	$\pi_M = \dfrac{V}{M}$
Investment	$\pi_I = \dfrac{V}{I}$
Fixed investment	$\pi_{I_F} = \dfrac{V}{I_F}$
Working capital	$\pi_{I_W} = \dfrac{V}{I_W}$

Thus productivity ratios may be stated for labor, materials, fixed investment and working capital, as shown in Table 2.4. A network of relationships between them, such as Figs 2.6, 2.8 and 2.9, helps to highlight their inter-dependence and the way they are affected by the ratios of factor inputs. Labor productivity is by far the most discussed ratio, although an increase in labor productivity does not necessarily result in a reduction in unit costs, or increase in return on capital, and the effect of the other productivity ratios may be more crucial at times. An "overall productivity ratio" may be defined as either V/I or V/C, the latter being none other than the inverse unit cost. An integrated model of productivity ratios and unit cost is given in Fig. 2.7 (Gold's model) and in Fig. 2.10.

REFERENCES

[1] Eilon, S., Gold, B. and Soesan, J. (1976). "Applied Productivity Analysis for Industry". Pergamon Press.

[2] Eilon, S. (1979). Some useful ratios in evaluating performance. *Omega* **7**, 166–8.

[3] Eilon, S. and Teague, J. (1973). On measures of productivity. *Omega* **1**, 565–576.

[4] Gold, B. (1955). "Foundations of Productivity Analysis". University of Pittsburgh Press, Pittsburgh.

[5] Gold, B. (1971). "Explorations in Managerial Economics". Macmillan, London.

[6] Gold, B. (1973). Technology, productivity and economic analysis. *Omega* **1**, 5–24.

[7] Gold, B. (1973). The impact of technological innovation—concepts and measurement. *Omega* **1**, 181–191.

[8] London Gazette (26 August, 1966). Notice on productivity agreements with reference to the Prices and Incomes Act 1966.

[9] Mali, P. (1978). "Improving Total Productivity". John Wiley, New York.

[10] National Board for Prices and Incomes (1967). "Productivity Agreements". Cmnd 3311, HMSO, London.

[11] Teague, J. and Eilon, S. (1973). Productivity measurement—a brief survey. *Appl. Econ.* **5**, 133–145.

[12] Westwick, C. A. (1973). "How to Use Management Ratios". Gower Press, London.

EXERCISES

1. A plant has been working at full capacity for ten years and data for its two major inputs and for its output are available. How would you go about constructing the production possibilities curve and what difficulties do you envisage?
2. Suggest some performance measures for a sales department.

3. Define

   Added value
   Profit before tax
   Revenue and cost proportions
   Return on investment
   Unit profit
   Effective fixed investment
   Efficiency of fixed investment
   Capacity utilization

4. Describe the productivity network.
5. How would you define labor productivity?
6. How can the measure of return on investment be disaggregated?
7. Comment on the following performance ratios

   Sales lost/salesman
   Sales lost/employee
   Revenue/customer
   Rework/production costs
   Machine time/total available time
   Meals served/customer
   Reports/administrator
   Advertising costs/employee

8. Assuming that the component of other expenses $E = 0$, show that materials productivity can be measured by

$$\pi_Q = \frac{1}{c}\left(\frac{W}{Q} + m\right)$$

3
Index Numbers

Themes in this chapter

The purpose and definition of an index number
Measuring the output of a single-product plant
The multi-product problem
Relative indices adjusted by volume, price or revenue
Aggregative volume and price indices
The Laspeyres, Paasche, Fisher and Edgeworth indices and their application
Examples to compare results for various indices
Methods for coping with gaps in the historical record of a product range
The base-transfer and time-reversal properties

THE INDEX SCALE AS A COMPARATIVE YARDSTICK

In attempting to make a judgement about phenomena in a given system, we naturally resort to comparisons, and in particular to changes that take place over a period of time. We compare the value of a given criterion or variable at one point in time to its corresponding value at another. Such a comparison, for a series of values, is conveniently carried out by the construction of *an index*, which provides a yardstick for *relative values instead of the use of absolute quantities*. Many examples may be cited for index scales; perhaps the most well-known are based on government and business statistics and are frequently quoted in the financial press, such as the Consumer (or Retail) Price Index, the Wholesale Price Index, the Index of Industrial Production, the House Price Index, the Average Wage Index, the Dow Jones Share Price Index, the FT (Financial Times) Index, and so on. In the case of individual enterprises we often need to measure the output of a system in one period and compare it with the output prevailing during another period, and here too the comparison can be done by means of an index number.

An index is simply a measure of a given variable related to its value at a specified time period, known as the *base period* or *the baseline*, when this value is usually taken as 100. Thus, when the index at a subsequent period rises to

110, it is because the value of the variable, which the index is designed to measure, has increased by 10%, compared with its absolute value in the base period.

AN INDEX FOR A SINGLE VARIABLE

When a single variable is involved, the construction of an index is a simple affair, which is best illustrated by an example. Suppose that we have statistics for the sale of cement for house building in a given area for several successive years, as shown in Table 3.1. If we take year 1 as the base and designate its sales as 100, the index figure for year 2 becomes

$$\frac{280}{250} \times 100 = 112$$

and for year 3

$$\frac{175}{250} \times 100 = 70$$

and so on, so that in terms of this index the sales are summarized in Table 3.2. The information in this table is not fundamentally different from that given in the previous table, except that it is presented in a *non-dimensional* form.

Notice that in converting the data from Table 3.1 to Table 3.2 we have, in fact, lost some information in that we are unable to tell from Table 3.2 what actual tonnage was sold in each year, but the presentation in index form helps us to make judgements about the changes that have taken place. For example, if the sales in another geographical area, calculated in a similar way, show a series of rising figures culminating with 160 in year 8, we may conclude that either the house building program or the market share, or both have been growing at a much faster rate than in the area depicted in Table 3.2. Another example is the

TABLE 3.1
An example of cement sales (in 10^3 tons).

Year	1	2	3	4	5	6	7	8
Sales	250	280	175	275	290	310	315	325

TABLE 3.2
Cement sales index (year 1 = 100).

Year	1	2	3	4	5	6	7	8
Sales	100	112	70	110	116	124	126	130

breakdown of expenditure by the population in a given area into such items as: housing, household goods, food and drink, entertainment, clothing, etc. *The rate of growth* in expenditure in each of these categories may be of greater importance to businessmen in assessing market potential for their corporate plans than the absolute values in monetary terms (though these cannot, of course, be entirely ignored).

Selection of a base period

In the case of a single variable it does not strictly matter which period is selected as the base, since the index values are merely scaled up or down, depending on the period chosen. It is desirable, however, to avoid the choice of an "abnormal" period as a base. For example, if for the data in Table 3.1 the base is taken as year 3, we would infer that over the five years from year 3 to year 8 sales almost doubled, whereas they increased by only 30% over the eight years shown in Table 3.2. Both statements are correct, but as sales in year 3 were exceptionally low for some reason, there seems no merit in selecting it as a yardstick, which could lead to misleading impressions. Many apparently conflicting arguments, particularly between politicians quoting statistics to prove a point, can be traced to a convenient choice of a base period. While one politician cites figures to show that unemployment and inflation have gone up, another is adamant that at least one, if not both, have gone down. Both assertions may, in fact, be correct and the difference between them due solely to different baselines.

In order to convert a given index to one related to a new base period we need not go back to the raw data, but merely divide the old index figures by the index value of the new base period. For example, if period 2 in Table 3.2 is to be taken as the base, we divide all the figures in Table 3.2 by 112 and multiply by 100 (since the baseline is always taken as 100), and if period 4 becomes the base, then the figures need to be divided by 110 and multiplied by 100. The results for alternative base years, shown in Table 3.3 and Fig. 3.1, follow the same pattern,

TABLE 3.3
Cement sales index.

Year	1	2	3	4	5	6	7	8
Base year								
Year 1 = 100	100	112	70	110	116	124	126	130
2 = 100	89.3	100	62.5	98.2	103.6	110.7	112.5	116.1
3 = 100	142.9	160	100	157.1	165.7	177.1	180	185.7
4 = 100	90.9	101.8	63.6	100	105.5	112.7	114.5	118.2
8 = 100	76.9	86.9	53.8	84.6	89.2	95.4	96.9	100

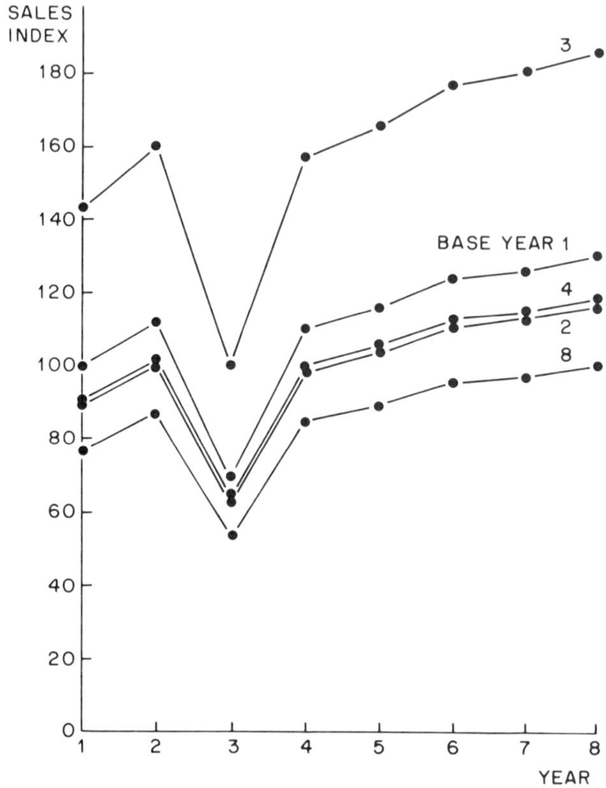

FIG. 3.1. Cement sales for different base years. (See Table 3.3.)

revealing the scaling up or scaling down effect; the low base figure in year 3 accentuates the amplitudes of changes over the eight year span, while the high base figure in year 8 tends to dampen the fluctuations.

This is why the index number is occasionally adjusted in cases of a rising series of statistics, such as those associated with inflation, so that after several years the old baseline is abandoned and a more recent period is taken as the base at 100. In this way the changes that take place are related to a recent starting point rather than to the more distant past.

Properties of the index number

From the above discussion it follows that, in comparing one year with another in the single product case, it is immaterial which baseline is used. If we denote

Y_{nb} as the value of the index Y in period n compared with 100 in the base period b, then this index is derived as

$$Y_{nb} = \frac{V_n}{V_b} 100 \tag{3.1}$$

where V_n = the value of the variable (in our case, the cement sales) in period n
V_b = the value of the variable in the base period b

Alternatively, the ratio

$$y_{nb} = \frac{V_n}{V_b} \tag{3.2}$$

measures the volume in year n when the base period is taken as 1.0. In most applications of index numbers, the base period is set at 100, so that

$$Y_{nb} = y_{nb} 100 \tag{3.3}$$

From these definitions we find that the *time reversal property*

$$y_{nb} y_{bn} = 1 \tag{3.4}$$

namely: *the value of the index of a variable in period n compared with its value in the base period b is the inverse of the index in year b compared with the base n* (when in both cases the index is taken as 1.0 in the base period).

More generally, if the base period is changed from b_1 to b_2, then when b_1 is taken as 1.0:

$$y_{nb_1} = \frac{V_n}{V_{b_1}}$$

and similarly for b_2:

$$y_{nb_2} = \frac{V_n}{V_{b_2}}$$

Or

$$y_{nb_2} = \frac{V_n}{V_{b_2}} = \frac{V_n}{V_{b_1}} \frac{V_{b_1}}{V_{b_2}}$$

Substituting the definitions

$$\frac{V_n}{V_{b_1}} = y_{nb_1}$$

$$\frac{V_n}{V_{b_2}} = y_{n_1 b_2}$$

and the *base transfer property* is then obtained:

$$y_{nb_2} = y_{nb_1} y_{b_1 b_2} \tag{3.5}$$

And when the base year is taken as 100, the application of Equation (3.3) yields

$$Y_{nb_2} = Y_{nb_1} y_{b_1 b_2}$$

$$= Y_{nb_1} \frac{Y_{b_1 b_2}}{100}$$

which is precisely how the computations for a new baseline in Table 3.3 were carried out, namely: conversion of the index by transfer from base b_1 to base b_2 is achieved by multiplying the original index based on b_1 by the factor $(Y_{b_1 b_2}/100)$, which is the index for the old base b_1 compared with the new base b_2 divided by 100.

Example

If we now wish to compare the volume of sales in year n_2 with year n_1, we can either use the raw data and find the value of $V_{n_2 n_1}/100$ by the ratio V_{n_2}/V_{n_1} or use the index figures for a given base b in the ratio $Y_{n_2 b}/Y_{n_1 b}$. For example, let us compare the cement sales in year 5 with year 4:

(1) Use the sales data in Table 3.1:

$$V_5 = 290$$

$$V_4 = 275$$

hence

$$\frac{V_5}{V_4} = \frac{290}{275} = 1.055, \text{ i.e. an increase of } 5.5\,\%$$

(2) Use the index figures based on year 1 (Table 3.2):

$$Y_{51} = 116$$

$$Y_{41} = 110$$

hence

$$\frac{Y_{51}}{Y_{54}} = \frac{116}{110} = 1.055$$

(3) Use the index figures based on year 4 (Table 3.3):

$$Y_{54} = 105.5$$

$$Y_{44} = 100 \; (= \text{the base year})$$

hence the ratio is 1.055

(4) Use the index figures based on year 8 (Table 3.3):

$$Y_{58} = 89.2$$

$$Y_{48} = 84.6$$

the ratio is 1.054, the small discrepancy compared with the previous results being due to rounding errors (if computed to three decimal places, then $Y_5(8) = 89.231$ and $Y_4(8) = 84.615$, so that the ratio becomes 1.055).

Comments on the base-transfer property

The various examples of the single product case cited above illustrate that the *time-reversal* and the *base-transfer properties*, as defined earlier, hold throughout. It should be pointed out that in general when the base-transfer property applies, the time-reversal property can be derived as a special case, as is clear from the following simple arguments: Given the relationship in Equation (3.5)

$$y_{nb_2} = y_{nb_1} y_{b_1 b_2}$$

we may then deliberately choose b_2 to coincide with n and b_1 is then denoted as b; the substitution

$$b_2 = n$$
$$b_1 = b$$

in Equation (3.5) yields

$$y_{nn} = y_{nb} y_{bn}$$

and since by definition $y_{nn} = 1$, the result for the time reversal property in Equation (3.4) is obtained. It should be stressed, though, that a base transfer is more general than time reversal. Hence, if the general case of base transfer holds, the particular case of time reversal must hold as well. But the converse does not apply, so that when the time-reversal property holds, it does not necessarily require that the more general base-transfer property must hold as well.

Other implications of the base-transfer property may be stated by reformulating it as follows:

1. *If three index numbers are defined, the first for period n_1 to base n_2, the second for period n_2 to base n_3 and the third for period n_3 to base n_1, then the product of the three is* 1 (when in each case the index is 1.0 in its base period). This result is easily derived from Equation (3.5) after substituting the time reversal property $y_{nb_2} = 1/y_{b_2 n}$ to obtain

$$y_{nb_1} y_{b_1 b_2} y_{b_2 n} = 1$$

With a new notation

$$n = n_1$$
$$b_1 = n_2$$
$$b_2 = n_3$$

we find that

$$y_{n_1 n_2} y_{n_2 n_3} y_{n_3 n_1} = 1 \qquad (3.6)$$

2. *The ratio of the index for period n_2 to the index for period n_1, both to a given baseline, is the same as the index for period n_2 taken to the base period n_1.* This is derived from the base transfer property by designating the periods n, b_1 and b_2 as follows
$$n = n_2$$
$$b_1 = n_1$$
$$b_2 = b$$
Substituting this notation into Equation (3.5), we get
$$y_{n_2 b} = y_{n_2 n_1} \, y_{n_1 b}$$
or
$$\frac{Y_{n_2 b}}{Y_{n_1 b}} = y_{n_2 n_1} \tag{3.7}$$

3. *The ratio of the index for period n_2 to that for period n_1, remains the same irrespective of the base period.* This result is directly obtained from Equation (3.7), so that when two alternative baselines b_1 and b_2 are considered, we get
$$\frac{y_{n_2 b_2}}{y_{n_1 b_2}} = \frac{y_{n_2 b_1}}{y_{n_1 b_1}}$$

All these results apply to the situation when a single product is considered, but the multi-product case is a different story altogether, as the following discussion will reveal.

AN INDEX FOR A COMPOUND VARIABLE

The case of a single output variable is rarely encountered in practice and we often need to consider a *compound variable*, which is an aggregation of several components, each of which is subject to change over time. A good example of a compound variable is a measure of output of an industrial enterprise, being the aggregate of various products manufactured at several plants, or sold in different markets, each of the products involving a certain level of cost, or a different price tag.

How do we, then, measure the total output of the enterprise, and how do we compare this output for one year with another? Similar questions may be posed about the total input to the system, consisting of a variety of resources, each with its own cost. With the passage of time the individual streams of inputs and outputs change, both with respect to quantities and prices (or costs), and this is where a suitably constructed index can help in the evaluation of the total change that takes place.

These problems become particularly acute when the level of remuneration (such as wages or pensions) and other payments are "indexed", namely when they are adjusted in line with the movement of an agreed index number. One example of indexation is the automatic adjustment of wages and salaries in countries which experience high rates of inflation; another is the comprehensive study on indexation of prices of commodities produced by developing countries, as called for by the General Assembly of the United Nations in 1973 (see [2]).

An example of a two-product output

Unlike the single variable case, where the construction of the index is quite straightforward, there are numerous ways in which an index for a compound variable can be designed. It is best to illustrate some of the alternatives through an example. Table 3.4 gives details of an output consisting of two products over a five-year period.

Figure 3.2 gives a simple representation of the output of this enterprise: physical quantities are manufactured by the production facilities and then converted into revenues through price factors. It is tempting, therefore, to regard the output in terms of the physical volume, or the price, or the revenue, but complications arise because of the existence of the two distinct products. Consider first the physical volume: if we take the total volume as the sum of the two product volumes, then the total becomes a single variable, for which an index may easily be constructed, as shown earlier, and the results are given in Table 3.5, with year 1 taken as the base year. But if the products are very different in character (for example, tables and chairs produced in a furniture factory), it may be meaningless to add the physical volumes of the two products. This is often referred to as "the apples and pears problem": if you

TABLE 3.4
An example for output consisting of two products.

Year	1	2	3	4	5
Physical quantity					
product I	75	60	80	100	120
product II	30	40	60	45	50
Price per product					
product I	4.0	4.4	4.0	5.0	5.2
product II	2.0	3.0	3.0	4.0	3.6
Revenue					
product I	300	264	320	500	624
product II	60	120	180	180	180

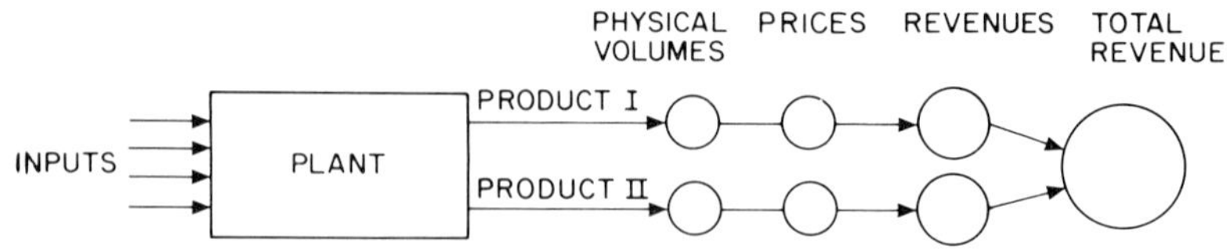

FIG. 3.2. Two-product output.

TABLE 3.5
Total output in terms of physical volume.

Year	1	2	3	4	5
Physical quantity					
product I	75	60	80	100	120
product II	30	40	60	45	50
total	105	100	140	145	170
index	100	95.2	133.3	138.1	161.9
Individual indices					
product I index	100	80	106.7	133.3	160
product II index	100	133.3	200	150	166.7
average index	100	106.7	153.3	141.7	163.3

have five apples and four pears, it is true that you have a total of nine pieces of fruit, but it is questionable whether it is better than having eight pieces of fruit of a different composition. The sum total of apples and pears is hardly of any use when trying to evaluate the change that takes place over time.

An alternative approach to constructing a multi-product quantity index is first to compute an index for each of the individual products (so that each is taken as 100 in the base year) and then finding the average index for each year. The results for the two individual indices and their average are given in Table 3.5, from which we see that they are different from those based on the total physical quantity. But this average index is also unsatisfactory, because an unweighted average gives great prominence to changes in lesser products ("lesser" with respect to their physical volumes). To illustrate this point, consider the example of product I with sales of 1000 tons in each of two consecutive years, while product II has sales of 1 ton and 10 tons respectively. In such a case the index of product I remains at 100 for the two years and the index of product II rises from 100 to 1000, so that the average index in year 2 is 550 (namely 5.5 times its value in year 1), giving quite a misleading impression about the total increase in volume.

Similar comments may be extended to an index based solely on price. The two alternatives corresponding to those shown in Table 3.5 are summarized in Table 3.6: the first is based on the average price for each year and then takes year 1 as 100; the second develops a price index for each product and then records the average index. Again, the results are different for the two alternatives (and different from those shown in Table 3.5), and highlight the fact that they are based on different weightings.

Clearly, neither approach of relying solely on physical volumes or on price is satisfactory as a measure of output of a multi-product system. We therefore turn to revenue, which takes account of both quantity and price. Of the two alternatives depicted in Table 3.7—one based on the total revenue and the other on individual revenue indices—the latter is open to the criticism, already stated, regarding the method of averaging individual indices. The total revenue

TABLE 3.6
Output in terms of a price index.

Year	1	2	3	4	5
Unit price					
product I	4.0	4.4	4.0	5.0	5.2
product II	2.0	3.0	3.0	4.0	3.6
average	3.0	3.7	3.5	4.5	4.4
index	100	123.3	116.7	150	146.7
Individual indices					
product I	100	110	110	125	130
product II	100	150	150	200	180
average	100	130	125	162.5	155

TABLE 3.7
Output in terms of revenue.

Year	1	2	3	4	5
Revenue					
product I	300	264	320	500	624
product II	60	120	180	180	180
total	360	384	500	680	804
index	100	106.7	138.9	188.9	223.3
Individual indices					
product I	100	88	106.7	166.7	208
product II	100	200	300	300	300
average	100	144	203.3	233.3	254

index is a good measure of changes in the total revenue, but not necessarily of *output*. It is quite possible for physical volume to decline and yet for revenue to remain the same or even to increase, owing to a sufficient rise in unit prices. Even in the event of both physical volumes and prices increasing, so that revenue also increases (and we have no doubt then about the qualitative statement that output has increased), there is still the problem of determining by *how much* it has changed, since in revenue terms some of the change is attributable to volume and some to price.

ALTERNATIVE METHODS OF CONSTRUCTING AN INDEX

Many computational procedures can be devised to take account of both price and volume in their effect on output. Two possible approaches are outlined in Fig. 3.3: the first is aimed at constructing *an adjusted relative index* for volume, price or revenue, in order to overcome the shortcomings of the average index computed in Tables 3.5–3.7. The means of adjusting the volume index may rest either on price or revenue (as elaborated later); similarly, the price index may be adjusted by volume or revenue and the revenue index by volume or price. This approach in Fig. 3.3 gives us six alternative indices. In addition, it is possible to relate the computations either to the base year or to the current year, or to both, so that the number of alternative indices rises to 18. The second approach in Fig. 3.3 is to modify *the aggregative revenue measure*, either by reference to volume or to price, in each case relating to the base year, or to the current year, or to both, or through an average value, thereby yielding eight further indices. We shall now proceed to examine how these indices can be computed in practice with reference to our example in Table 3.4.

Relative indices

The 18 alternatives for the first approach in Fig. 3.3 are summarized in Table 3.8. Section (A) lists six indices with reference to the base year b, Section (B) with reference to the current year n, and Section (C) takes the average of the corresponding indices in (A) and (B). The symbol Σ denotes summation over all the products and the 100 is inserted as the base figure with which the index is compared. For example, the index for 1 (A) for year n, denoted as Y_{nb} (1A), or Y_n (1A) for short (when the base year is self understood), is

$$Y_n(1\text{A}) = \frac{\dfrac{V_n(1)}{V_b(1)}\,p_b(1) + \dfrac{V_n(2)}{V_b(2)}\,p_b(2) + \dfrac{V_n(3)}{V_b(3)}\,p_b(3) + \ldots}{p_b(1) + p_b(2) + p_b(3) + \ldots}\,100$$

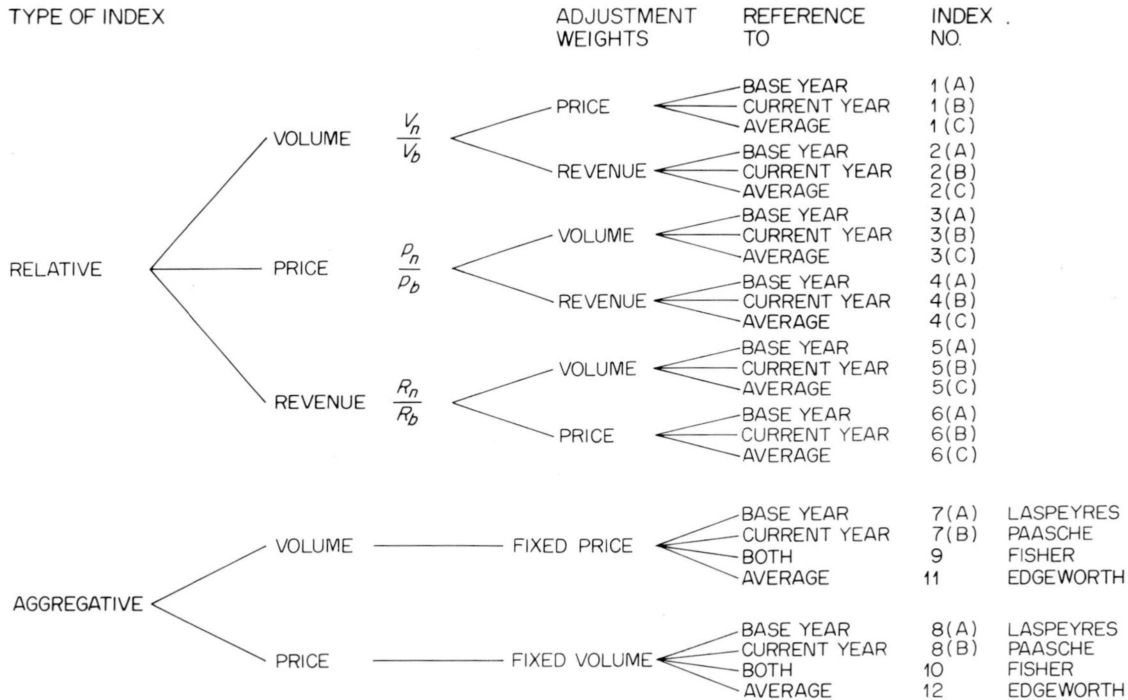

FIG. 3.3. Various types of index.

TABLE 3.8
Adjusted relative indices for year n.

	Volume adjusted	Price adjusted	Revenue adjusted
(A) Reference to base year			
Volume index		$1(A) = \frac{\Sigma \frac{V_n}{V_b} p_b}{\Sigma p_b} 100$	$2(A) = \frac{\Sigma \frac{V_n}{V_b} R_b}{\Sigma R_b} 100$
Price index	$3(A) = \frac{\Sigma \frac{p_n}{p_b} V_b}{\Sigma V_b} 100$		$4(A) = \frac{\Sigma \frac{p_n}{p_b} R_b}{\Sigma R_b} 100$
Revenue index	$5(A) = \frac{\Sigma \frac{R_n}{R_b} V_b}{\Sigma V_b} 100$	$6(A) = \frac{\Sigma \frac{R_n}{R_b} p_b}{\Sigma p_b} 100$	
(B) Reference to current year			
Volume index		$1(B) = \frac{\Sigma \frac{V_n}{V_b} p_n}{\Sigma p_n} 100$	$2(B) = \frac{\Sigma \frac{V_n}{V_b} R_n}{\Sigma R_n} 100$
Price index	$3(B) = \frac{\Sigma \frac{p_n}{p_b} V_n}{\Sigma V_n} 100$		$4(B) = \frac{\Sigma \frac{p_n}{p_b} R_n}{\Sigma R_n} 100$
Revenue index	$5(B) = \frac{\Sigma \frac{R_n}{R_b} V_n}{\Sigma V_n} 100$	$6(B) = \frac{\Sigma \frac{R_n}{R_b} p_n}{\Sigma p_n} 100$	
(C) Average of (A) and (B)			
Volume index		$1(C) = \frac{1}{2}[1(A) + 1(B)]$	$2(C) = \frac{1}{2}[2(A) + 2(B)]$
Price index	$3(C) = \frac{1}{2}[3(A) + 3(B)]$		$4(C) = \frac{1}{2}[4(A) + 4(B)]$
Revenue index	$5(C) = \frac{1}{2}[5(A) + 5(B)]$	$6(C) = \frac{1}{2}[6(A) + 6(B)]$	

where $V_n(1)$, $V_n(2)$, $V_n(3)$, etc., are the volumes in the current year n for products 1, 2, 3, etc.
$V_b(1)$, $V_b(2)$, etc., are the corresponding values in the base year
$p_b(1)$, $p_b(2)$, $p_b(3)$, etc., are the unit prices in the base year for products 1, 2, 3, etc.

Notice that if the volumes do not change from year b to year n (i.e. if $V_n = V_b$ for all the products) then the volume index becomes 100, irrespective of whether it is price adjusted or revenue adjusted, i.e. all the indices 1(A), 2(A), 1(B), 2(B), 1(C) and 2(C) yield the same result. It is when $V_n \neq V_b$ that the manner of adjustment can affect the outcome. Notice also the difference between 1(A) and 1(B): the former relies entirely on base year price adjustments, while the latter relies on current prices and ignores the base prices. As a compromise, 1(C) takes the average between 1(A) and 1(B). Similar observations apply to the other indices.

We can now compute these indices for our two product example in Table 3.4 as shown below.

Volume index, price adjusted, 1 (A)

The individual index values (V_2/V_1) 100 for each product and those for subsequent years (V_3/V_1) 100, (V_4/V_1) 100, etc., are given in Table 3.5. Hence for year 2 the value of this index is

$$Y_2\,(1\text{A}) = \frac{\Sigma \dfrac{V_2}{V_1} p_1}{\Sigma p_1}\, 100 = \frac{(80 \times 4) + (133.3 \times 2)}{4+2} = 97.8$$

and similarly for subsequent years:

$$Y_3\,(1\text{A}) = \frac{(106.7 \times 4) + (200 \times 2)}{6} = 137.8$$

$$Y_4\,(1\text{A}) = \frac{(133.3 \times 4) + (150 \times 2)}{6} = 138.9$$

$$Y_5\,(1\text{A}) = \frac{(160 \times 4) + (166.7 \times 2)}{6} = 162.2$$

Volume index, price adjusted, 1 (B)

In 1(A) the prices of the base year prevail, whereas here we need to account for the current prices:

$$Y_2(1B) = \frac{\Sigma \frac{V_2}{V_1} p_2}{\Sigma p_2} = \frac{(80 \times 4.4) + (133.3 \times 3.0)}{4.4 + 3.0} = 101.6$$

and for subsequent years:

$$Y_3(1B) = \frac{(106.7 \times 4) + (200 \times 3)}{4 + 3} = 146.7$$

$$Y_4(1B) = \frac{(133.3 \times 5) + (150 \times 4)}{5 + 4} = 140.7$$

$$Y_5(1B) = \frac{(160 \times 5.2) + (166.7 \times 3.6)}{5.2 + 3.6} = 162.7$$

Volume index, price adjusted, 1 (C)

This is simply the average between 1(A) and 1(B):

$$Y_2(1C) = \tfrac{1}{2}[Y_2(1A) + Y_2(1B)] = \tfrac{1}{2}(97.8 + 101.6) = 99.7$$

$$Y_3(1C) = \tfrac{1}{2}(137.8 + 146.7) = 142.2$$

$$Y_4(1C) = \tfrac{1}{2}(138.9 + 140.7) = 139.8$$

$$Y_5(1C) = \tfrac{1}{2}(162.2 + 162.7) = 162.5$$

Similarly, the other indices can be computed, using the formulae in Table 3.8, and the results are given in Table 3.10.

Aggregative indices

The second approach in Fig. 3.3 is summarized in Table 3.9. To start with, we have the two index numbers 7(A) and 7(B), both are volume based, so that the current volumes V_n appear in the numerator, while the base year volumes are in the denominator. Both (A) and (B) attempt to relate the revenue to constant prices. In (A) the constant prices are taken as those prevailing in the *base year*, so that the denominator in 7(A) represents the revenue in the base year, while the numerator stands for the revenue that would have obtained in the current year, had the base year prices prevailed during the current year as well; this method is called after its originator, the economist Etienne Laspeyres (1834–1913), the *Laspeyres Index*. In (B) the constant prices are taken as the *current prices*, and the index is called after the economist Hermann Paasche (1851–1925), the *Paasche Index*; here the numerator is the current revenue,

TABLE 3.9
Aggregative indices.

	Formula	(A) Reference to base year Laspeyres	(B) Reference to current year Paasche
Volume based	7	$\frac{\Sigma p_b V_n}{\Sigma p_b V_b} 100$	$\frac{\Sigma p_n V_n}{\Sigma p_n V_b} 100$
Price based	8	$\frac{\Sigma p_n V_b}{\Sigma p_b V_b} 100$	$\frac{\Sigma p_n V_n}{\Sigma p_b V_n} 100$
		Fisher	
Volume based	9	$100 \left(\frac{\Sigma p_b V_n}{\Sigma p_b V_b} \frac{\Sigma p_n V_n}{\Sigma p_n V_b} \right)^{1/2}$	
Price based	10	$100 \left(\frac{\Sigma p_n V_b}{\Sigma p_b V_b} \frac{\Sigma p_n V_n}{\Sigma p_b V_n} \right)^{1/2}$	
		Edgeworth	
Volume based	11	$\frac{\Sigma p V_n}{\Sigma p V_b} 100$ where $p = \frac{1}{2}(p_b + p_n)$	
Price based	12	$\frac{\Sigma p_n V}{\Sigma p_b V} 100$ where $V = \frac{1}{2}(V_b + V_n)$	

while the denominator represents the revenue that would have obtained in the base year with current price levels. As we see, 7(A) completely ignores current prices, while 7(B) ignores base year prices. It is similarly possible to devise price-based Laspeyres and Paasche indices, as shown by the formulae 8(A) and 8(B) in Table 3.9.

Of the Laspeyres and Paasche indices, the former is often favored because the weights relating to the base year remain constant when the current year moves forward and hence they are known in advance, whereas weights attributed to the current year may not always be readily available, so that computations of the index then have to be delayed until all the statistical data have been assembled. For example, the problem of measuring the level of a multi-product output for a plant is mirrored at the national level, when we seek to determine the degree to which total production volume rises or declines from year to year. An index of industrial production is therefore needed as an indicator of the general state of production in the national economy, though the make-up of the various industrial sectors and the product-mix may undergo drastic changes from one year to another. For this output index the Laspeyres quantity index is commonly used, where it is convenient to have the denominator in formula 7(A) constant, irrespective of which year n is compared with the base year.

One way to combine the Laspeyres and Paasche volume-based approaches is to take their arithmetic average value, similar to the approach in index 1(C) in Table 3.8, namely

$$Y_n(7C) = \tfrac{1}{2}[Y_n(7A) + Y_n(7B)]$$

and

$$Y_n(8C) = \tfrac{1}{2}[Y_n(8A) + Y_n(8B)]$$

Another alternative is due to the economist Irving Fisher (1867–1947), who proposed his "ideal index", known as the *Fisher Index* and provided by the formula 9 in Table 3.9. This index is merely the geometric mean of Laspeyres and Paasche and yields a value between the two. A similar price-based compromise is given in formula 10 as an alternative to the average index 8(C). But because the Fisher Index is cumbersome to compute, a reasonable approximation is provided by the *Edgeworth Index*, named after Francis Edgeworth (1845–1926), where for the volume index (in formula 11) the average price is taken as the constant price, both in the numerator and in the denominator*, and a similar approach using average volumes is adopted for the price index (in formula 12).

The volume-based indices 7, 9 and 11 for our two-product example are computed as follows:

7(*A*) *Laspeyres*

For year 2:

$$Y_2(7A) = \frac{\Sigma p_1 V_2}{R_1} 100 = \frac{(4 \times 60) + (2 \times 40)}{360} 100 = 88.9$$

and for subsequent years:

$$Y_3(7A) = \frac{(4 \times 80) + (2 \times 60)}{360} 100 = 122.2$$

$$Y_4(7A) = \frac{(4 \times 100) + (2 \times 45)}{360} 100 = 136.1$$

$$Y_5(7A) = \frac{(4 \times 120) + (2 \times 50)}{360} 100 = 161.1$$

* S. Fabricant followed the same approach in his numerous studies of industrial output in the US, for example in his publication "The Output of Manufacturing Industries, 1898–1937", National Bureau of Economic Research, 1940 (see also [4]).

These results are exactly the same as those obtained for the volume index 2(A). The reason for this is that for year n the index 2(A), as stated in Table 3.8, is

$$Y_n(2\text{A}) = \frac{\Sigma \dfrac{V_n}{V_1} R_1}{\Sigma R_1} 100$$

but $\dfrac{R_1}{V_1} = p_1$, hence

$$Y_n 2(\text{A}) = \frac{\Sigma p_1 V_n}{\Sigma R_1} 100 = Y_n(7\text{A})$$

7(*B*) *Paasche*

For year 2:

$$Y_2(7\text{B}) = \frac{R_2}{\Sigma p_2 V_1} 100 = \frac{384}{(4.4 \times 75) + (3 \times 30)} 100 = 91.4$$

and similarly,

$$Y_3(7\text{B}) = \frac{500}{(4 \times 75) + (3 \times 30)} 100 = 128.2$$

$$Y_4(7\text{B}) = \frac{680}{(5 \times 75) + (4 \times 30)} 100 = 137.4$$

$$Y_5(7\text{B}) = \frac{804}{(5.2 \times 75) + (3.6 \times 30)} 100 = 161.4$$

9 *Fisher*

For year 2:

$$Y_2(9) = [Y_2(7\text{A}) Y_2(7\text{B})]^{1/2} = (88.9 \times 91.4)^{1/2} = 90.2$$

and for subsequent years:

$$Y_3(9) = (122.2 \times 128.2)^{1/2} = 125.2$$

$$Y_4(9) = (136.1 \times 137.4)^{1/2} = 136.7$$

$$Y_5(9) = (161.1 \times 161.4)^{1/2} = 161.3$$

11 *Edgeworth*

For year 2 the average prices for products I and II respectively are

$$p_{\text{I}} = \tfrac{1}{2}(4.0 + 4.4) = 4.2$$

$$p_{\text{II}} = \tfrac{1}{2}(2.0 + 3.0) = 2.5$$

Hence

$$Y_2(11) = \frac{(4.2 \times 60) + (2.5 \times 40)}{(4.2 \times 75) + (2.5 \times 30)} 100 = 90.3$$

Similarly, for subsequent years:

$$Y_3(11) = \frac{(4 \times 80) + (2.5 \times 60)}{(4 \times 75) + (2.5 \times 30)} 100 = 125.3$$

$$Y_4(11) = \frac{(4.5 \times 100) + (3 \times 45)}{(4.5 \times 75) + (3 \times 30)} 100 = 136.8$$

$$Y_5(11) = \frac{(4.6 \times 120) + (2.8 \times 50)}{(4.6 \times 75) + (2.8 \times 30)} 100 = 161.3$$

These results are very close to those obtained from the Fisher formula.

Comments

All these results (as well as the price-based indices 8, 10, 12) are summarized in Table 3.10, where 32 indices are listed. This formidable array of alternatives demonstrates the wide variety of choice available to analysts (needless to say, many other formulae can be devised) and hence the care with which such choice must be exercised, depending on the purpose for which the index is to be used. The following are some general comments about the indices in Table 3.10.

1. Three types

In order to emphasize their purpose, the indices in Table 3.10 are grouped under three headings: volume, price and revenue. Volume indices are meant for the measurement of changes in physical output for a multi-product system, while price and revenue indices attempt to measure the price and revenue levels respectively of the output. In each case the index purports to compare like with like and the different formulae reflect the various interpretations of what "like with like" means.

The first index of each type in Table 3.10 is the result of a simple summation—total physical volume, average price (which is the total price divided by the number of products), total revenue—and of these only the latter has a meaning in economic terms. Both total physical volume and average price suffer from the "apples and pears syndrome" and are meaningless for most analytical purposes, although it must be stated that the total physical volume is often found in company statements and is quoted by commentators as a general indication of the total level of activity. As we have seen, however, this aggregate measure can be very misleading. For example, if the total tonnage produced by a steel company declines in conjunction with a drastic change in the product-mix in favor of specialized and high-value steels, then the reduction in total tonnage may give the impression of a contraction, whereas the total revenue and profitability may have risen. Similarly, a change in the total mileage flown by an airline may not tell us much about its performance, if the balance between short and long journeys (again, the product-mix) changes as well. In short, the total physical volume is a useful measure of output when the product-mix remains unchanged, but can be quite meaningless otherwise.

The second index in each group, computed as the average of individual indices, is also open to the charge of not distinguishing between "apples and pears" by allowing products with a trivial contribution to the business to have the same status as the major products in the composition of an index. This is why there are hardly any circumstances in which the average of individual indices of quantity, price or revenue can be justified.

2. The revenue indices

As already stated, the total revenue index is a measure of the total output in financial terms (see Fig. 3.2), but is a poor measure of physical output or price levels when prices and quantities are subject to change. It is thus possible for physical output to decline and yet for the aggregate revenue to rise (even when the product-mix does not change), and vice versa. Even when the analyst's aim is to measure total revenue, the revenue index in Table 3.10 is often regarded as inadequate, if the change in prices is largely the result of inflation, since the value of money in year 5 is depreciated compared with year 1.

Rather than rely on historical figures, many analysts prefer to compare them with an inflation index. Suppose that the general price index, as published by government agencies, is known for the five-year period; we can first normalize it, so that it reads 100 in year 1 (this is obtained by dividing the index by its value in year 1 and multiplying by 100) and then dividing the revenue index by the normalized price index and multiplying by 100. This process is shown in Table 3.11. In relation to the price index (i.e. in year 1 price terms) the adjusted revenue shows a decline of 3 % in year 2 (going down from 100 to 97) and improving by 39.3 % in year 5, whereas in historical terms the rise in year 5 amounted to 123.3 %.

The exercise in Table 3.11 immediately raises the question of which inflation index to apply, since it must be obvious from the foregoing discussion that, when it comes to the use of an index, there are many alternatives to choose

TABLE 3.10

Various indices for the two-product example.

	Derived from	Year 1	2	3	4	5
Volume indices						
total physical volume	Table 3.5	100	95.2	133.3	138.1	161.9
average of individual indices	Table 3.5	100	106.7	153.3	141.7	163.3
price adjusted (A)	1(A), Table 3.8	100	97.8	137.8	138.9	162.2
price adjusted (B)	1(B), Table 3.8	100	101.6	146.7	140.7	162.7
price adjusted (C)	1(C), Table 3.8	100	99.7	142.2	139.8	162.5
revenue adjusted (A)	2(A), Table 3.8	100	88.9	122.2	136.1	161.1
(B)	2(B), Table 3.8	100	96.7	140.3	137.7	161.5
(C)	2(C), Table 3.8	100	92.8	131.2	136.9	161.3
Laspeyres	7(A), Table 3.9	100	88.9	122.2	136.1	161.1
Paasche	7(B), Table 3.9	100	91.4	128.2	137.4	161.4
Fisher	9, Table 3.9	100	90.2	125.2	136.7	161.3
Edgeworth	11, Table 3.9	100	90.3	125.3	136.8	161.3
Price indices						
average price	Table 3.6	100	123.3	116.7	150.0	146.7
average of individual indices	Table 3.6	100	130.0	125.0	162.5	155.0
volume adjusted (A)	3(A), Table 3.8	100	121.4	114.3	146.4	144.3
volume adjusted (B)	3(B), Table 3.8	100	126.0	121.4	148.3	144.7
volume adjusted (C)	3(C), Table 3.8	100	123.7	117.9	147.4	144.5

revenue adjusted (A)	4(A), Table 3.8	100	116.7	108.3	137.5	138.3
revenue adjusted (B)	4(B), Table 3.8	100	122.5	118.0	144.9	141.2
revenue adjusted (C)	4(C), Table 3.8	100	119.6	113.2	141.2	139.8
Laspeyres	8(A), Table 3.9	100	116.7	108.3	137.5	138.3
Paasche	8(B), Table 3.9	100	120.0	113.6	138.8	138.6
Fisher	10, Table 3.9	100	118.3	111.0	138.1	138.5
Edgeworth	12, Table 3.9	100	118.2	111.2	138.2	138.5
Revenue indices						
total revenue	Table 3.7	100	106.7	138.9	188.9	223.3
average of individual indices	Table 3.7	100	144.0	203.3	233.3	254.0
volume adjusted (A)	5(A), Table 3.8	100	120.0	161.9	204.8	234.3
volume adjusted (B)	5(B), Table 3.8	100	132.8	189.5	208.0	235.1
volume adjusted (C)	5(C), Table 3.8	100	126.4	175.7	206.4	234.7
price adjusted (A)	6(A), Table 3.8	100	125.3	171.1	211.1	238.7
price adjusted (B)	6(B), Table 3.8	100	133.4	189.5	225.9	245.6
price adjusted (C)	6(C), Table 3.8	100	129.4	180.3	218.5	242.2

TABLE 3.11
Taking account of inflation in the revenue index.

	Year				
	1	2	3	4	5
Revenue	360	384	500	680	804
Revenue index	100	106.7	138.9	188.9	223.3
Price index	136	149.6	167.5	192	218
Normalized price index	100	110	123.2	141.2	160.3
Adjusted revenue index	100	97	112.8	133.8	139.3

from: the retail price index (the RPI, which attempts to measure the general level of prices for a typical basket of goods purchased by the general public), the wholesale price index, the price index of goods produced by the industry in which the company is operating, the price index of goods purchased by that industry to maintain its operations, the wages and salaries index (in the economy as a whole, or in the particular industry), and so on. While the retail price index is the one often chosen for comparative purposes, there are clearly advantages and disadvantages associated with any price index in current use.

The other six revenue indices listed at the bottom of Table 3.10 attempt to adjust the total revenue index by volume or price weights, although their procedures would be regarded by many as unacceptable for reasons that are further discussed in the next section. It should also be observed from the results in Table 3.10 that in our example the values for indices 5 and 6 consistently exceed those of the total revenue index, so that any notion that the volume or price weights in indices 5 and 6 can adjust for inflation are obviously erroneous. Another point to note is that the categorization of indices in Table 3.10 is somewhat arbitrary in that all the aggregative indices in Table 3.9 may be regarded as revenue adjusted indices (as discussed later), but they are labeled as volume or price indices in Tables 3.9 and 3.10 to emphasize the purpose which they aim to measure.

3. Volume and price adjusted indices

These adjusted indices (i.e. the four indices 1, 3, 5 and 6 described in the first two columns in Table 3.8) have either the sum of physical volumes or prices in their denominators; as we have already seen this "mixed bag aggregation" is a dubious proposition, although the criticism cannot be leveled with equal force at all the four indices. Let us examine them in further detail.

The volume index 1(A) with year 1 as the base can be written as

$$Y_n(1A) = \frac{\Sigma \frac{V_n}{V_b} p_b}{\Sigma p_b} 100$$

$$= \Sigma \frac{V_n p_b}{V_b \Sigma p_b} 100$$

Substitute $\Sigma p_b = m \bar{p}_b$
where $\bar{p}_b$ = average unit price for all the products in the base year
m = the number of products in the product mix
Hence,

$$Y_n(1A) = \frac{1}{m} \Sigma \frac{p_b V_n}{\bar{p}_b V_b} 100$$

This index may, therefore, be regarded as the average of a series of m indices. Each of the terms in the series corresponds to a particular product. The numerator denotes the revenue that would have obtained for that product in year n at the base year price, while the denominator stands for the revenue in the base year for that product, had the average price $\bar{p}_b$ applied. The meaning of each index in the series is highly questionable, and we have already referred to the weakness of any method that relies on averaging a series of individual indices. The shortcomings of the price adjusted index 1(A) stems from the sum of prices in the denominator, and these criticisms may, therefore, be extended to index 1(B), as well as to 6(A) and 6(B), and hence to the average indices 1(C) and 6(C).

The volume adjusted indices in the first column of Table 3.8 fare not much better. Index 3(A) for year n with year b as the base is

$$Y_n(3A) = \frac{\Sigma \frac{p_n}{p_b} V_b}{\Sigma V_b} 100$$

Now, $p_n V_b$ denotes the revenue in year n at constant volumes V_b, and when that revenue is divided by the price p_b we get the notional volume in the base year resulting from the effect of change in price from p_b to p_n. The numerator in the index is the sum of these notional volumes and it is compared with the actual total physical volume in the base year. The objection to the use of the undifferentiated sum of physical volumes in the denominator has already been alluded to. Similarly, the index 3(B) can be seen as

$$Y_n(3B) = \frac{\Sigma \frac{p_n}{p_b} V_n}{\Sigma V_n} 100 = \frac{\Sigma \frac{R_n}{p_b}}{\Sigma V_n} 100$$

Here R_n/p_b is the notional volume in year n had the base price p_b been in operation, and again a comparison is made between the sum of notional volumes with the actual total volume, except that now the point of reference is the current year.

If we now turn to the revenue index 5(A), we find that

$$Y_n(5A) = \frac{\Sigma \frac{R_n}{R_b} V_b}{\Sigma V_b} 100 = \frac{\Sigma \frac{R_n}{p_b}}{\Sigma V_b} 100$$

The numerator may be interpreted as the volumes that would have obtained for the various products in year n at base year prices, so that the index compares the sum of these notional volumes with the actual total physical volume in the base year. We also note that after substituting $R_n = p_n V_n$ this index can be written as

$$Y_n(5A) = \frac{\Sigma \frac{p_n}{p_b} V_n}{\Sigma V_b} 100$$

where the numerator is identical to that of 3(B), so that while 5(A) takes the total physical volume in year b as the baseline, index 3(B) compares the same notional volume with the total in year n.

As for index 5(B), we have

$$Y_n(5B) = \frac{\Sigma \frac{R_n}{R_b} V_n}{\Sigma V_n} 100 = \frac{\Sigma \frac{R_n}{R_b/V_n}}{\Sigma V_n} 100$$

Here R_b/V_n represents the notional price in the base year if the volume V_n had applied. Dividing R_n by this notional price gives a notional volume for each product in year n and the numerator in this index is the sum of all these notional volumes, compared with the total volume in year n in the denominator.

It is interesting to note that while in Tables 3.8 and 3.10 index 3 is called a price index and index 5 a revenue index, they are both guided by the idea of comparing a sum of notional volumes with the actual total volume, either for the base year or for the current year. In a sense, therefore, both may be regarded as volume indices and the classification adopted in Table 3.8 may be accepted as a matter of semantic convenience.

4. The reference year

The proliferation of indices in Table 3.8 is exacerbated by the two computational methods specified in Tables 3.8 and 3.9: method (A) has its reference point in the base year, while (B) refers to the current year. The results for our example in Table 3.10 show that for any given index the differences between the numerical values of (A) and (B) are often relatively small, but there are cases where the differences are quite substantial. For example, indices 1(A) and 1(B) do not even agree on whether the output increased or decreased in year 2, and indices 2(A) and 2(B) differ greatly on the result in year 3, as do indices 3(A) and 3(B). These differences reflect the different contributions made by the adjusting factors and the changes that they undergo from the base year to the current year.

5. The revenue adjusted indices

The index 2(A) is a volume index, revenue adjusted (see Table 3.8). It should be noted that the adjustment is, in fact, guided by the revenue proportions of the various products, since

$$Y_n\,(2\text{A}) = \frac{\Sigma \dfrac{V_n}{V_b} R_b}{\Sigma R_b} 100$$

$$= \Sigma\left(\frac{V_n}{V_b} 100\right)\left(\frac{R_b}{\Sigma R_b}\right)$$

Thus $(V_n/V_b)\,100$ is the volume index for each individual product, and this value is weighted by $R_b/\Sigma R_b$ which is the proportion of the total revenue in the base year generated by a given product. The sum of all the revenue proportions is 1. In our example in Table 3.7 the revenue proportions in the base year were:

$$\text{product I} \quad \frac{R_b}{\Sigma R_b} = \frac{300}{360} = 0.833$$

$$\text{product II} \quad \frac{R_b}{\Sigma R_b} = \frac{60}{360} = 0.167$$

Total 1.000

It may, therefore, be said that the method behind index 2(A) is an attempt to overcome the weakness of averaging the individual indices for the various products by applying weights to these indices, in order to take account of the relative importance of the products in the product-mix. This relative importance is signified by the revenue proportions. In our example, the revenue for product I in the base year was five times higher than that of product II and it is therefore proper that volume changes of product I should be given greater prominence. Notice that as the weights in 2(A) relate to the

TABLE 3.12
Index 2(A) for the two-product example.

	Year 1	2	3	4	5
Individual volume indices V_n/V_1					
product I	100	80	106.7	133.3	160
product II	100	133.3	200	150	166.7
Revenue weights $R_1/(\Sigma R_1)$ in year 1					
product I	0.833				
product II	0.167				
Hence Y (2A)	100	88.9	122.3	136.1	161.1

base year, they remain constant throughout the time horizon, and the computation can follow the simple procedure in Table 3.12.

Index 2(B) is very similar in concept, namely

$$Y_n(2B) = \frac{\Sigma \frac{V_n}{V_b} R_n}{\Sigma R_n}$$

$$= \Sigma \left(\frac{V_n}{V_b} 100 \right) \left(\frac{R_n}{\Sigma R_n} \right)$$

so that the revenue proportions pertain to the current year. The computational procedure is similar to that in Table 3.12, except that the revenue weights no longer remain constant but need to be determined afresh for each year. As we can see from the results in Table 3.10, the difference between the two indices 2(A) and 2(B) can sometimes be significant, and the average index 2(C) is a compromise between the two, to take account of revenue proportions both in the base year and in the current year.

Price indices with revenue proportions as weights are similarly constructed in 4(A) for the base-year revenue proportions and in 4(B) for the proportions in the current year, while 4(C) is the average between the two.

6. The Laspeyres, Paasche, Fisher and Edgeworth indices

The indices in Table 3.8 are somewhat cumbersome to compute, particularly when a large number of products is involved. By contrast, the Laspeyres and Paasche indices in Table 3.9 are much simpler and therefore more commonly found. We note that the Laspeyres volume index 7(A) is identical to the volume index (revenue adjusted) 2(A) and that the Laspeyres price index 8(A) is identical to the price index (revenue adjusted) 4(A).

"The standard practice in constructing index numbers of quantity (or price) is to relate each quantity (or price) to a base reference value" [1, p. 702]. The basic idea is that of *fixed weight aggregation*: the Laspeyres index compares a weighted revenue (or expenditure) figure to actual revenue (or expenditure) in the base year, whereas Paasche compares actual current revenue (or expenditure) with a weighted figure. In either case, the volume indices are used to measure changes in the physical output, while the price indices attempt to measure changes in the price level. It is interesting to note the symmetry in the definitions in Table 3.9: The denominator in the Laspeyres volume index is the same as that in the price index (both being $\Sigma p_b V_b$ or R_b), and similarly the numerator for both the Paasche indices is the same (i.e. $\Sigma p_n V_n$ or R_n). The weighted figures in the numerator for the Laspeyres index are produced by weighting—for the volume index—the current volumes by the base prices, and similarly—for the price index—the current prices by the base volumes. For Paasche the procedure is reversed, so that the denominator for the Paasche volume (or price) index is the same as the numerator in the Laspeyres price (or volume) index. We thus have two pair of aggregate values:

Actual revenues (or expenditures)

base year $\Sigma p_b V_b = R_b$
current year $\Sigma p_n V_n = R_n$

Weighted revenues (or expenditures)

base prices and current volumes $\Sigma p_b V_n$
base volumes and current prices $\Sigma p_n V_b$

When each of the first pair is compared with each of the second, four indices emerge, as follows (see Table 3.9):

Actual revenues	$\Sigma p_b V_b = R_b$ (denominator)	$\Sigma p_n V_n = R_n$ (numerator)
Weighted revenues	(numerator)	(denominator)
$\Sigma p_b V_n$	Laspeyres volume 7(A)	Paasche price 8(B)
$\Sigma p_n V_b$	Laspeyres price 8(A)	Paasche volume 7(B)

From this table it is clear that

$$Y(7\text{A})Y(8\text{B}) = Y(7\text{B})Y(8\text{A}) = \frac{R_n}{R_b}$$

The generally accepted way to reconcile Laspeyres and Paasche is through the Fisher index in Table 3.9, but because it is somewhat cumbersome to compute, the Edgeworth method is recognized as a very good approximation [3]. The logic of the Edgeworth formula is based on a comparison of weighted current revenue with weighted base revenue: the volume index (formula 11 in Table 3.9) relies on constant prices (taken as the average between the current price and the base year price), and the price index (formula 12 in Table 3.9) relies on constant volumes (also taken as the average). Objections may, of course, be raised as to the efficacy of using the average price or average volume in these formulae, particularly when a long period of time has elapsed since the base year and when prices (in the case of formula 11) or volumes (in the case of formula 12) have greatly changed; nevertheless, the Edgeworth volume method is generally regarded as a good measure for "an economically oriented concept of physical output", because—as Gold puts it—"it weights the quantity of each product by the economic value commanded by it" [3, p. 53], and similar comments may be made about the Edgeworth price index.

7. Cost of living index

As stated by Stigler, "the percentage change in the cost of living (index of consumer prices) of a family is defined as the percentage change in expenditure (from one year to another, or from one place to another) necessary to maintain the same standard of living" [6, p. 87]. While it is natural to think in terms of what the family buys, in trying to evaluate its standard of living, it is clear that what it buys must relate to its hierarchy of tastes and preferences, to the price level, to its ability to afford the range of goods and services on offer, and to the perception of their utility to the consumer.

The many arguments about the cost of living index, and what it should measure or purport to measure, are exemplified by the view of Franklin Fisher and Karl Shell that it is impossible to ascertain the "level of true utility (whatever that may mean) . . . one never steps into the same river twice, and the comparison between a man's utility and his utility yesterday stands precisely on the same lack of footing as the comparison of two different men" [5, p. 18]. They go on to say "The true cost-of-living index does not answer the question 'how much income would it take today to make me as well off as I was yesterday with yesterday's budget constraint?' That question is unanswerable. A similar-seeming question which *can* be answered is 'how much income is required *today* to make me just as indifferent between facing yesterday's budget constraint and a budget constraint defined by today's prices and the income in question?' The latter question refers to a choice which can in principle be posed; the former does not" [5, p. 19]. Whether such questions can be posed or not, many will doubt the meaning and validity of the answers.

As it is difficult to take account of all these considerations even for a given family, let alone for the population as a whole, the problem is commonly reduced to that of constructing a price index, based on a basket of goods that the "average" family purchases. This basket is determined by a census or a survey and, if it is assumed to remain constant, then a price index such as Laspeyres' reveals the expenditure involved in purchasing the basket in any given year compared with the expenditure for the same basket in the base year. This price index is interpreted as a cost of living index. In principle, it is again possible to carry out the calculations with reference to the base year (Laspeyres) or to the current year (Paasche), though "almost all cost of living indexes are computed by the Laspeyres formula, because this is cheaper than making continuing studies of family budgets" [6, p. 90], and as prices tend to increase, the index has an upward trend and is larger than the true index of the cost of living. Fisher and Shell also note that the general practice is to use the Laspeyres index with base-period volume weights, but add the comment "In the case of no taste change, a frequently encountered proposition is that a Laspeyres index overstates price rises and a Paasche index understates them. If tastes change, however, and if we agree that it is the current-taste cost-of-living in which we are interested, a Laspeyres index loses much of its meaning" [5, pp. 21–2]. Thus, Laspeyres is often regarded as an upper bound for a true cost-of-living index, while Paasche remains a lower bound for current tastes, but not for base-year tastes.

Another point to bear in mind with the Laspeyres index is that as the base year recedes into the distant past, questions are bound to arise about the validity of a basket of goods defined long ago and it is common to find that account is taken of changing purchasing habits by amending the basket every year, for example by using the moving average quantity for each item in the basket over a two- or three-year period. In this sense the index begins to incorporate an element of the Edgeworth approach and ceases to be a pure Laspeyres index.

8. The effect of quality change

One of the major difficulties encountered in the construction of an index (and this is perhaps at the root of major criticisms voiced by economists and statisticians against many indices in current use) is how to account for changes in quality. With the passage of time, a given product is modified in specifications, in methods of manufacture, in the properties of materials used.

A good example is found in the automobile industry, where the introduction of product improvement is the name of the game and where, apart from some discrete major innovations that take place from time to time (such as automatic transmissions), there is a constant stream of design modifications and new gadgets to improve the appearance, performance and appeal of the automobile. Some accessories, which at one time are regarded as extra, may subsequently become part of the standard model, so that the definition of what is an automobile and what is its price needs to be adjusted (for example, by deflating the observed price by the estimated degree of quality change, or by incorporating a later improvement in the base-year model), thereby attempting to compare "like with like". In a study under the delightful title "hedonic price indexes for automobiles", Griliches concludes that "over the whole period since 1937, the CPI (Consumer Price Index) may be overestimating the rise in automobile prices by at least a third. Since the CPI is a Laspeyres index, the appropriate quality adjustment should also be based on 'base' period weights. If this is done, about three-fourths of the rise in automobile prices in the CPI since 1937 could be attributed to quality improvements" [5, p. 57]. He goes on to highlight the fact that the choice of a base year in such circumstances can have a significant effect on the results and that the Laspeyres and Paasche indices may then diverge sharply [5, p. 59].

9. Discontinuities in the product-mix

One of the problems encountered in practice is the computation of an index number when there is a discontinuity in the product-mix over a given time period. In one case study of a steel plant, for example, it was found that the production of some products had been discontinued after the base year and that new lines had been introduced [3, p. 117]. The problem may be demonstrated by the example in Table 3.13, where the product-mix in year 1

TABLE 3.13
An example of discontinuities in the product-mix.

	Year			
	1	2	3	4
Physical quantities				
product I	100	80	40	–
product II	50	60	65	64
product III	–	–	20	30
Prices				
product I	2.0	2.2	2.5	–
product II	4.0	4.5	4.8	5.0
product III	–	–	6.0	6.4
Average prices $p = \frac{1}{2}(p_1 + p_n)$				
product I		2.1	2.25	?
product II		4.25	4.4	4.5
product III		?	?	?
Total revenue	400	446	532	512
Revenue index	100	111.5	133	128

consisted of two products, but while production of product I declined over the years and ceased in year 4, a new product was introduced in year 3. If we wish to apply the volume Edgeworth formula as a measure of output, we need to compute the average prices, and we find that there are gaps in the average price table owing to discontinuities in the data record (see Table 3.13). The problem can be resolved in one of the following ways:

(1) ***Assume that the missing price is identical to that prevailing at the nearest point in time***. In the case of product I this means that $p_4(\mathrm{I}) = p_3(\mathrm{I}) = 2.5$ where $p_4(\mathrm{I})$ and $p_3(\mathrm{I})$ are the prices for product I for years 4 and 3 respectively. Hence the average price for product I for year 4 is equal to that for year 3. Similarly, we assume that $p_1(\mathrm{III}) = p_2(\mathrm{III}) = p_3(\mathrm{III}) = 6.0$, so that the average prices for product III for years 2, 3 and 4 become 6.0, 6.0 and 6.2 respectively. The disadvantage of this method is that it relies on hypothetical prices that may not have any basis in reality, as we do not know whether the unit price 6.0 for product III would have been a fair and realistic price to quote, had this product been available for sale in years 1 and 2. In particular, this method is oblivious to the effect of inflation. As we can see from Table 3.13, the price trends are upwards throughout, presumably under the influence of inflation, and such price trends should not be ignored in filling the gaps in the price table.

(2) ***Account for product price trends***. To overcome the shortcomings just alluded to, an assessment can be made of trends in the price data. When the prices in Table 3.13 are plotted (see Fig. 3.4), the upward trend becomes evident and by extrapolation we can estimate for product I that $p_4(\mathrm{I}) = 2.7$, and for product III that $p_1(\mathrm{III}) = 5.2$ and $p_2(\mathrm{III}) = 5.6$. The missing average prices in Table 3.13 can now be computed and the Edgeworth formula applied. The assumption in this method is that the trends extend beyond the known data, so that extrapolation is valid. Also, sufficient information is required to allow the trends to be established. In our case, we have only two datum points for product III, a rather shaky foundation on which to postulate a trend. The situation is even worse for newly introduced products, when only one reference point (the current price) is known. The problem of extrapolation is further exacerbated with the passage of time: when we get to the tenth year, for example, the linear extrapolation for the price of product I would be 4.2 and it has to be asked whether such estimation is valid, based as it is on data that belong to the distant past. For product III there is the added nuisance that as we progress into the future, and as more price data become available, the trend line in Fig. 3.4 would have to be modified, giving fresh estimates for past prices in years 1 and 2, thereby affecting earlier calculations of output; thus, whatever output figure is computed for the current year, it is conditional

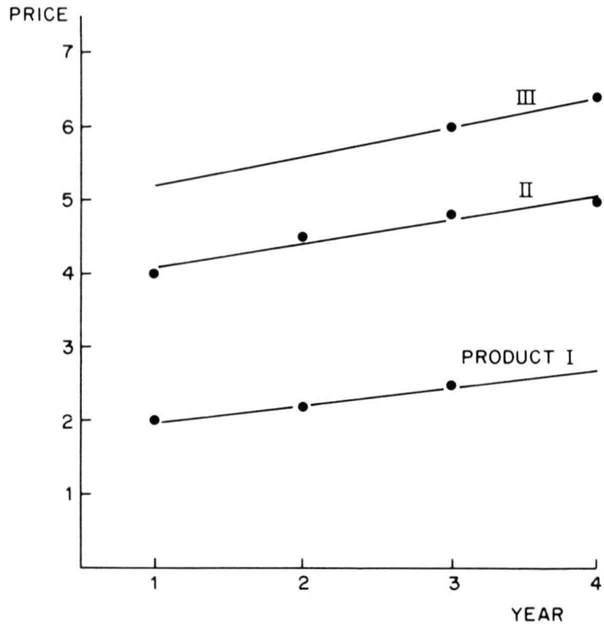

FIG. 3.4. Unit price extrapolation.

on future data not disturbing the estimated price for the base year.

(3) ***Assume market prices for the missing data.*** It may be possible to determine the price charged by competitors for similar products at the time when our products I and III were not available for sale, or use any other market information, to estimate surrogate prices for the gaps in Table 3.13.

(4) ***Use a price index as an adjustor.*** As in the case of adjusting historical revenue data to take account of inflation (see Table 3.11), we can apply a price index to estimate the missing price levels in Table 3.13. For example, if the normalized price index in Table 3.11 holds for the four-year span in Table 3.13, then $p_4(\mathrm{I})$ may be derived by applying the change in the price index between years 3 and 4 (namely, $\frac{Y_4}{Y_3} = \frac{141.2}{123.2}$ where Y_3 and Y_4 are the price index values for years 3 and 4 respectively) to the known price in year 3, i.e.

$$p_4(\mathrm{I}) = p_3(\mathrm{I})\,\frac{Y_4}{Y_3} = 2.5 \times \frac{141.2}{123.2} = 2.87$$

Since the price trend for product I may not necessarily follow the price index (and in our case it does not), the estimate for $p_4(\mathrm{I})$ may well depend on which year is selected for the calculation. If we select the price in year 2, then

$$p_4(\mathrm{I}) = p_2(\mathrm{I})\,\frac{Y_4}{Y_2} = 2.2 \times \frac{141.2}{110} = 2.82$$

and if year 1 is taken,

$$p_4(\mathrm{I}) = p_1(\mathrm{I})\,\frac{Y_4}{Y_1} = 2.0 \times \frac{141.2}{100} = 2.82$$

so that the last two estimates coincide because in our case the change in price from 2.0 in year 1 to 2.2 in year 2 does exactly follow the change in the normalized index from 100 to 110. Of the alternative estimates it is usual to prefer the first, since it is based on the latest available price information, unless there are grounds for suspecting that the information is fickle or unrepresentative of the general trend. Similarly, we can estimate the missing prices for product III for the first two years:

$$p_2(\mathrm{III}) = p_3(\mathrm{III})\,\frac{Y_2}{Y_3} = 6.0\,\frac{110}{123.2} = 5.36$$

$$p_1(\mathrm{III}) = p_3(\mathrm{III})\,\frac{Y_1}{Y_3} = 6.0\,\frac{100}{123.2} = 4.87$$

As in the case of the inflation-adjusted revenue computations, there are several price indices to choose from, and the results may well depend on which index is selected.

(5) ***Amalgamate product categories***. Since the specification of product groups is a matter of definition, it may often be possible (particularly when a large number of products is involved) to combine discontinued and new lines with other products to present a continuous historical record without the gaps in Table 3.13*. This process of amalgamating products is a matter of judgement regarding their technical similarity, their unit price and the degree of substitution in the market. In our example we do not have many product lines or sufficient information about the products for this method to be properly demonstrated, but if for illustrative purposes we assume that products I and III can be combined† into a new group labeled product IV, then for the period when an overlap occurred (in our case, year 3) the data for the combined new product group are reconstituted (here for year

* An example is cited in the case study in [3, pp. 117–8].

† In practice this would not be very likely, in view of the significantly different unit prices.

3) as follows:

$$\text{Revenue} = 40 \times 2.5 + 20 \times 6.0 = 220$$

$$\text{Physical volume} = 40 + 20 = 60$$

$$\text{Average unit price} = \frac{220}{60} = 3.67$$

It must be reiterated that there are circumstances when such an approach would be undesirable, for example when the products are rather different in their technical specification, in market appeal, in price, or in the type and mix of resources required for their manufacture [3, p. 118].

Comparison of the estimating methods To illustrate the effect of these five estimating methods on the value of the output, the Edgeworth index was computed for each and the results are summarized in Table 3.14. First, we find that the results are identical for year 2, since the estimated prices do not affect the Edgeworth computation for that year (in that respect p_2(III) becomes irrelevant). Secondly, it transpires that while at times the choice of an estimating procedure has little effect on the result (such as procedures (1)–(4) in year 3), this is certainly not always the case, as we can see from the index values for year 4. The risk involved in procedure (5) in amalgamating products of vastly different prices is also clearly demonstrated, where the high average price in year 4 for the combined product IV distorts the value of the output in the base year to such an extent that the resultant index is markedly different from the values derived by other methods. The moral of this story is that great care needs to be exercised in selecting an appropriate estimating procedure, and that it may well be advisable to discard a distant base year in favor of a more recent baseline, in order to mute the effect of rapidly changing circumstances on the computational results.

10. Base-transfer and time reversal properties

The base transfer property is clearly a convenient feature to maintain, so that the ratios between index values corresponding to given years remain unchanged with a change in the base year. As we have seen, these relationships easily obtain in the case of a single-product output, but when a compound variable is involved, the outcome may depend on which index is chosen and on the circumstances for which it is applied.

To illustrate this point, take the example of the two-product output in Table 3.4, for which the volume indices Laspeyres 7 (A), Paasche 7 (B), Fisher and Edgeworth are computed, with the base year being changed from year 1 (for which the results in Table 3.10 apply) to year 2, then year 3, etc. The index

TABLE 3.14
The Edgeworth index for four estimating procedures.

	Method	Year 1	2	3	4
Average prices					
product I	(1)		2.1	2.25	2.25
	(2)		2.1	2.25	2.35
	(3)		2.1	2.25	2.6
	(4)		2.1	2.25	2.43
product III	(1)		6.0	6.0	6.2
	(2)		5.4	5.6	5.8
	(3)		5.6	5.7	5.9
	(4)		5.11	5.68	5.88
product IV	(5)		2.1	2.83	4.2
Edgeworth index	(1)	100	100.1	111.5	105.3
	(2)	100	100.1	109.7	100.4
	(3)	100	100.1	110.1	95.9
	(4)	100	100.1	110.0	99.2
	(5)	100	100.1	90.6	64.2
Revenue index	Table 12	100	111.5	133	128
Inflation adjusted revenue index		100	101.4	108	90.7

NOTE: The following prices are assumed:

		p_4(I)	p_1(III)	p_2(III)	p_3(IV)
Procedure	(1)	2.5	6.0	6.0	
	(2)	2.7	5.2	5.6	
	(3)	3.2	5.4	5.8	
	(4)	2.87	5.36	4.87	
	(5)				3.67

Prices for (3) are based on market data.

values for each of the five years, taken to each of the five baselines, are summarized in Table 3.15. The last two columns in the table record two ratios (the first is the ratio of the index in year 3 to its value in year 1, and the second is the ratio for year 5 to year 1). For the base transfer property to hold, these ratios need to remain constant when a shift in the base year takes place, in order to satisfy Equation (3.8), but it is clear from the table that they do not. In some cases the effect is fairly minor (for example, the ratio of year 5 to year 1 for Laspeyres), but there are many instances where a change in the base year can affect the ratio by several percentage points (for example, the output in year 3 compared with year 1 according to Laspeyres rose by 22.2% if year 1 is taken as a base, but by 29.3% if year 4 becomes the baseline). Hence the base-transfer property does not hold (as commented upon in the application of the

TABLE 3.15

Effect of choice of base year.

	Year 1	2	3	4	5	Index yr 3 / Index yr 1	Index yr 5 / Index yr 1
Laspeyres—Index 7(A)							
Base year 1	100	88.9	122.22	136.11	161.11	1.222	1.611
2	109.38	100	138.54	149.74	176.56	1.267	1.614
3	78.00	72.00	100	107.00	126.00	1.282	1.615
4	72.79	67.65	94.12	100	117.65	1.293	1.616
5	61.94	56.72	78.61	84.83	100	1.269	1.614
Paasche—Index 7(B)							
Base year 1	100	91.43	128.21	137.37	161.45	1.282	1.614
2	112.50	100	138.89	147.83	176.32	1.235	1.567
3	81.82	72.18	100	106.25	127.22	1.222	1.555
4	73.47	66.78	93.46	100	117.89	1.273	1.605
5	62.07	56.64	79.37	85.00	100	1.279	1.611
Fisher—Index 9							
Base year 1	100	90.15	125.18	136.74	161.28	1.252	1.613
2	110.93	100	138.71	148.78	176.44	1.250	1.591
3	79.89	72.09	100	106.62	126.61	1.252	1.585
4	73.13	67.21	93.79	100	117.77	1.283	1.610
5	62.00	56.68	78.99	84.91	100	1.274	1.613
Edgeworth—Index 11							
Base year 1	100	90.26	125.33	136.84	161.31	1.253	1.613
2	110.80	100	138.71	148.70	176.43	1.252	1.592
3	79.79	72.09	100	106.58	126.68	1.253	1.588
4	73.08	67.25	93.83	100	117.77	1.284	1.611
5	61.99	56.68	78.94	84.91	100	1.273	1.613

Edgeworth index to a case-study in a steel plant [3, pp. 118–9]), and the effect of the base year may well be exacerbated when discontinuities in the product-mix occur.

The reason for the failure of the base-transfer property lies in the definition of each index. For example, if the Laspeyres index for year n_2 is compared with that in year n_1, both to base b_1, then we find that

$$y_{n_2 b_1} = \frac{\Sigma p_{b_1} V_{n_2}}{R_{b_1}}$$

$$y_{n_1 b_1} = \frac{\Sigma p_{b_1} V_{n_1}}{R_{b_1}}$$

so that

$$\frac{y_{n_2 b_1}}{y_{n_1 b_1}} = \frac{\Sigma p_{b_1} V_{n_2}}{\Sigma p_{b_1} V_{n_1}} \neq y_{n_2 n_1}$$

Similarly, in relation to base b_2,

$$\frac{y_{n_2 b_2}}{y_{n_1 b_2}} = \frac{\Sigma p_{b_2} V_{n_2}}{\Sigma p_{b_2} V_{n_1}} \neq y_{n_2 n_1}$$

and clearly

$$\frac{y_{n_2 b_1}}{y_{n_1 b_1}} \neq \frac{y_{n_2 b_2}}{y_{n_1 b_2}}$$

Similar considerations apply to the other indices in Table 3.15. On the other hand, the time reversal property does apply both to the Fisher and Edgeworth indices. In the case of Edgeworth the index value in year n to base b is

$$Y_{nb} = \frac{\Sigma p\, V_n}{\Sigma p\, V_b} 100$$

and when the years are reversed it becomes

$$Y_{bn} = \frac{\Sigma p\, V_b}{\Sigma p\, V_n} 100$$

so that the time reversal condition holds, whereas the more general base transfer property does not.

CONCLUSION

The discussion in this chapter first suggests that for the multi-product case there is no universal "true" index, and that the numerical value of different indices may vary, depending on their definitions. For the most part, indices are constructed to measure changes in revenue, or volume or price. The most common revenue indices are the pure index (which usually compares the single variable of revenue in a given year with that in the base year), or the pure index adjusted for inflation (by relating the pure index to a price index, such as the cost of living or the retail price index). Other adjustments to the revenue index by price or volume weights are often found wanting.

Of the many possible volume and price indices, the most common are those of Laspeyres, Paasche, Fisher and Edgeworth. Of the first two, Laspeyres is more commonly employed. The Fisher Index is the geometric average of the first two, and its value is usually very close to that of the Edgeworth formula, though the latter is much simpler to use and is generally to be preferred, particularly in conjunction with an analysis of plant or company performance. It should be borne in mind, however, that discontinuities in the product-mix and the selection of the base year for the chosen index may affect the results and have implications on their interpretation.

The base-transfer property is defined in Equation (3.5) and is rephrased in Equations (3.6), (3.7) and (3.8). When it applies, the time reversal property as defined in Equation (3.4) automatically holds as well. The base-transfer condition is met in the single product case but fails for a multi-product output, because of the definitions that are employed in the construction of the index numbers, though in the case of the Fisher and Edgeworth indices the time reversal property is maintained.

REFERENCES

[1] Allen, R. G. D. (1963). "Mathematical Economics". Macmillan, London (published in USA by St. Martin's Press).
[2] Cuddy, J. D. A. (1976). "International Price Indexation". Saxon House, D.C. Heath or Lexington Books.
[3] Eilon, S., Gold, B. and Soesan, J. (1976). "Applied Productivity Analysis for Industry". Pergamon Press.
[4] Gold, B. (1971). "Explorations in Managerial Economics". Macmillan, London.
[5] Griliches, Z. (ed.) (1971). "Price Indexes and Quality Change". Harvard University Press.
[6] Stigler, G. J. (1952). "The Theory of Price". Macmillan, London.

EXERCISES

1. For the single product case, prove the following relationships:

(a) $\dfrac{y_{n_2 b}}{y_{n_1 b}} = \dfrac{Y_{n_2 n_1}}{100}$

(b) $\dfrac{Y_{n_2 b_2}}{Y_{n_1 b_1}} = \dfrac{Y_{n_2 n_1}}{Y_{b_2 b_1}}$

(c) $Y_{n b_2} Y_{b_2 b_1} = Y_{n b_3} Y_{b_3 b_1}$

(d) $Y_{n n_1} Y_{n_1 n_2} Y_{n_2 n_3} Y_{n_3 n_4} \ldots Y_{n_m b} = Y_{n b}\, 100^m$.

2. Compute the index numbers for the data in Table 3.4 for the range of indices shown in Table 3.8 and compare your results with those shown in Table 3.10.
3. For the same data, compute the price indices 8 (A), 8 (B), 10 and 12 (see Table 3.8).
4. If for a multi-product case the base transfer property holds, prove that
   (a) the time reversal property holds
   (b) equation (3.7) holds.
5. Show that the time reversal property applies to the Fisher Index.
6. The following examples are taken from Stigler [6, pp. 88–90]:
   (a) A family buys the following combination of goods in two years:

	Quantity		Price	
	Year 1	Year 2	Year 1	Year 2
Bread	500	550	0.10	0.09
Beef	100	95	0.30	0.40

   What is the standard of living and the cost of living of the family in year 2 compared with year 1, following the Laspeyres Index?
   (b) If the expenditure figures for a second example are

	Quantity		Price	
	Year 1	Year 2	Year 1	Year 2
Bread	500	510	0.10	0.08
Beef	100	90	0.30	0.45

   how would you interpret the results following Paasche?
7. For the data in Table 3.13, complete the computations for the Edgeworth volume index for procedures (2) and (5) and check your results against those given in Table 3.14.

4
The Rules of Incremental Calculus

Themes in this chapter

Definition of the relative increment
The addition rule
Subtraction
Multiplication
Division
Sensitivity analysis
Iso-cost-increase maps
Numerous examples

ADDITION

Example

Consider the case of a company that generates revenues R_1 and R_2 in the home and export markets respectively. The total revenue R is

$$R = R_1 + R_2$$

Now, what happens to the total revenue when the home market expands by δR_1? The total revenue increases by δR, which in this case is the same amount as δR_1, provided the export market remains unchanged, but in percentage terms the increase in the total revenue R will be less than the percentage increase in R_1, the magnitude of the increase depending on how much of the total revenue R is contributed by the home market. The three cases in Table 4.1 illustrate this point, where the increase in R_1 in each case is 10 %. In case 1 the two markets yield equal revenues and the effect of increasing the home market by 10 % is to increase the total by 5 %. In case 2 the home market is relatively stronger and the effect of a 10 % increase in R_1 leads to an 8 % increase in total, while in case 3 the home market is the sole contributor to revenue and the 10 % increase in R_1 is fully reflected as a 10 % increase in R (since in this case R_1 and R are identical).

TABLE 4.1
Three examples of revenue from two markets.

	Home market R_1	Export market R_2	Total R	Increase in R_1	New total revenue	Increase in total revenue %
Case 1	80	80	160	8	168	5.0
2	80	20	100	8	108	8.0
3	80	0	80	8	88	10.0

The reason for these different effects obviously lies in the value of the total revenue: since the absolute increase in revenue in all three cases is the same ($\delta R_1 = 8$), the relative increase is smaller when the value of R is high than when it is small. We further observe from the three examples that the relative increase in R depends on the ratio R_1/R, and that when this ratio is multiplied by the relative increase in R_1 then the relative increase in the total is obtained. Specifically, in our case we find that

$$\frac{\delta R}{R} = \frac{R_1}{R}\frac{\delta R_1}{R_1}$$

where $\frac{\delta R}{R}$ = relative increase in the total

$\frac{R_1}{R}$ = proportion of the home market in the total, prior to the change

$\frac{\delta R_1}{R_1}$ = relative change in R_1.

The relative incremental change

Notice that our interest centers on the *relative changes that take place*, namely on the ratio of the absolute change, or increment, in a given variable to the absolute original value of that variable. When several variables are involved, we seek to establish the relationships between their relative increments, such as in the example cited above, where this relationship is expressed in the equation for $\delta R/R$. One reason for concentrating our attention on *relative increments*, rather than on absolute increments, emanates from the fact that relative changes (shown in percentage terms) are always quoted in various managerial reports and in published accounts (as already mentioned in Chapter 3). But a second, and more compelling, reason is that many relationships of a more general nature can thereby be established than those that depend on the absolute values of the variables involved.

The idea of relying on ratios is akin to the method of dimensionless analysis in engineering, which allows inferences to be made about the behavior of engineering systems on a large scale (such as machines, structures, even whole plants) from available data on their behavior on a smaller scale (obtained, for example, from experimental results in the laboratory, or from the operation of a pilot plant). Thus, the relationship between $\delta R/R$ and $\delta R_1/R_1$ in our example will remain valid irrespective of the absolute value of the total revenue R or its component R_1 (for example, in all cases that R_1 contributes half the total revenue, the relative increase in R will be half the relative increase in R_1). We are, therefore, interested in *incremental calculus* (strictly, it should be called the *calculus of relative increments*, but the shorter term incremental calculus will be used henceforth), the rules of which will help us establish the relationships between relative increments in a variety of circumstances*.

The addition rule

The *rule for addition* may be developed as follows: Let x_1 and x_2 be two variables that add up to form the sum y, so that

$$y = x_1 + x_2 \tag{4.1}$$

similar to the example of $R = R_1 + R_2$ mentioned above. Let the variable x_1 be subject to an incremental change δx_1 and x_2 to an incremental change δx_2, thereby changing the total y by δy. The relative changes that have taken place are denoted by starred variables as follows:

$$\left.\begin{aligned} x_1^* &= \frac{\delta x_1}{x_1} \\ x_2^* &= \frac{\delta x_2}{x_2} \\ y^* &= \frac{\delta y}{y} \end{aligned}\right\} \tag{4.2}$$

The following should be noted:

(1) The starred variables may be positive or negative, depending on the original values of the basic variables and on the increments. There are

* The development of the basic concepts of incremental calculus by the author date back to the early 1970s, see [1, 2]

many cases where the basic variables are always positive, for example when we consider data of costs or revenues. If a given variable x is always positive, then the sign of x^* corresponds to the sign of δx, i.e. $x^* > 0$ when $\delta x > 0$ and $x^* < 0$ when $\delta x < 0$. But when the basic variable can become negative, for example in the case of profit denoted by Z, then $Z^* > 0$ implies an increase in profit only if its original value was positive.

(2) No constraint is implied regarding the manner in which a change in a given variable takes place. The change may be gradual or abrupt. In other words, the analysis that we wish to pursue is not conditional on x in Fig. 4.1 being a continuous function with the passage of time. All we are concerned with here is comparing the values of x at two points in time and we note that x changes from x to $x+\delta x$, so that although the two cases in Fig. 4.1 obviously differ in the interim values for x (and the fact that x is continuous in one case but is subject to a discrete jump in the other), the value of x^* is the same for both. The only way in which the difference between the two cases can be revealed is by recording the values of x at interim points in time, for which the corresponding increments δx will no longer be identical.

Returning to the Equations (4.1) and (4.2), the situation before and after the changes is summarized below:

Before the change **After the change**

$$x_1 \rightarrow \qquad x_1+\delta x_1 = x_1\left(1+\frac{\delta x_1}{x_1}\right) = x_1(1+x_1^*)$$

$$x_2 \rightarrow \qquad x_2+\delta x_2 = x_2\left(1+\frac{\delta x_2}{x_2}\right) = x_2(1+x_2^*)$$

$$y \rightarrow \qquad y+\delta y = y\left(1+\frac{\delta y}{y}\right) = y(1+y^*)$$

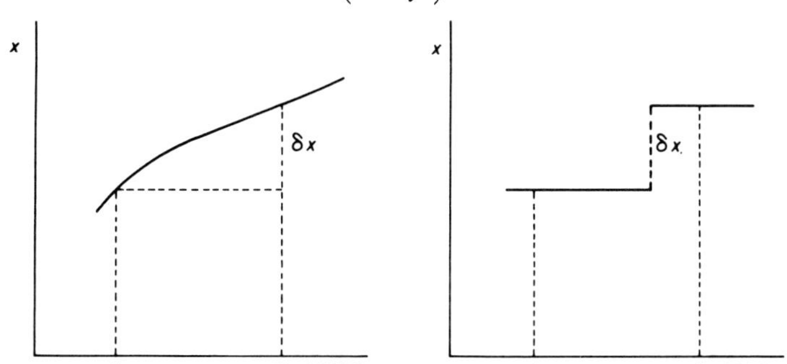

FIG. 4.1. A continuous and discrete change to variable x.

Now, the absolute increment in y is

$$\delta y = \delta x_1 + \delta x_2$$

and the relative increment

$$y^* = \frac{\delta y}{y} = \frac{\delta x_1 + \delta x_2}{x_1 + x_2} = \frac{x_1}{x_1 + x_2}\frac{\delta x_1}{x_1} + \frac{x_2}{x_1 + x_2}\frac{\delta x_2}{x_2}$$

If we substitute the definitions of x_1^* and x_2^* from (4.2), and also

$$\left.\begin{aligned} k_1 &= \frac{x_1}{x_1 + x_2} = \text{the proportion of } x_1 \text{ in the total prior to the change} \\ k_2 &= \frac{x_2}{x_1 + x_2} = \text{the proportion of } x_2 \text{ in the total prior to the change} \end{aligned}\right\} \quad (4.3)$$

where $$k_1 + k_2 = 1, \quad (4.4)$$

we obtain the result for the addition rule:

$$y^* = (x_1 + x_2)^* = k_1 x_1^* + k_2 x_2^* \quad (4.5)$$

where $(x_1 + x_2)^*$ denotes the relative change in the sum $x_1 + x_2$ (namely, y). Applying this result to the example for the total revenue given earlier, we obtain

$$R^* = \rho_1 R_1^* + \rho_2 R_2^* \quad (4.6)$$

Where

$$\left.\begin{aligned} \rho_1 &= \frac{R_1}{R} = \text{the revenue proportion from the home market} \\ \rho_2 &= \frac{R_2}{R} = \text{the revenue proportion from the export market} \end{aligned}\right\} \quad (4.7)$$

In case 2, for example, $\rho_1 = 0.8$ and $\rho_2 = 0.2$. For the special case where the export market remains unchanged, we have $R_2^* = 0$, so that

$$R^* = \rho_1 R_1^*$$

which is precisely the expression derived earlier for $\delta R/R$. Thus, in general terms the relative change of a sum $a + x$, where a is a constant (namely, $a^* = 0$), is given by

$$(a + x)^* = ka^* \qquad \text{where } k = \frac{x}{a + x} \quad (4.8)$$

Another notable special case is when the variables x_1 and x_2 change by the same relative amount, namely $x_1^* = x_2^*$, in which case the relative increment of the sum is

$$(x_1 + x_2)^* = (k_1 + k_2)x_1^* = x_1^*$$

since $k_1 + k_2 = 1$, and this result is independent of the value of k_1.

The way these special cases relate to the general case can be illustrated by rewriting Equation (4.5) as

$$\frac{(x_1 + x_2)^*}{x_1^*} = k_1 + (1 - k_1)\frac{x_2^*}{x_1^*} \tag{4.9}$$

The left-hand side is the ratio of the relative increment of the total $x_1 + x_2$ to the relative increment of the component x_1, this ratio being a linear function of the ratio x_2^*/x_1^* and shown in Fig. 4.2 for several values of x_2^*/x_1^*. The two

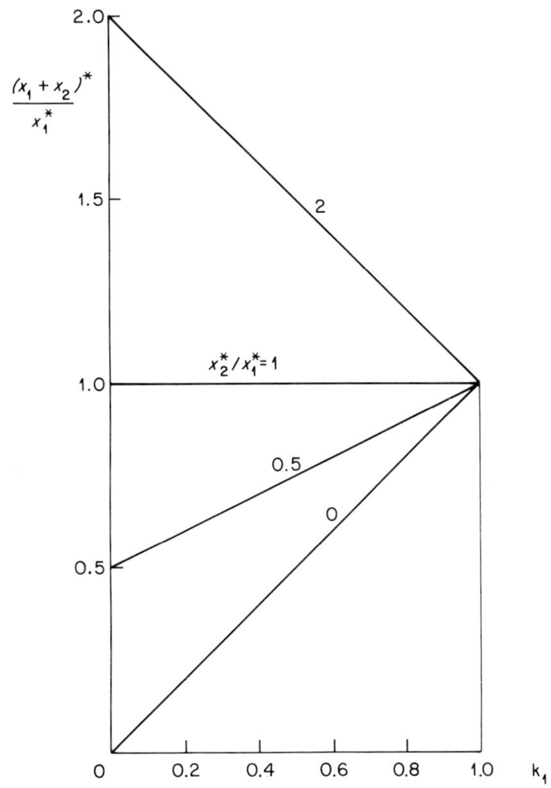

FIG. 4.2. The function $\dfrac{(x_1 + x_2)^*}{x_1^*}$.

special cases referred to above are represented by the lines $x_2^*/x_1^* = 0$ (which corresponds to $x_2^* = 0$) and $x_2^*/x_1^* = 1$ (where the result is not affected by the value of k_1) respectively.

It is not difficult to extend the addition rule to three variables and to show that

$$y = (x_1 + x_2 + x_3)^* = k_1 x_1^* + k_2 x_2^* + k_3 x_3^* \tag{4.10}$$

where

$$k_1 = \frac{x_1}{y}, \quad k_2 = \frac{x_2}{y}, \quad k_3 = \frac{x_3}{y} \quad \text{and} \quad k_1 + k_2 + k_3 = 1 \tag{4.11}$$

In general terms, the sum of n variables denoted by x_1, x_2, x_3, etc., gives rise to a relative increment

$$y^* = (\Sigma x_i)^* = k_1 x_1^* + k_2 x_2^* + \ldots + k_n x_n^* \tag{4.12}$$

where Σ stands for $\sum_{i=1}^{n}$ and

$$\left.\begin{aligned} k_i &= \frac{x_i}{y} \text{ for } i = 1, 2, \ldots, n \\ \text{so that} \quad \Sigma k_i &= 1. \end{aligned}\right\} \tag{4.13}$$

SUBTRACTION

When we are interested in the difference between two variables, each subjected to a given relative change, the *rule for subtraction* applies. For example, profit Z may be expressed as

$$Z = R - C \tag{4.14}$$

where R is the revenue from sales and C is the cost of the products or services sold.

In general, if $y = x_1 - x_2$, we can follow the method used earlier:

$$\begin{aligned} y^* = (x_1 - x_2)^* &= \frac{\delta x_1 - \delta x_2}{x_1 - x_2} \\ &= \frac{x_1}{x_1 - x_2}\frac{\delta x_1}{x_1} - \frac{x_2}{x_1 - x_2}\frac{\delta x_2}{x_2} \\ &= k_1 x_1^* - k_2 x_2^* \end{aligned} \tag{4.15}$$

where, as before,

$$\left.\begin{aligned} k_1 &= \frac{x_1}{x_1 - x_2} \\ k_2 &= \frac{x_2}{x_1 - x_2} \\ \text{but now} \qquad k_1 - k_2 &= 1 \end{aligned}\right\} \qquad (4.16)$$

Subtraction can, of course, be derived as a special case of addition, simply by allowing x_1 or x_2 to become negative, but it is generally more convenient to regard x_1 and x_2 as positive entities, thereby making the coefficients k_1 and k_2 (which represent the component proportions) positive in all cases. This means that in the case of addition the k values are between 0 and 1, whereas in the case of subtraction they remain positive but $k_1 > 1$ while k_2 may be either greater or smaller than 1 (and clearly $k_1 > k_2$ if $x_1 - x_2 > 0$).

Alternatively, we can convert an expression involving subtraction to one involving addition by simply rearranging. For example, Equation (4.14) can be rewritten as

$$R = Z + C$$

i.e. revenue equals profit plus cost, and Equation (4.5) may then be applied. If instead of k_1 in that equation we substitute the commonly used concept of profit margin, defined as

$$\alpha = \frac{Z}{R} \qquad (4.17)$$

then

$$R^* = \alpha Z^* + (1 - \alpha)C^*$$

from which we obtain

$$Z^* = \frac{R^*}{\alpha} - \frac{1-\alpha}{\alpha} C^* \qquad (4.18)$$

It is not difficult to verify that the same result is derived by applying the subtraction rule in (4.15).

MULTIPLICATION

There are many cases in which a quantity is derived from, or can be expressed as, a product of two variables. Some examples to illustrate this point are:

(1) The total labor cost W may be expressed as

$$W = wN \qquad (4.19)$$

where w = average wage or salary or cost per employee
N = number of employees

(2) The revenue R (or sales turnover) may be described as

$$R = pV \tag{4.20}$$

where p = unit price (i.e. average price per unit of product sold)
V = physical volume

(3) The total cost C is

$$C = cV \tag{4.21}$$

where c = unit cost (i.e. average cost per unit)

(4) The profit Z is

$$Z = zV \tag{4.22}$$

where z = unit profit (i.e. average profit obtained per unit).

The multiplication rule

All these cases follow the same pattern of being depicted as the product of two variables, say, x_1 and x_2, which may be subjected to the increments δx_1 and δx_2 respectively. A geometrical presentation of the product $x_1 x_2$ can be given by a rectangle y determined by its two dimensions x_1 and x_2, as shown in Fig. 4.3. The original area is

$$y = x_1 x_2 \tag{4.23}$$

and the new area

$$y + \delta y = (x_1 + \delta x_1)(x_2 + \delta x_2)$$

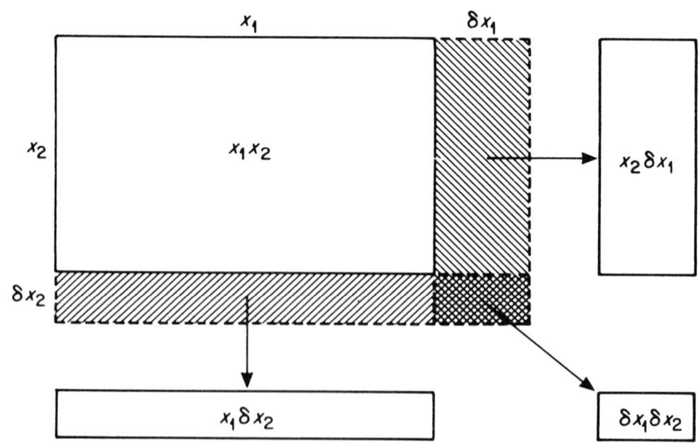

FIG. 4.3. A rectangle with dimensions x_1 and x_2 subject to changes δx_1 and δx_2.

so that the change in area is

$$\begin{aligned}\delta y &= (x_1+\delta x_1)(x_2+\delta x_2)-x_1x_2\\ &= x_2\delta x_1+x_1\delta x_2+\delta x_1\delta x_2\end{aligned}$$

The three terms are shown by the shaded rectangles in Fig. 4.3 and it should be noted that:

$x_2\delta x_1$ is the incremental area caused by the change in x_1;
$x_1\delta x_2$ is the incremental area due to a change in x_2;
$\delta x_1\delta x_2$ is a further incremental area of a second order magnitude, resulting from the interaction between the change in x_1 and the change in x_2.

The presentation in Fig. 4.3 is based on the assumption that the increments δx_1 and δx_2 are positive, but the results for δy in the last equation are equally valid when x_1 and/or x_2 decline in value (see Fig. 4.4).

The relative increment in area is

$$\begin{aligned}y^* = \frac{\delta y}{y} &= \frac{1}{x_1x_2}(x_2\delta x_1+x_1\delta x_2+\delta x_1\delta x_2)\\ &= \frac{\delta x_1}{x_1}+\frac{\delta x_2}{x_2}+\frac{\delta x_1}{x_1}\frac{\delta x_2}{x_2}\end{aligned}$$

hence the result may be expressed in several ways:

$$\left.\begin{aligned}(x_1x_2)^* &= x_1^*+x_2^*+x_1^*x_2^*\\ &= x_1^*+x_2^*(1+x_1^*)\\ &= x_1^*(1+x_2^*)+x_2^*\\ &= (1+x_1^*)(1+x_2^*)-1\end{aligned}\right\}\qquad(4.24)$$

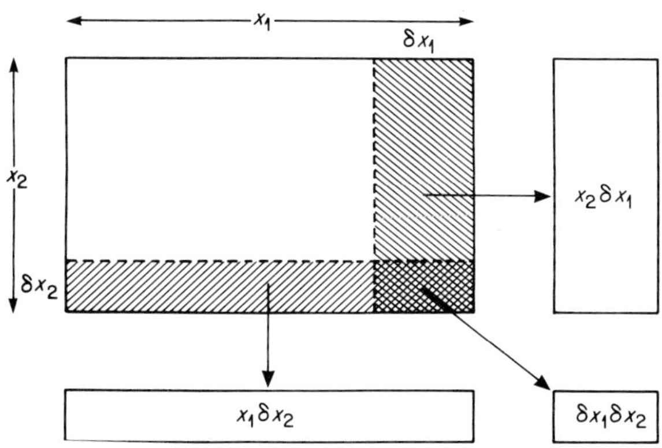

FIG. 4.4. A rectangle with declining dimensions.

An approximation

When either the increment δx_1 or δx_2 is relatively small (and particularly when both are small), the third term in Equation (4.24) becomes negligible, so that a good approximation is given by

$$(x_1 x_2)^* \simeq x_1^* + x_2^* \qquad \text{for } |x_1^*| \text{ and/or } |x_2^*| \ll 1 \tag{4.25}$$

The effect of the third term is illustrated by the first two examples in Table 4.2. In the first example x_1 and x_2 change by 20 % and 30 % respectively, so that $x_1^* x_2^*$ amount to 0.06, or 6 %, whereas in the second example x_1 and x_2 change by 1 % and 4 % respectively, so that the term $x_1^* x_2^*$ contributes very little to the total change. Clearly, therefore, the approximation (4.25) is not adequate when relatively large increments occur.

TABLE 4.2
Three examples for the multiplication rule.

Example	x_1^*	x_2^*	$x_1^* + x_2^*$ approx (4.25)	$x_1^* x_2^*$	$x_1^* + x_2^* + x_1^* x_2^*$ Eq. (4.24)	% error
1	0.2	0.3	0.5	0.06	0.56	56.0
2	0.01	0.04	0.05	0.0004	0.0504	5.0
3	−0.2	−0.3	−0.5	0.06	−0.44	−44.0

Notice that when x_1 and x_2 are subject to a decrease, namely when δx_1 and δx_2 are negative, then the total relative decline caused by $x_1^* + x_2^*$ is tempered by the effect of $x_1^* x_2^*$, which is then positive. A graphical presentation of this case is shown in Fig. 4.4, from which we see that the declining segments $x_2 \delta x_1$ and $x_1 \delta x_2$ overlap, so that the small area $\delta x_1 \delta x_2$ is included in both (unlike Fig. 4.3, where it is included in neither) and is thereby counted twice, so to speak, in the first two terms of Equation (4.24), so that the third term with its positive value may be interpreted as a means of adjusting this "double counting". The third example in Table 4.2 illustrates the contrast with the first example. In both cases x_1 and x_2 change by 20 % and 30 % respectively, but in example 1 the changes are positive while in example 3 they are negative. The major effect $x_1^* + x_2^*$ adds up to +50 % for example 1 and −50 % for example 3. The secondary effect $x_1^* x_2^*$ amounts to +6 % in both cases, so that the total result is 56 % in example 1 but −44 % in example 3. This asymmetry is entirely due to the effect of the third term in equation (4.24).

Note: In order to avoid possible errors in computations it is safest to express the variables in decimal form. In the case of addition it is obviously immaterial whether percentages or the decimal form is used; for example, if $x_1^* = 5\,\%$,

$x_2^* = 15\%$, $k_1 = 0.4$ and $k_2 = 0.6$, then

$$(x_1 + x_2)^* = 0.4\, x_1^* + 0.6\, x_2^*$$

In percentages: $(x_1 + x_2)^* = 0.4 \times 5 + 0.6 \times 15 = 11\%$
In decimal form: $(x_1 + x_2)^* = 0.4 \times 0.05 + 0.6 \times 0.15 = 0.11$
However, in the case of multiplication the use of percentages in the formula is not applicable because of the term $x_1^* x_2^*$, hence for $x_1^* = 5\%$ and $x_2^* = 15\%$:

$$\begin{aligned}(x_1 x_2)^* &= x_1^* + x_2^* + x_1^* x_2^* \\ &= 0.05 + 0.15 + 0.05 \times 0.15 \\ &= 0.2075 \qquad \text{i.e. } 20.75\%.\end{aligned}$$

Three variables

In the case of three variables, the result (4.24) can easily be extended to the following:

$$\begin{aligned}(x_1 x_2 x_3)^* &= (1 + x_1^*)\,(1 + x_2^*)\,(1 + x_3^*) - 1 \\ &= x_1^* + x_2^* + x_3^* + x_1^* x_2^* + x_2^* x_3^* + x_3^* x_1^* + x_1^* x_2^* x_3^* \end{aligned} \tag{4.26}$$

The first three terms represent the first order of magnitude; the next three are the secondary terms, expressing the effect of the interaction between changes in pairs of variables; the last term is of a third order, corresponding to the interaction of all three changes. If $|x_1^*|, |x_2^*|, |x_3^*| < 1$ then the first group $x_1^* + x_2^* + x_3^*$ contributes most to the total change, while the last term $x_1^* x_2^* x_3^*$ contributes the least; when $|x_1^*|$, $|x_2^*|$ and $|x_3^*|$ are relatively small the last term becomes negligible, and for practical purposes even the second group may be ignored, leaving us with the reasonable approximation

$$(x_1 x_2 x_3)^* \simeq x_1^* + x_2^* + x_3^* \qquad \text{for } |x_1^*|, |x_2^*|, |x_3^*| \ll 1 \tag{4.27}$$

A special case of Equation (4.24) is obtained when one of the variables becomes constant. Thus, if a = constant then $a^* = 0$, so that

$$(ax)^* = a^* + x^*\,(1 + a^*)$$

Substituting $a^* = 0$ we get

$$(ax)^* = x^*$$

and similarly it is easy to verify that

$$(ax_1 x_2)^* = (x_1 x_2)^*$$

Examples

1. Analysis of the wages bill

Analysis of the total wage cost in one company showed an increase of 29.1 % in a given period compared with the previous period. The chief executive was puzzled by this figure, because he had been assured that the wage negotiators had agreed to a 26.9 % increase.

A further analysis revealed the following data:

		Previous period	Given period	Increase %
Remuneration/person	w	5546	7040	26.9
Employees	N	7326	7453	1.7
Total wages	W	40630	52649	29.1

As stated in Equation (4.19), $W = wN$ and applying Equation (4.24)

$$W^* = w^* + N^* + w^* N^*$$

corresponding to

$$29.1 = 26.9 + 1.7 + 0.5$$

w^* accounts for the increase in wage rates (26.9 %), N^* for the increase in the labor force (1.7 %) and $w^* N^*$ for the secondary effect of the interaction between them (i.e. for the increase in wage for the increased number of employees), adding another 0.5 % to the cost.

2. Tradeoff between two variables

As part of its wage negotiations and planning process management wishes to put a limit on the amount of the total wage cost increase, i.e. on the value of W^*. Let us assume that this limit is stated as 10 %. The equation for W^* provides the perceived tradeoff between w^* and N^* which management is prepared to tolerate in order to ensure that W^* does not exceed 10 %. For example, if an increase in wage rates of 15 % is considered, then from the equation for W^* we find that

$$N^* = \frac{W^* - w^*}{1 + w^*} = \frac{0.10 - 0.15}{1.15} = -0.043$$

i.e. a reduction of 4.3% in the workforce will have to follow, whereas if no reduction in the workforce is acceptable then $N^* = 0$ leading to

$$w^* = \frac{W^* - N^*}{1 + N^*} = \frac{0.1 - 0}{1} = 0.1$$

or a maximum increase of only 10% in the wage rates.

In general terms, when we have a case which conforms to the format in Equation (4.24), the tradeoff between x_1^* and x_2^* can be easily tabulated for any given value of $(x_1 x_2)^*$, such as in Table 4.3, which is constructed for $(x_1 x_2)^* = 0.1$. x_1^* and x_2^* in this table are interchangeable, so that if $x_1^* = 15\%$ we read from the table that $x_2^* = -4.3\%$, and if $x_2^* = 15\%$ then $x_1^* = -4.3\%$. This tradeoff table can be expanded for various values of $(x_1 x_2)^*$, as shown in Table 4.4, or as presented in Fig. 4.5, and thus become available for a variety of applications.

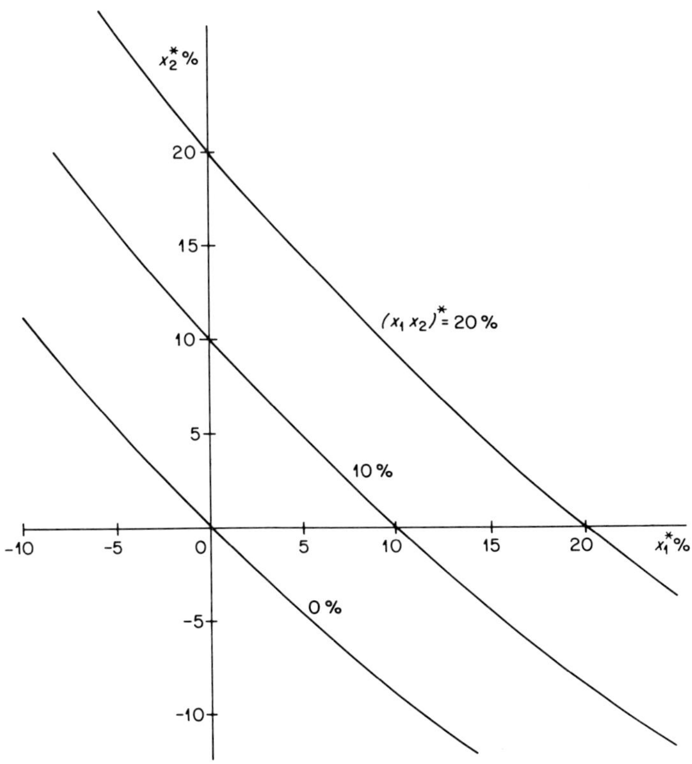

FIG. 4.5. Trade-off between x_1^* and x_2^* for given values of $(x_1 x_2)^*$.

TABLE 4.3

Tradeoff between x_1^* and x_2^* for $(x_1 x_2)^* = 10\%$.

$x_1^*(\%) =$	−10	−8	−6	−4	−2	0	2	4	6	8	10	15	20	25
$x_2^*(\%) =$	22.2	19.6	17.0	14.6	12.2	10.0	7.8	5.8	3.8	1.9	0	−4.3	−8.3	−12.0

TABLE 4.4

Values of x_2^* (in %) for given x_1^* (in %) and $(x_1 x_2)^*$.

x_1^* (%)	−10	−8	−6	−4	−2	0	2	4	6	8	10	15	20	25
$(x_1 x_2)^* =$ −20%	−11.1	−13.0	−14.9	−16.7	−18.4	−20.0	−21.6	−23.1	−24.5	−25.9	−27.3	−30.3	−33.3	−36.0
−10%	0	−2.2	−4.3	−6.3	−8.2	−10.0	−11.8	−13.5	−15.1	−16.7	−18.2	−21.7	−25.0	−28.0
−5%	5.6	3.3	1.1	−1.0	−3.1	−5.0	−6.9	−8.7	−10.4	−12.0	−13.6	−17.4	−20.8	−24.0
0	11.1	8.7	6.7	4.2	2.0	0	−2.0	−3.8	−5.7	−7.4	−9.1	−13.0	−16.7	−20.0
5%	16.7	14.1	11.7	9.4	7.1	5.0	2.9	1.0	−0.9	−2.8	−4.6	−8.7	−12.5	−16.0
10%	22.2	19.6	17.0	14.6	12.2	10.0	7.8	5.8	3.8	1.9	0	−4.3	−8.3	−12.0
15%	27.8	25.0	22.3	19.8	17.3	15.0	12.7	10.6	8.5	6.5	4.5	0	−4.2	−8.0
20%	33.3	30.4	27.7	25.0	22.4	20.0	17.6	15.4	13.2	11.1	9.1	4.3	0	−4.0
25%	38.9	35.9	33.0	30.2	27.6	25.0	22.5	20.2	17.9	15.7	13.6	8.7	4.2	0

TABLE 4.5
A case study in a steel plant [2, p. 124] (year **1** taken as 100, when the absolute figures were all positive).

Year	Return on investment r	Unit profit z	Capacity utilization e	Capacity/ investment k
1	100	100	100	100
2	55.6	55.4	103.4	96.9
3	46.2	51.3	98.5	91.4
4	43.0	54.2	101.8	78.0
5	7.7	11.8	88.4	73.9
6	−57.5	−81.0	91.5	77.5
7	28.5	35.8	104.3	76.2
8	18.1	22.2	99.3	82.4

3. The function of return on investment

As discussed in Chapter 2, the return on total investment, denoted as r, may be expressed as follows [2, p. 122]:

$$
\begin{aligned}
r &= \frac{\text{Profit}}{\text{Investment}} \\
&= \left(\frac{\text{Profit}}{\text{Output}}\right)\left(\frac{\text{Output}}{\text{Capacity}}\right)\left(\frac{\text{Capacity}}{\text{Investment}}\right) \\
&= zek
\end{aligned}
$$

where z = unit profit = Profit/Output
e = capacity utilization of the plant = Output/Capacity
k = provision of capacity by total investment = Capacity/Investment

An examination of the causes for changes in the rate of return may thus be related to the three factors z, e and k. A case study of a steel plant revealed that over a period of eight years the rate of return fluctuated wildly, suffering a serious setback in the second, fifth and sixth year, as shown in Table 4.5, in which year 1 is taken as the base year with performance denoted as 100. The three factors z, e and k are also recorded, again in relation to year 1 taken as 100 in each case.

Using Equation (4.26) the relative change in rate of return is given by

$$r^* = z^* + e^* + k^* + z^* e^* + e^* k^* + k^* z^* + z^* e^* k^*$$

and various comparisons are made in Table 4.6: Year 2 is compared with year 1, year 3 compared with year 2, and so on.

Table 4.6

Inter-year comparisons for the steel plant [2, p. 124] (all figures in %).

Comparison	r^*	z^*	e^*	k^*	$z^* e^*$	$e^* k^*$	$k^* z^*$	$z^* e^* k^*$	Primary terms $z^* + e^* + k^*$	Secondary terms $z^* e^* + e^* k^* + k^* z^*$
2 vs 1	−44.4	−44.6	3.4	−3.1	−1.52	−0.11	1.38	0.05	−44.3	−0.2
3 vs 2	−16.9	−7.4	−4.7	−5.7	0.35	0.28	0.42	−0.02	−17.8	1.0
4 vs 3	−6.9	5.7	3.4	−14.7	0.19	−0.50	−0.84	−0.03	−5.6	−1.2
5 vs 4	−82.1	−78.2	−13.2	−5.3	10.32	0.70	4.14	−0.55	−96.7	15.2
8 vs 7	−36.5	−38.0	−4.8	8.1	1.82	−0.39	−3.08	0.15	−34.7	−1.7
8 vs 1	−81.9	−77.8	−0.7	−17.6	0.54	0.12	13.69	−0.10	−96.1	14.3

Note:
2 vs 1 means comparison of year 2 with year 1.
In some rows the sum of the various changes does not add up to the value of r^*. The reason for this lies in rounding errors, since the figures in the previous table have been rounded off to one decimal place.

The computations are carried out as follows: In comparing year 2 with year 1 we find r^* by

$$r^* = \frac{r_2 - r_1}{r_1}$$

where r_1 and r_2 refer to years 1 and 2 respectively, hence

$$r^* = \frac{55.6 - 100}{100} = -0.444 \text{ or } -44.4\%$$

and similarly when year 3 is compared with year 2

$$r^* = \frac{r_3 - r_2}{r_2} = \frac{46.2 - 55.6}{55.6} = -0.169 \text{ or } -16.9\%.$$

We thus find that the decline in r in year 2 compared with year 1 was 44.4%, largely due to a decline in the unit profit, with e^* and k^* almost cancelling each other, while the total effect of the secondary terms was negligible. At times, however, the secondary terms may not be ignored. For example, in comparing year 5 with year 4 we find that the effect of the secondary term z^*e^* is greater than the effect of the primary term k^*, and similarly in comparing year 8 with year 1 the term k^*z^* is relatively large. Also, whenever all the three primary relative increments z^*, e^* and k^* are negative (as in the second, fourth and sixth row in Table 4.6), the secondary terms are all positive, thereby ameliorating to some extent the total effect of the primary terms. In all the cases shown here, the effect of the tertiary term $z^*e^*k^*$ is, as expected, negligible. Thus, an examination of the eight year period, through the examples cited in Table 4.6, clearly indicates that the major reason for the deterioration in the rate of return is attributable to a decline in the unit profit, suggesting either a decline in the unit price or an increase in the unit cost, or both, thereby outlining the direction which further analysis should take. In this way, the presentation of the information in Table 4.6 provides a more vivid picture, and hence gives a better insight, than the more basic data in Table 4.5.

DIVISION

If we have a function of the form

$$y = \frac{x_1}{x_2} \tag{4.28}$$

then we can find y^* as follows:

$$yx_2 = x_1$$

hence

$$(yx_2)^* = x_1^*$$

substitute

$$(yx_2)^* = y^*(1+x_2^*)+x_2^*$$

hence

$$y^*(1+x_2^*)+x_2^* = x_1^*$$

or

$$\left(\frac{x_1}{x_2}\right)^* = \frac{x_1^* - x_2^*}{1+x_2^*} \tag{4.29}$$

and when the absolute value x_2^* is small in relation to 1, this expression is reduced to the approximation

$$\left(\frac{x_1}{x_2}\right)^* \simeq x_1^* - x_2^* \qquad \text{for } |x_2^*| \ll 1 \tag{4.30}$$

Example

The average unit cost is derived from Equation (4.28)

$$c = \frac{C}{V}$$

Hence

$$c^* = \frac{C^* - V^*}{1+V^*}$$

Suppose that the total cost increases by 10%. The effect that volume has on unit cost is seen from Table 4.4 along the row corresponding to $(x_1 x_2)^* = 10\%$, while the volume corresponds to x_1^* at the top of the table (alternatively, Fig. 4.5 can be used). Thus, no increase in volume obviously causes the unit cost to increase by the same amount as the total cost, namely by the full 10%, a 2% increase in volume means 7.8% increase in unit cost, and for the unit cost not to increase at all the volume must increase by 10%. This result is also evident from the equation for c^*, from which we see that $c^* = 0$ provided $C^* = V^*$.

The effect of the approximation (4.30) is demonstrated in the two examples given in Table 4.7, the first for $C^* = 10\%$ and the second for $C^* = 30\%$. The figures suggest that the quality of the approximation depends only on V^* and

TABLE 4.7
Examples to illustrate the approximation (4.30).

V^* (%)	0	1	2	5	10	20	30	−20	−30
$C^* = 10\%$									
c^*	10.0	8.9	7.8	4.8	0	−8.3	−15.4	37.5	57.1
c^* appr.	10.0	9.0	8.0	5.0	0	−10.0	−20.0	30.0	40.0
$C^* = 30\%$									
c^*	30.0	28.7	27.5	23.8	18.2	8.3	0	62.5	85.7
c^* appr.	30.0	29.0	28.0	25.0	20.0	10.0	0	50.0	60.0

$$c^* = \frac{C^* - V^*}{1 + V^*}$$

c^* appr. $= C^* - V^*$.

not on C^*. This conclusion is verified by looking at the value of c^* appr. in relation to c^* (provided $C^* \neq V^*$):

$$\frac{c^* \text{ appr.}}{c^*} = 1 + V^*$$

Hence, the approximation always overestimates c^* when the volume increases and underestimates c^* when the volume decreases (except for the case $C^* = V^*$, when the result is $c^* = 0$).

SENSITIVITY ANALYSIS

Managers often need to know whether a given measure of performance is likely to be greatly affected when several of the variables in the system are liable to change. If, for example, the measure of performance is y, which has a functional relationship with a single variable x in the form

$$y = f(x)$$

then a useful measure of sensitivity may be expressed as

$$\frac{y^*}{x^*}$$

which shows the degree to which y changes to a given relative change in x.

In the simple example

$$y = a + bx$$

where a and b are constant, we find that

$$y^* = kx^*$$

where $k = bx/y$, so that

$$\frac{y^*}{x^*} = k$$

and if $k < 1$ it follows that the absolute relative change in y will always be smaller than the absolute relative change in x, and more specifically that the smaller the value of k the less sensitive y is to a change in x.

When we have two variables x_1 and x_2 the sensitivity of y^* may be measured by y^*/x_1^* and y^*/x_2^* respectively. Thus, for the case

$$y = x_1 + x_2$$

Equation (4.9) yields

$$\frac{y^*}{x_1^*} = k_1 + k_2\frac{x_2^*}{x_1^*} \tag{4.31}$$

and we see that apart from the effect of k_1 the sensitivity depends not only on x_1^* but on x_2^* as well. In the special case of changing one variable at a time, namely when $x_2^* = 0$, the sensitivity relationship is reduced to

$$\frac{y^*}{x_1^*} = k_1 \qquad (\text{for } x_2^* = 0)$$

Similarly, the sensitivity in relation to x_2^* is

$$\frac{y^*}{x_2^*} = k_1\frac{x_1^*}{x_2^*} + k_2$$

For the case

$$y = x_1 x_2$$

the sensitivity in relation to x_1^* is

$$\frac{y^*}{x_1^*} = \frac{x_1^* + x_2^* + x_1^* x_2^*}{x_1^*} = 1 + \frac{x_2^*}{x_1^*} + x_2^* \tag{4.32}$$

so that the sensitivity of y^* with respect to x_1^* depends here not only on x_2^*/x_1^*, but on the value of x_2^* as well, although for small relative changes the effect of the latter becomes marginal. For example, if $x_2^* = 3\%$ then

$$\frac{y^*}{x_1^*} = 1 + \frac{x_2^*}{x_1^*} + 0.03 \simeq 1 + \frac{x_2^*}{x_1^*}$$

A similar result can be obtained for the sensitivity measure $\dfrac{y^*}{x_2^*}$.

APPLICATIONS TO VARIOUS FUNCTIONAL RELATIONSHIPS

The basic operations developed in this chapter can be used to examine a variety of functions encountered by the analyst. Some examples are explored below.

(1) Suppose we have a quadratic relationship of the form

$$y = x^2$$

We can find y^* by using Equation (4.24), namely

$$y^* = (xx)^* = 2x^* + (x^*)^2$$

hence

$$(x^2)^* = (x^* + 1)^2 - 1$$

(2) Similarly, it is not difficult to show that

$$(x^n)^* = (x^* + 1)^n - 1$$

and

$$\left(\frac{1}{x^n}\right)^* = \frac{1}{(x^* + 1)^n} - 1 \qquad (4.33)$$

(3) In examining the cost of a manufacturing plant, two types of costs are often identified (as discussed in greater detail in the next chapter): the variable cost, which increases with output, and the fixed cost, which is independent of the level of output. Typically, the cost function C takes the linear form

$$C = sV + F$$

where s = marginal cost per unit of output
V = output
F = fixed cost

At any given time the variable and fixed cost elements constitute the proportions f_1 and f_2 of the total cost, namely

$$f_1 = \frac{sV}{C}$$

and

$$f_2 = \frac{F}{C}$$

so that

$$f_1 + f_2 = 1$$

We can now determine the composition of the relative change in cost C^* as follows:

$$C^* = f_1\,(sV)^* + f_2 F^*$$

If the marginal cost s per unit remains unchanged, then $s^* = 0$, hence

$$(sV)^* = V^*$$

so that

$$C^* = f_1 V^* + f_2 F^*$$

Notice that this result is independent of the actual value of the cost parameters s and F. Also, it should be pointed out that if the fixed cost F is unchanged ($F^* = 0$), the total cost C does not increase (or decrease) by the same proportion as the increase (or decrease) in the volume, but only by a fraction of that change (since $f_1 < 1$ and the case $f_1 = 1$ only applies if the fixed cost F is zero), and similarly if $V^* = 0$ then the relative change in C is a fraction of the relative change in F.

(4) If x has an initial value $x_{(0)}$, and then $x_{(1)}$, and $x_{(2)}$ at the end of the two periods 1 and 2 respectively, as shown in Fig. 4.6, then the total increment in x during the two periods is

$$\delta x = \delta x_{(1)} + \delta x_{(2)}$$

where $\delta x_{(1)}$ is the increment during the first period, and $\delta x_{(2)}$ during the second. The total relative change is associated with the initial level $x_{(0)}$, namely

$$x^* = \frac{\delta x}{x_{(0)}}$$

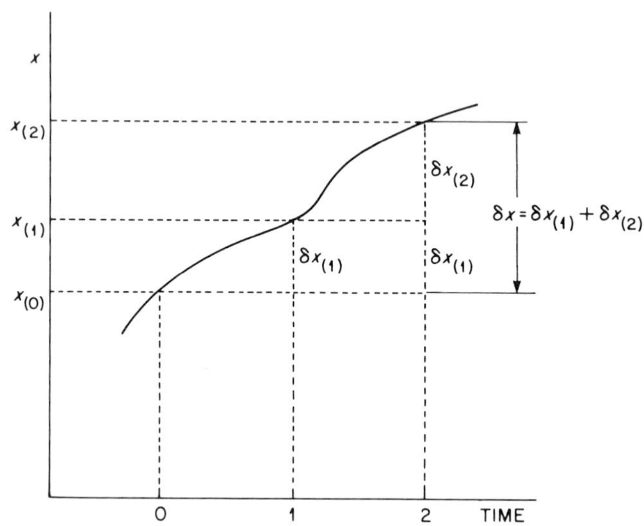

FIG. 4.6. The variable x for three points in time.

When disaggregated into changes related to each period, this total relative change becomes

$$\begin{aligned} x^* &= \frac{\delta x}{x_{(0)}} = \frac{\delta x_{(1)} + \delta x_{(2)}}{x_{(0)}} \\ &= \frac{\delta x_{(1)}}{x_{(0)}} + \frac{\delta x_{(2)}}{x_{(1)}} \frac{x_{(1)}}{x_{(0)}} \\ &= x^*_{(1)} + x^*_{(2)}(1 + x^*_{(1)}) \\ &= (1 + x^*_{(1)})(1 + x^*_{(2)}) - 1 \end{aligned}$$

which is the result in the multiplication rule (4.24). Thus, if in the first period there is an increase of $x^*_{(1)} = 10\%$ and in the second period $x^*_{(2)} = 16\%$, then the total relative increase during the two periods amounts to

$$x^* = 1.10 \times 1.16 - 1 = 1.276, \text{ i.e. } 27.6\%.$$

This example illustrates well the "compounding" effect, namely the fact that the increase of 16% in the second period applies not only to the base value $x_{(0)}$ but also to the increase of 10% in the first period, as we get in compound interest. Needless to say, if the time span is divided into n periods, the result can be expanded to:

$$x^* = (1 + x^*_{(1)})(1 + x^*_{(2)}) \ldots (1 + x^*_{(n)}) - 1$$

(5) In inventory control, the economic batch quantity for ordering replenishment stock from a supplier, is derived from the juxtaposition of two types of costs:

- A fixed cost per order, for processing the order and the consignment when it arrives; if the fixed cost is a, the cost per unit is a/x, where x is the size of the batch, and this cost per unit decreases as x increases.
- A carrying cost per unit of stock held in the store, equivalent to bx, where b is a constant parameter independent of the batch size x.

Thus the cost per unit y for replenishment of the stock is given by

$$y = \frac{a}{x} + bx$$

Let us define the cost proportions as

$$k_1 = \frac{a/x}{y}$$

$$k_2 = \frac{bx}{y}$$

and

$$k_1 + k_2 = 1$$

The economic batch quantity is defined as the value of x that minimizes the unit cost y. It is easily seen that if y is a continuous function of x, the optimal value is derived by equating the derivative of y to zero, i.e.

$$\frac{\mathrm{d}y}{\mathrm{d}x} = -\frac{a}{x^2} + b = 0$$

from which the well-known classical formula for the economic batch size is obtained (it is readily verified from the second derivative that this is a point of minimum)

$$x_{\text{opt}} = \left(\frac{a}{b}\right)^{1/2}$$

The two components of the cost function y are shown in Fig. 4.7, and it

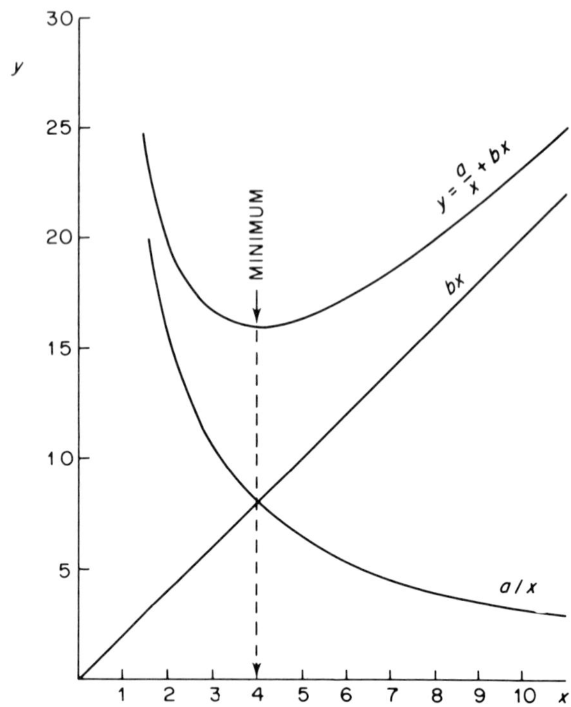

FIG. 4.7. The economic reorder quantity.

can easily be verified that at the optimum point x_{opt} we have

$$\frac{a}{x_{\text{opt}}} = bx_{\text{opt}}$$

which means that the optimum corresponds to the point where the two cost curves intersect.

Another vivid way of presenting the relationship between the two cost components of y is to plot a/x against bx when x is taken as a variable. We obtain a graph, shown in Fig. 4.8, on which every point corresponds to a particular value of x. As we move along the curve (starting from the left) the value of a/x decreases, but that of bx increases, and the last equation simply implies that the two components are equal where the curve is intersected by a straight line drawn from the origin at 45°.

Now, to determine the effect of a change in the batch size x on the cost y we can compute the value of y^*

$$y^* = k_1 \left(\frac{a}{x}\right)^* + k_2\,(bx)^*$$

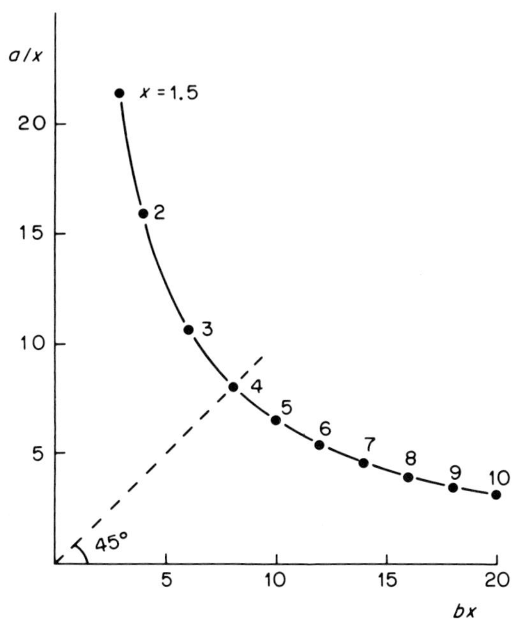

FIG. 4.8. The cost components a/x and bx.

Since a and b are constant (i.e. $a^* = b^* = 0$), it follows that

$$y^* = -k_1 \frac{x^*}{1+x^*} + k_2 x^*$$

or, after substituting $k_2 = 1 - k_1$,

$$y^* = x^*\left(1 - k_1 \frac{2+x^*}{1+x^*}\right)$$

If the relative change in x is positive ($x^* > 0$), then y declines ($y^* < 0$) if

$$1 - k_1 \frac{2+x^*}{1+x^*} < 0$$

or when

$$k_1 > \frac{1+x^*}{2+x^*}$$

but when x^* is negative the inequality sign in this condition is reversed. The function $(1+x^*)/(2+x^*)$ is plotted in Fig. 4.9, so that the domain above the line for $x^* > 0$ corresponds to a decline in the cost y and below the line to an increase, while for $x^* < 0$ the position is reversed. For example, if $k_1 = 0.55$ we see from the graph that the cost will decrease provided x is allowed to increase by less than 22 % ($0 < x^* < 0.22$); if x increases by 22 % the value of y is expected to remain unchanged, while an increase in x by more than 22 % is expected to lead to a rise in y. We further see from Fig. 4.9 that along the vertical axis $y^* = 0$ and that the curve crosses the vertical axis (i.e. when $x^* = 0$) at $k_1 = 0.5$ (i.e. where $a/x = bx$). If we deviate from this point to the right or to the left (namely if x^* is either positive or negative) we enter the domain where y will increase, and that means that the point $k_1 = 0.5$ corresponds to the optimum value of y, thus verifying the result obtained earlier. This means that if $k_1 > 0.5$ it would be necessary to increase x (i.e. to have $x^* > 0$) in order to reduce y, and if $k_1 < 0.5$ then x would have to be reduced, a conclusion compatible with Figs 4.7 and 4.8.

A more detailed presentation of these results is given in Fig. 4.10, in which *iso-cost-increase* lines are drawn, namely lines on which the relative cost increase is the same. The line $y^* = 0$ is as shown in Fig. 4.9; $y^* = 5$ % is the locus of points for various combinations of values of k_1 and x^* for which the cost increase is 5 %, and so on. The diagram is reminiscent of a contour map in geography, so that the height of any point in the "terrain" can be estimated from its position in relation to the contour lines shown. For example, if $k_1 = 0.4$ and if $x^* = 20$ % is considered, we see from the diagram that the expected rise in the cost y would be just over 5 %. We

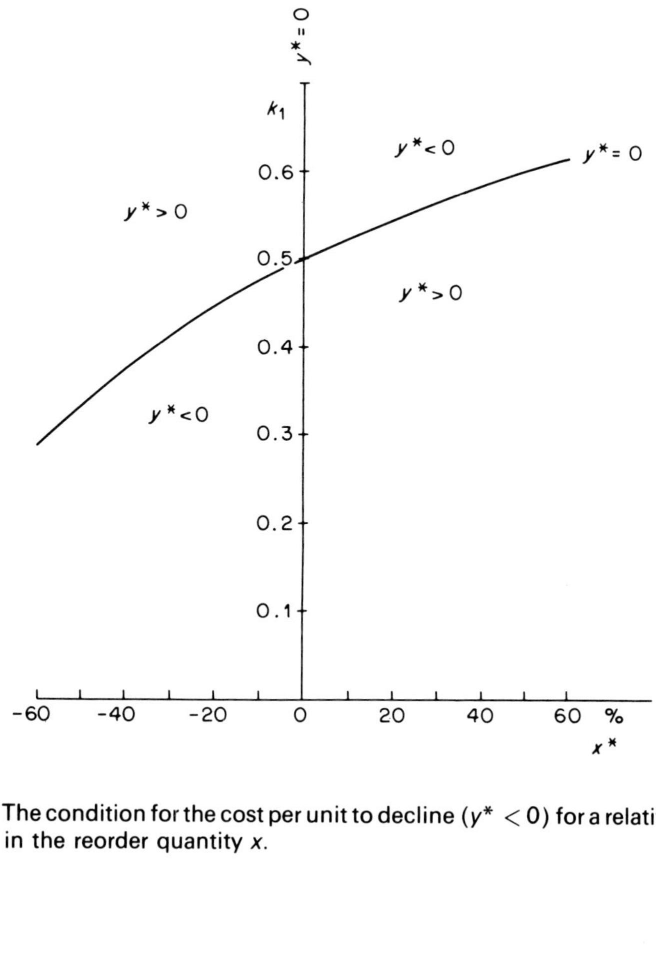

FIG. 4.9. The condition for the cost per unit to decline ($y^* < 0$) for a relative change in the reorder quantity x.

further see that the "terrain" in the vicinity of the optimum point corresponding to $x^* = 0$ and $k_1 = 0.5$ is flat, leaving a large area to cover the range $-5\% < y^* < 5\%$ and this is an indication of the insensitivity of the cost curve to deviations of x from the optimum point, so that x would have to be reduced by as much as 27 % or increased by 37 % before the cost y increases by as little as 5 %. Notice that these results are general and do not depend on the specific values of the coefficients in the cost function y (needless to say, these conclusions can be derived from the classical inventory model for economic batch quantities [3]).

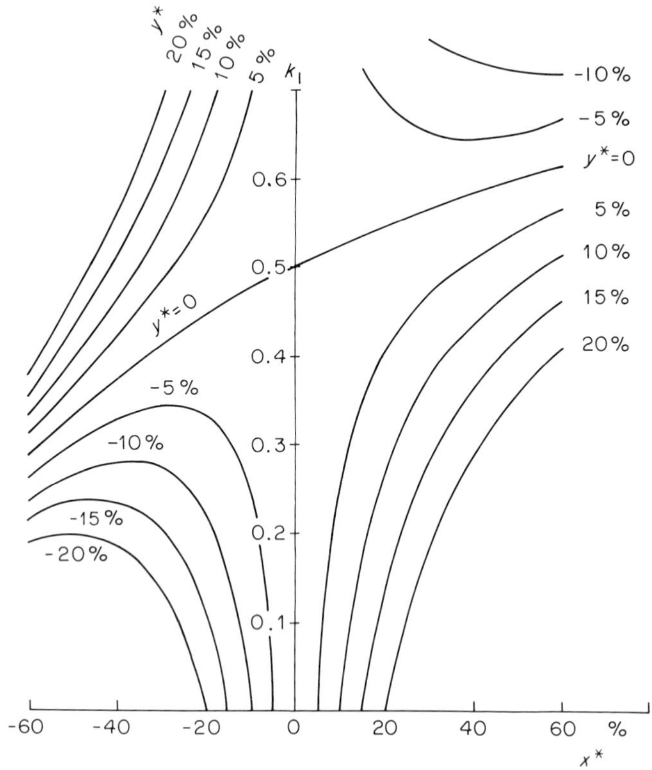

FIG. 4.10. Iso-cost-increase lines.

CONCLUSION

The main rules of incremental calculus are summarized in the following table. The rules can be used to explore a wide variety of functional relationships $y = f(x)$. The relative increments x^* and y^* are non-dimensional, so that if results can be derived to express y^* in terms of x^*, general conclusions may often be drawn which would be more difficult to arrive at without the help of incremental calculus. The ratio y^*/x^* is a good indicator of sensitivity and can be useful when the robustness of any given measure of effectiveness needs to be investigated.

TABLE 4.8
The basic rules of incremental calculus.

(1) $(x_1+x_2)^* = k_1 x_1^* + k_2 x_2^*$

where $k_1 = \dfrac{x_1}{x_1+x_2}$

$k_2 = \dfrac{x_2}{x_1+x_2}$

$k_1+k_2 = 1$

(2) $(x_1-x_2)^* = k_1 x_1^* - k_2 x_2^*$

where $k_1 = \dfrac{x_1}{x_1-x_2}$

$k_2 = \dfrac{x_2}{x_1-x_2}$

$k_1-k_2 = 1$

(3) $(x_1 x_2)^* = x_1^* + x_2^* + x_1^* x_2^*$

(4) $\left(\dfrac{x_1}{x_2}\right)^* = \dfrac{x_1^* - x_2^*}{1+x_2^*}$

REFERENCES

[1] EILON, S. (1975). Changes in profitability components. *Omega* **3**, 353–354.

[2] EILON, S., GOLD, B. and SOESAN, J. (1976). "Applied Productivity Analysis for Industry". Chapter 7. Pergamon Press, Oxford.

[3] EILON, S. (1962). "Elements of Production Planning and Control". Chapter 10. Macmillan, New York.

[4] EILON, S., WATSON-GANDY, C. D. T. and CHRISTOFIDES, N. (1971). "Distribution Management—Mathematical Modelling and Practical Analysis". Charles Griffin, London.

EXERCISES

1. Prove Equation (4.10).
2. Find the value of $(a+x_1+x_2)^*$ where a = constant.
3. For given values of x_1^* and x_2^*, under what conditions will the ratio $(x_1+x_2)^*/x_1^*$ in Equation (4.9) rise with k_1?
4. We aim at increasing the total revenue R in case 2 of Table 4.1 by 10%. Show by means of a graph what combinations of R_1^* and R_2^* will yield that result.

5. Derive Equation (4.18) from the subtraction rule (4.15).
6. If $y = a_1 x_1 + a_2 x_2$
   (a) Find x_2^* as a function of y^* and x_1^* and examine its sensitivity to k_1, where $k_1 = a_1 x_1 / y$.
   (b) State the conditions for $x_2^* > 0$.
7. Derive Equation (4.26)
8. In Equation (4.26) state the conditions under which the tertiary effect can be greater than one of the secondary effects, and similarly the condition for one of the secondary effects to be greater than one of the primary effects (consider only positive values for the starred variables).
9. Add to Table 4.6 a comparison of year 7 with year 6 based on the data in Table 4.5.
10. Determine the value of $\left(\dfrac{a}{x}\right)^*$ where $a =$ constant.
11. Prove Equation (4.33).
12. Figure 4.2 shows the ratio $(x_1 + x_2)^*/x_1^*$. Suggest a way for presenting the ratio $(x_1 x_2)^*/x_1^*$. State the conditions for x_1^* to exceed the value of $(x_1 x_2)^*$ and for the two to be equal in value.
13. If $y = x_1/x_2$, state the conditions $y^* < x_1^*$ using an analysis similar to that undertaken for Exercise 12.
14. For the function $y = x + x^2$ find an approximate value of y^* when x^* is relatively small and investigate how good your approximation is.
15. Plot Figs 4.6 and 4.7 for the case $y = 32/x + 2x$.
16. Given a continuous cost function per unit $y = a/x + bx$ and the current operation at $k_1 = \dfrac{a/x}{y}$, find the value of x^* that results in the maximum reduction in unit cost.
17. If $y = x_1 + x_2$ when is the relative sensitivity of y^* with respect to x_1^* the same as it is with respect to x_2^*?
18. The corporate plan for a company calls for increase in revenue $R^* = 24\%$ in one year.
   (a) If the relative monthly increases are to be the same, denoted as $R_{(1)}^*$, find this monthly value.
   (b) If the monthly increases for six months are to be $R_{(1)}^*$ per month and for the other six months $R_{(2)}^*$ per month, and if the management aims at $R_{(2)}^* = 1.2\, R_{(1)}^*$, find the value of $R_{(1)}^*$.
19. If $ay = bx^{\frac{1}{2}}$ where a and b are constant, find y^*.
20. It has been suggested that the cost of running a warehouse may follow the concave function [4, p. 76]

$$y = a_1 + a_2 x + a_3 x^{\frac{1}{2}}$$

where x = average volume throughput (tons) per day.
In a particular case it was found that

$a_1 = 500$
$a_2 = 3$
$a_3 = 100$

and that the warehouse was operating at $x = 9044$. Find the value of y^* as a function of x^*.

21. (a) In the two product case, show that the index numbers 2(A) and 4(A) for volume and price respectively are equivalent to the following two respective expressions:

$$V^* = \rho_1 V_1^* + \rho_2 V_2^*$$
$$p^* = \rho_1 p_1^* + \rho_2 p_2^*$$

where

V^* = overall relative change in the volume of the product-mix

$\rho_1 = R_1/R$ = revenue proportion of product 1 in the base year

$\rho_2 = R_2/R$ = revenue proportion of product 2 in the base year etc.

(b) Find the error incurred in computing $R^* = (pV)^*$ using these expressions.

(c) Use your results for the following example:

	Product 1		Product 2	
	p	V	p	V
Base year	1.0	100	3.0	100
Current year	1.1	120	3.6	90

(d) Compare the results in (a) with those obtained from the use of the Edgeworth method.

5
Total Cost and Unit Cost

Themes in this chapter

Definitions of:
fixed cost and variable cost
average unit cost
average unit fixed cost
average unit variable cost
marginal cost
Conditions leading the relative incremental unit cost to decline
Optimum output to minimize unit cost
Types of cost models, including:
the factor inputs model
the linear cost model
a hybrid model incorporating both

INCREMENTAL CHANGE IN UNIT COST

As indicated in previous chapters, the two commonly used measures of performance—profit Z and return on capital r—depend on the level of total cost C, since

$$Z = R - C$$

and

$$r = \frac{R - C}{I}$$

While the revenue R is determined by what happens in the market as a result of the interaction between unit price and volume, the total cost C is the sum of all the cost items associated with resource inputs and is dependent on the total unit cost and on the output. In the management of industrial operations it is not surprising, therefore, that special attention is devoted to the analysis and control of the total cost and unit cost.

As shown in Chapter 2, the unit cost can be disaggregated into its constituent elements to allow the appropriate managerial functions to monitor specific ratios, including productivity ratios and factor input proportions, so that an adequate understanding of the behavior of these ratios can provide a sound basis for managerial action.

The total cost C is given by

$$C = cV \tag{5.1}$$

where c is the unit cost and V the volume. The function C is cited in Chapter 4 as an example of how the multiplication rule can be used in incremental analysis, leading to

$$C^* = c^* + V^* + c^* V^* \tag{5.2}$$

Table 4.4 and Fig. 4.5 in Chapter 4 refer to the general case $(x_1 x_2)^*$ and these results apply here with c and V denoting the variables x_1 and x_2 respectively. Following Equation (4.24), C^* can also be written as

$$\begin{aligned} C^* &= c^* + V^*(1 + c^*) \\ &= c^*(1 + V^*) + V^* \end{aligned}$$

so that when c^* and/or V^* are relatively small (i.e. if $|c^*|, |V^*| \ll 1$) then the following approximation applies

$$C^* \simeq c^* + V^* \tag{5.3}$$

These equations underline the obvious comment that an increase in total cost, viewed in isolation, is not necessarily unwelcome, since it depends on where the increase emanates from. An increase in total cost due to an increase in volume ($V^* > 0$) is a perfectly natural outcome, and provided the profit margin is positive (i.e. $p > c$), the resultant increase in revenue will exceed the increase in cost and hence improve total profit. But an increase in the unit cost ($c^* > 0$) has an undesirable effect, both on its own and in its interaction with the increase in volume (as expressed by the term $c^* V^*$ in Equation (5.2)). This is why action is often sought to reduce c or, when an increase is inevitable, at least to keep the increase to a minimum. Thus, in a managerial control context we find that attention tends to be focused on the unit cost rather than on the total cost. From Equation (5.1),

$$c^* = \frac{C^* - V^*}{1 + V^*} \tag{5.4}$$

Now, by definition $V^* > -1$ (only in the extreme case, when production ceases altogether, volume declines by 100%, resulting in $V^* = -1$), hence the denominator $1 + V^*$ is positive, and it follows that

$$\begin{aligned} c^* &> 0 \quad \text{if } C^* > V^* \\ c^* &= 0 \quad \text{if } C^* = V^* \\ c^* &< 0 \quad \text{if } C^* < V^* \end{aligned}$$

These three statements may be combined into one by using the symbol $\gtrless$:

$$c^* \gtrless 0 \qquad \text{if } C^* \gtrless V^* \tag{5.5}$$

where the > sign in the first expression corresponds to the > sign in the second expression, and similarly there is a one to one correspondence between the = signs and the < signs respectively in the two expressions. Hence, if both C^* and V^* are positive, the relative increase in volume must exceed that of the total cost, and if both are negative, the relative reduction in volume should be less than that of the total cost, in order for the unit cost to decline.

COST FUNCTIONS

As emphasized in Chapter 2, costs depend on the array of resource inputs and on factor prices of these inputs. But when the concept of a *cost function* is used, the major variable under consideration is that of *output*, so that attention is directed to the functional relationship between total cost and volume V. Specifically, management is interested in the effect of change of volume on cost, both in the short term, when fluctuations in demand are bound to occur, and in the long term, when a plant is faced with the prospect of expansion or contraction.

Examples of cost functions are shown in Figs 5.1 to 5.5:

In Fig. 5.1—the cost increases linearly with volume, corresponding to *constant return to scale.*

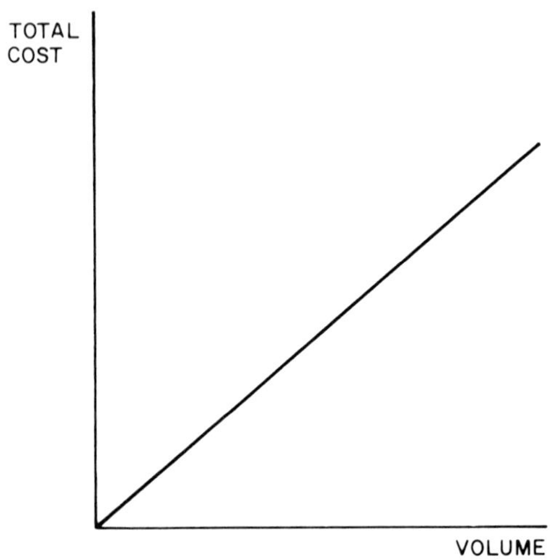

FIG. 5.1. A linear cost function with constant return to scale.

In Fig. 5.2—the cost function is again linear, but a fixed cost element is incurred, even before manufacture begins.

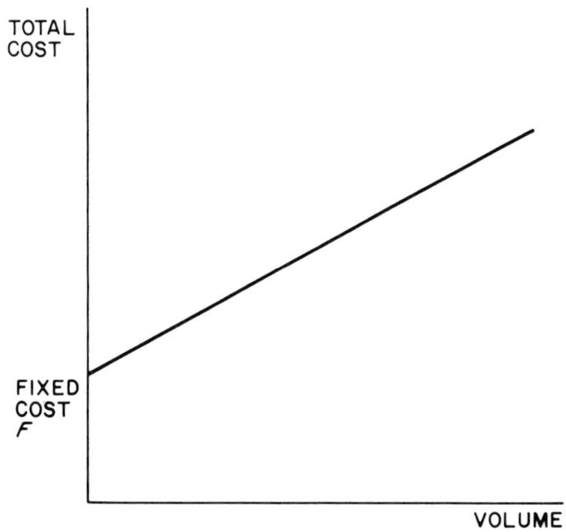

FIG. 5.2. A linear cost function with a fixed cost element.

In Fig. 5.3—the cost function is concave, namely the incremental cost per unit declines as output increases, and an operation with this characteristic is known as being subject to *economies of scale.* An additional fixed cost element may apply here, as in Fig. 5.2.

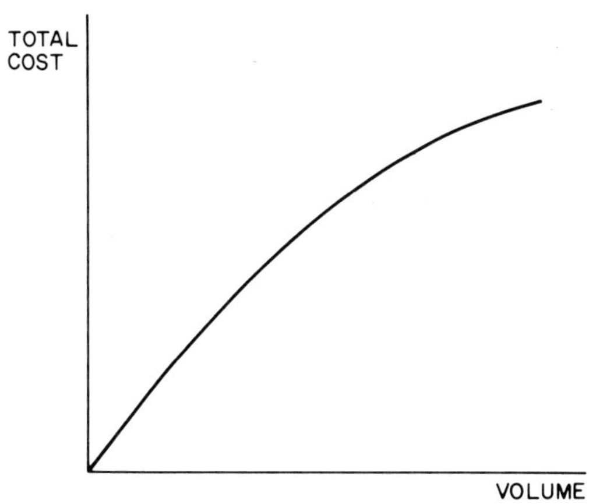

FIG. 5.3. A concave cost function yielding economies of scale.

In Fig. 5.4—the cost function is convex, corresponding to *diseconomies of scale*, i.e. the production cost for each additional unit increases as the size of the operation increases.

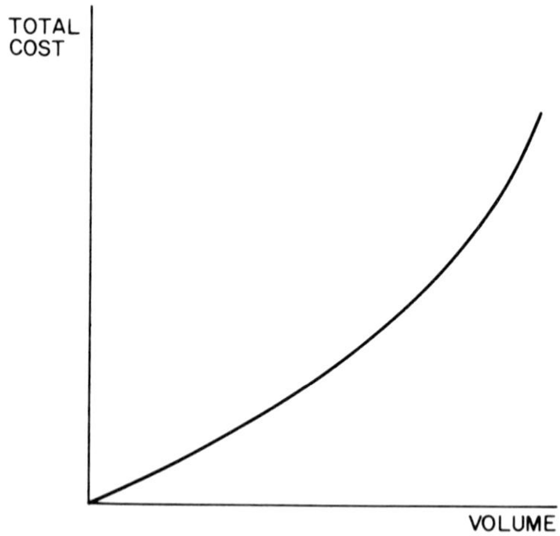

FIG. 5.4. A convex cost function involving diseconomies of scale.

In Fig. 5.5—the cost function starts off in a concave fashion with economies of scale, and then a point of inflexion is reached, beyond which diseconomies of scale set in.

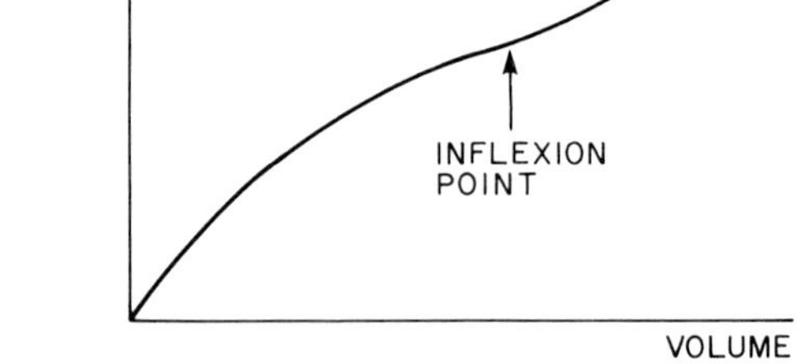

FIG. 5.5. A cost function with a point of inflexion.

For the purpose of analysis and decision making it is obviously important for management to know the characteristics of the cost function, both in the short term and in the long term (a point made by many economists, see for example [1]).

THE FIXED AND VARIABLE COST FUNCTION

It is convenient, for the sake of analyzing the effect of volume on cost, to differentiate between *fixed costs* and *variable costs*:

Fixed costs, denoted by F, include all the cost elements that remain unchanged, irrespective of a change in the level of output V. These costs often cover rent, rates and local taxes (where these are determined by the locality and type of premises used and not by turnover or profit), depreciation of machinery and plant, interest on borrowings to finance fixed assets, salaries and benefits of personnel who must be retained even when output declines, and any other fixed charges on services and facilities. When F is said to be fixed, it is fixed in relation to the output V, i.e. when V changes (all other things being equal), the value of F remains unaffected. There may, of course, be other factors that have an effect on F from one time period to another, such as a change in the level of rent, interest rates, fixed manning of machines, office staff, or changes in maintenance procedures, consequences of investment in fixed assets, etc. Any one or a combination of these changes may cause F to change to $F + \delta F$, the relative increment being $F^* = \delta F/F$.

Variable costs, denoted by $S(V)$, cover the cost items that are affected by a change in volume, such as direct labor and materials, maintenance of plant (except for maintenance which is needed even when the plant is idle), sales commissions, depreciation of fixed assets associated with their being in use (i.e. excluding depreciation attributed to the passage of time when the machinery is idle), fuel, etc.

This distinction between fixed and variable costs is not always easy to make. First, the segregation of cost elements into fixed and variable categories is often dictated by internal accounting conventions, which may be rooted in historical reasons and may seem arbitrary at times. Secondly, the fixed cost component tends to change by discrete amounts when the level of activity reaches certain levels.

In Fig. 5.6 we see that the initial fixed cost is F_0, but when output reaches V_1, there is a need to strengthen and improve the central facilities; the Chief Executive needs an assistant, another sales manager is appointed, more space is acquired, machines are replaced by faster and bigger ones, investment in

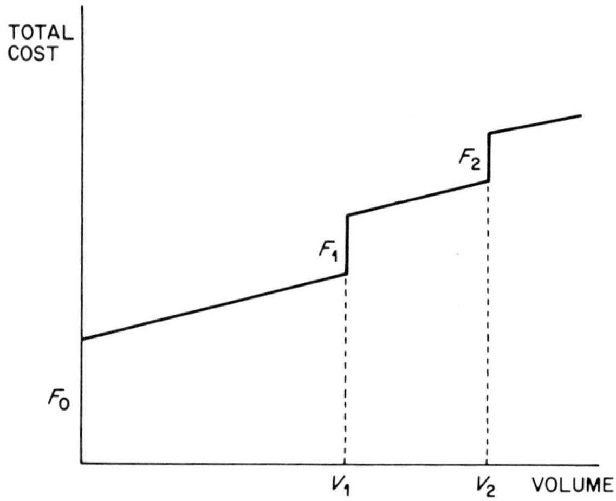

FIG. 5.6. Discrete increases in fixed cost at given output levels.

mechanization is made, the advertising budget is increased, and so on. Each of these additional cost items is justified by the larger scale of operations, but none can be attributed to increase in output from V to $V+1$. Nevertheless, the increase in the fixed cost from F_0 to F_0+F_1 as shown in Fig. 5.6 comes about when the volume reaches V_1, and a further discrete increase F_2 occurs at V_2, and so on. In other words, *the fixed costs F are not really fixed irrespective of V but in fact depend on V*. We choose to designate them as fixed costs because they *do not change at the margin*, i.e. when the volume increases or decreases by one unit. Needless to say, when numerous discrete increases take place and are spread along the output axis, as shown in Fig. 5.7, it may be preferable not to regard them as part of the fixed cost but incorporate them into the variable cost, represented by the fitted curve in Fig. 5.7. Usually, management is interested in the characteristics of the cost function for a limited range of output, for which the cost model can take the form

$$C(V) = F + S(V)$$

or

$$C = F + S \qquad (5.6)$$

for short, where S is the variable cost component (sometimes called *the cost of sales*) with the property that $S = 0$ when $V = 0$.

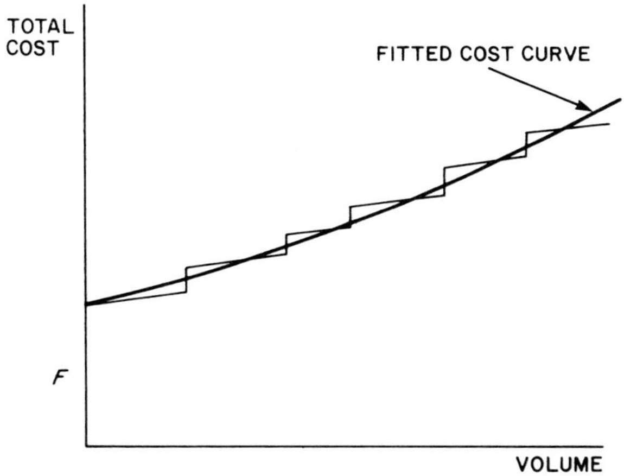

FIG. 5.7. Increments in fixed cost incorporated in the variable cost element by a fitted cost curve.

AVERAGE AND MARGINAL COST

The average cost of producing an output V is called the unit cost and is given by

$$c = \frac{C}{V} = \frac{F}{V} + \frac{S}{V} \tag{5.7}$$

The average unit fixed cost component is $c_F = \dfrac{F}{V}$

and the average variable cost component is $c_S = \dfrac{S}{V}$

hence the total unit cost is $c = c_F + c_S$

The marginal cost is defined as the change in cost when volume changes at the margin, and this is interpreted as a change by one unit from V to $V+1$ or from V to $V-1$. Strictly, the marginal total cost is

$$\frac{dC(V)}{dV}$$

but the usual interpretation is that

$$\frac{dC(V)}{dV} = C(V+1) - C(V) = C(V) - C(V-1)$$

From a mathematical point of view this interpretation is an approximation which holds good when V is relatively large compared with a postulated increment of one unit. The term *incremental cost* is sometimes used in the literature interchangeably with the *marginal cost*, though it is preferable to confine the latter to a change at the margin, leaving the former to refer to a change when the volume increment δV has an absolute value greater than one unit.

We can now proceed to determine the marginal costs as follows:
Marginal fixed cost—since F is independent of V it follows that

$$\frac{\mathrm{d}F}{\mathrm{d}V} = 0$$

Marginal variable cost $= \dfrac{\mathrm{d}S}{\mathrm{d}V}$

$$\textit{Marginal total cost} = \frac{\mathrm{d}C}{\mathrm{d}V} = \frac{\mathrm{d}S}{\mathrm{d}V} \tag{5.8}$$

Marginal unit fixed cost

$$\frac{\mathrm{d}c_F}{\mathrm{d}V} = \frac{\mathrm{d}}{\mathrm{d}V}\left(\frac{F}{V}\right) = -\frac{F}{V^2} \tag{5.9}$$

hence, if V increases, the marginal unit fixed cost declines.
Marginal unit variable cost

$$\begin{aligned}\frac{\mathrm{d}c_S}{\mathrm{d}V} &= \frac{d}{\mathrm{d}V}\left(\frac{S}{V}\right)\\ &= \frac{V\dfrac{\mathrm{d}S}{\mathrm{d}V} - S}{V^2}\\ &= \frac{1}{V}\left(\frac{\mathrm{d}S}{\mathrm{d}V} - c_S\right)\end{aligned} \tag{5.10}$$

from which it is clear that

$$\frac{\mathrm{d}c_S}{\mathrm{d}V} \gtrless 0 \qquad \text{if } \frac{\mathrm{d}S}{\mathrm{d}V} \gtrless c_S \tag{5.11}$$

Marginal total unit cost

Management is particularly interested in knowing whether the marginal total unit cost is positive or negative. From the above,

$$\frac{\mathrm{d}c}{\mathrm{d}V} = \frac{\mathrm{d}c_F}{\mathrm{d}V} + \frac{\mathrm{d}c_S}{\mathrm{d}V}$$

into which the results from Equations (5.9) and (5.10) may be substituted. Alternatively,

$$\frac{\mathrm{d}c}{\mathrm{d}V} = \frac{\mathrm{d}\left(\dfrac{C}{V}\right)}{\mathrm{d}V} = \frac{V\dfrac{\mathrm{d}C}{\mathrm{d}V} - C}{V^2}$$

$$= \frac{1}{V}\left(\frac{\mathrm{d}C}{\mathrm{d}V} - \frac{C}{V}\right)$$

$$= \frac{1}{V}\left(\frac{\mathrm{d}S}{\mathrm{d}V} - c\right) \tag{5.12}$$

The expression in the brackets is the difference between the marginal cost and the average unit cost; that difference divided by the output gives the rate at which the unit cost changes at the margin with a change in V. It follows that

$$\frac{\mathrm{d}c}{\mathrm{d}V} < 0 \qquad \text{if } \frac{\mathrm{d}S}{\mathrm{d}V} < c$$

i.e. *the unit cost declines with output when it exceeds the marginal cost per unit*; this is what was referred to earlier as the effect of economies of scale. The unit cost increases with output when the inequality sign reads the other way, so that similar to Expression (5.11) we have

$$\frac{\mathrm{d}c}{\mathrm{d}V} \gtrless 0 \qquad \text{if } \frac{\mathrm{d}S}{\mathrm{d}V} \gtrless c \tag{5.13}$$

The incremental change in unit cost is given in Equation (5.4). From the form in Equation (5.6) we have

$$C^* = f_1 S^* + f_2 F^*$$

where f_1 and f_2 are the cost proportions

$$f_1 = \frac{S}{C} \text{ and } f_2 = \frac{F}{C}$$

hence

$$c^* = \frac{f_1 S^* + f_2 F^* - V^*}{1 + V^*}$$

and if we assume that F is not affected by V (i.e. $F^* = 0$), the effect of a change in V is given by

$$c^* = \frac{f_1 S^* - V^*}{1 + V^*} \tag{5.14}$$

The change in the unit cost is, therefore, determined by the following condition:

$$c^* \gtrless 0 \qquad \text{if} \quad f_1 S^* \gtrless V^* \tag{5.15}$$

When volume increases ($V^* > 0$) and with it the variable cost ($S^* > 0$), then clearly the total cost rises ($C^* > 0$) but it is possible for the unit cost to decline, provided

$$S^* < \frac{V^*}{f_1} \tag{5.16}$$

Now, since $f_1 \leqslant 1$ it follows that

$$S^* < V^* \tag{5.17}$$

is a sufficient condition to ensure that the unit cost declines.

Example of a quadratic cost function

Consider the following quadratic cost function

$$C = F + a_1 V + a_2 V^2 \tag{5.18}$$

where $a_1, a_2 > 0$. Here the variable cost is

$$S = a_1 V + a_2 V^2$$

the marginal cost

$$\frac{\mathrm{d}C}{\mathrm{d}V} = \frac{\mathrm{d}S}{\mathrm{d}V} = a_1 + 2a_2 V > 0$$

The average unit cost

$$c = \frac{C}{V} = \frac{F}{V} + a_1 + a_2 V$$

and the marginal average unit cost

$$\frac{\mathrm{d}c}{\mathrm{d}V} = -\frac{F}{V^2} + a_2$$

Hence, c first declines with V and then rises, the point of minimum being when

$$\frac{\mathrm{d}c}{\mathrm{d}V} = -\frac{F}{V^2} + a_2 = 0$$

i.e. at

$$V_{\mathrm{opt}} = \left(\frac{F}{a_2}\right)^{1/2} \tag{5.19}$$

Since the unit profit is

$$z = p - c$$

it follows that if $p =$ const (i.e. the unit price is independent of volume) then the unit profit z is maximized when c is minimized, i.e. at V_{opt}. Graphs are shown in Fig. 5.8 for a numerical example to illustrate how various functions change with V. The graphs shown cover the total cost C, the fixed cost F, the variable cost S, the average unit cost c, the marginal total cost dC/dV and the marginal unit cost dc/dV. Note the following:

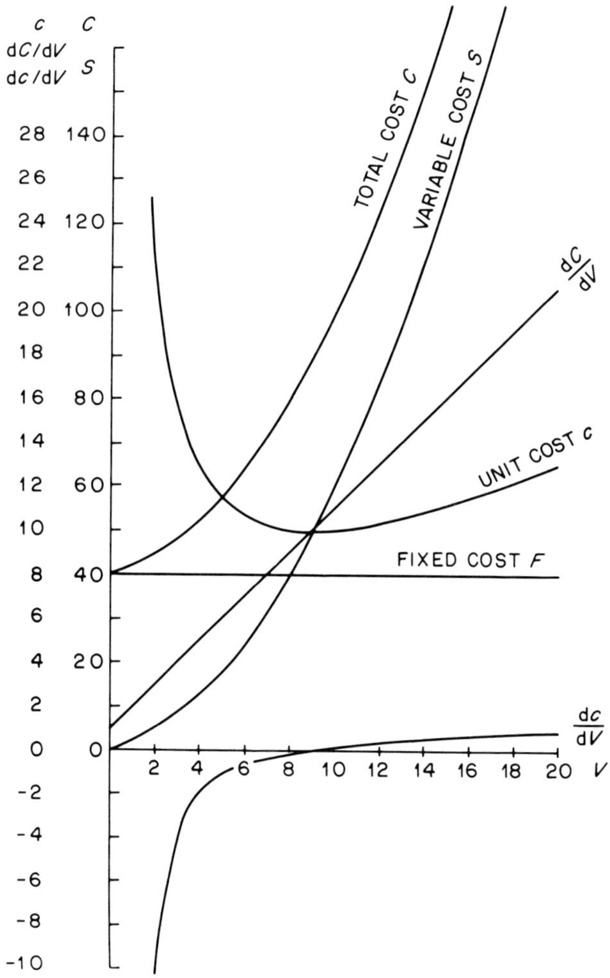

FIG. 5.8. An example to illustrate the properties of a cost function.

- The total cost C and variable cost S functions increase with V.
- The average unit cost c first declines, reaches a minimum at V_{opt} as shown in Equation (5.19) and then rises.
- The marginal unit cost $\mathrm{d}c/\mathrm{d}V$ is negative while $V < V_{\text{opt}}$ but becomes positive beyond that point.
- The slow rise in the value of the marginal unit cost beyond V_{opt} causes the unit cost c to be less sensitive to a change in volume beyond V_{opt} than below it.
- V_{opt} occurs at the point $c = \mathrm{d}S/\mathrm{d}V$ as suggested in Equation (5.12).
- The marginal cost function $\mathrm{d}C/\mathrm{d}V$ increases linearly with V.

The incremental relative change in unit cost for this example is given in (5.14), and the condition for c^* to be negative (i.e. unit cost to decline) is stated in (5.16). This condition is compatible with the conclusion derived from (5.12), as shown by the following: for relatively small increments the approximation

$$\frac{\delta S}{\delta V} \simeq \frac{\mathrm{d}S}{\mathrm{d}V}$$

may be used, or

$$\delta S \simeq \frac{\mathrm{d}S}{\mathrm{d}V}\,\delta V$$

so that

$$S^* = \frac{\delta S}{S} \simeq \frac{\mathrm{d}S}{\mathrm{d}V}\frac{\delta V}{V}\frac{V}{S} = \frac{VV^*}{S}\frac{\mathrm{d}S}{\mathrm{d}V}$$

Substituting into (5.16) the condition for unit cost to decline when V increases becomes

$$\frac{VV^*}{S}\frac{\mathrm{d}S}{\mathrm{d}V} < \frac{V^*}{f_1}$$

but by definition

$$S = f_1 C$$

hence the condition is

$$\frac{\mathrm{d}S}{\mathrm{d}V} < \frac{C}{V} = c$$

which is the same as that given following Equation (5.12).

A numerical example for the quadratic cost function

If in the cost function (5.18) the specific values are:

$$F = 40, \qquad a_1 = 1, \qquad a_2 = 0.5, \qquad V = 16$$

then

$$C = F + a_1 V + a_2 V^2 = 184$$

Here the cost proportions are defined as

$$f_1 = \frac{S}{C} = \frac{144}{184} = 0.783 = f'_1 + f''_1$$

$$f'_1 = \frac{a_1 V}{C} = \frac{16}{184} = 0.087$$

$$f''_1 = \frac{a_2 V^2}{C} = \frac{128}{184} = 0.696$$

$$f_2 = \frac{F}{C} = \frac{40}{184} = 0.217$$

Hence in this case

$$\begin{aligned} S^* &= \frac{f'_1}{f_1} V^* + \frac{f''_1}{f_1} (V^2)^* = 0.111\, V^* + 0.889\, (V^2)^* \\ &= (0.111 + 1.778)\, V^* + 0.889\, (V^*)^2 \\ &\simeq 1.89\, V^* \end{aligned}$$

so that the condition (5.16) cannot be satisfied for an increase in volume ($V^* > 0$). The same result is obtained by considering (5.12), where

$$c = \frac{C}{V} = \frac{184}{16} = 11.5$$

but

$$\frac{dS}{dV} = a_1 + 2a_2 V = 17.0$$

so that the marginal variable cost exceeds the average unit cost and hence any increase in volume will cause the unit cost to rise.

TYPES OF COST MODELS

Following the preceding discussion and the arguments in Chapter 2, analysis of costs needs often to be undertaken in some detail, and for that purpose a disaggregation of the cost function becomes necessary. It is often said that a properly managed accounting system should be able to provide the disaggregated information in a form amenable to analysis. However, there are many ways in which cost items can be classified, depending on the conventions associated with any given accounting system and on the prevailing managerial control mechanisms, designed to assist the relevant executives in their task. There is no attempt to review here all the possible types of cost models, but the following four should be mentioned as the most commonly found in practice:

1. The factor inputs model
2. The linear cost model
3. The divisional or product cost model
4. The functional cost model

1. The factor inputs cost model

The framework of this model was outlined in Chapter 2, the total cost takes the form [2]

$$C = M + W + E \tag{5.20}$$

to account for the expenditure incurred in materials (M), manpower (W) and other expenses (E). Each of the three terms can be disaggregated further, as appropriate in any particular case, for example into various categories of materials, or different inputs of labor, or the specific components that make up the "other expenses". The advantage of this model is that it is resource input oriented, so that the contribution of each input can be monitored and analyzed with the aid of the network of productivity ratios and the scheme of input proportions discussed in Chapter 2. The unit cost then becomes

$$c = c_M + c_W + c_E \tag{5.21}$$

where c_M = unit material cost = M/V
c_W = unit labor cost = W/V
c_E = unit cost incurred by other expenses = E/V

and the effect of each resource input on the total unit cost can then be more easily identified than in some other model. The factor input cost model is analyzed in some detail in the next chapter.

2. The linear cost model

This model is a special case of the general fixed and variable cost function in Equation (5.6) and follows the form shown in Fig. 5.2. Within a relatively limited range of output values, the variable cost S is assumed to be a linear function of V, an assumption which is often acceptable for practical purposes. Such a linear relationship is also commonly found for the revenue function. Starting with the revenue R, in Fig. 5.9, it is first decomposed into the profit Z

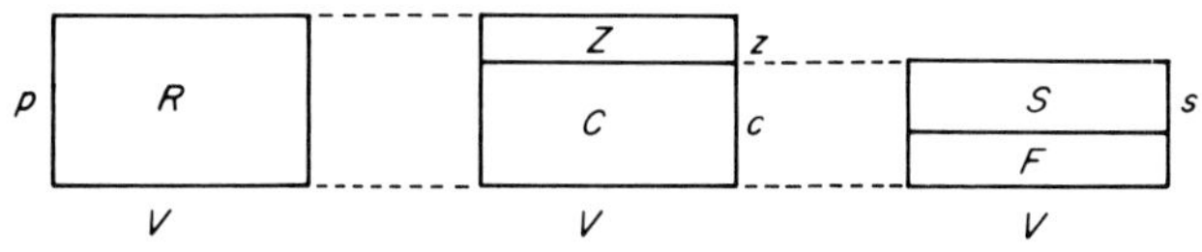

FIG. 5.9. Revenue and cost.

and the cost C, which in turn is decomposed into S and F. It is then convenient for all these, with the exception of the fixed element F, to be expressed as linear functions of V:

$R = pV$ where p is the average unit price
$Z = zV$ where z is the average unit profit
$C = cV$ where c is the average unit cost
$S = sV$ where s is the constant direct cost per unit of output

s is sometimes called *the incremental cost* or, preferably, *the marginal cost*, to indicate that the total cost C increases by the increment s when the volume increases by one unit, and similarly that C decreases by s when the volume decreases by one unit. Under these assumptions we have [3]:

The total cost function

$$C = F + sV \tag{5.22}$$

and the unit cost

$$c = \frac{F}{V} + s \tag{5.23}$$

where the fixed cost F can be subdivided into several categories, if needs be, such as

$$F = F_1 + F_2 + F_3$$

with each term in this sum reflecting particular cost items (for example, F_1 = rent, F_2 = depreciation charges, etc.), all being independent of the level of activity V. The linear total cost function, shown in Fig. 5.10, is convenient for break-even analysis. The total cost line intersects the revenue line at the point $R = C$, namely when

$$pV = F + sV$$

and the profit may then be expressed as

$$Z = \phi V - F \tag{5.24}$$

where $\phi = p - s$ is the marginal profit per unit, being the amount of money retained after the direct costs s are covered, so that ϕV is sometimes referred to as "the contribution of sales towards overheads". The break-even point, where

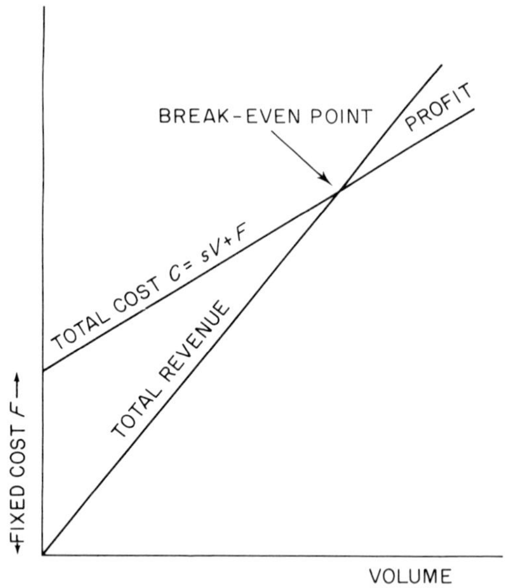

FIG. 5.10. The break-even point.

$Z = 0$, obtains when

$$V_{\text{bep}} = \frac{F}{\phi} \tag{5.25}$$

and the profit function becomes

$$Z = \phi(V - V_{\text{bep}}) \tag{5.26}$$

When $V > V_{\text{bep}}$ profit is made, and operational managers are constantly anxious to know the value of V_{bep} and to operate at a safe margin above it.

Other results for the linear cost function are summarized below:

- Marginal variable cost = marginal total cost = marginal cost
$$= \frac{\mathrm{d}S}{\mathrm{d}V} = \frac{\mathrm{d}C}{\mathrm{d}V} = s.$$
- Marginal unit variable cost $= \dfrac{\mathrm{d}s}{\mathrm{d}V} = 0.$
- Marginal total unit cost = marginal unit fixed cost $= \dfrac{\mathrm{d}c}{\mathrm{d}V} = -\dfrac{F}{V^2}.$
- Since $\mathrm{d}c/\mathrm{d}V < 0$ it follows that the unit cost continues to decline when V increases, so that economies of scale apply.

An example to illustrate the properties of the linear cost function is shown in Fig. 5.11. Note the following:

- The total cost line C and the variable cost S line are parallel.
- The vertical distance between them is F.
- The unit cost curve declines and converges towards the marginal value s.
- The marginal unit cost is always negative and converges to zero.

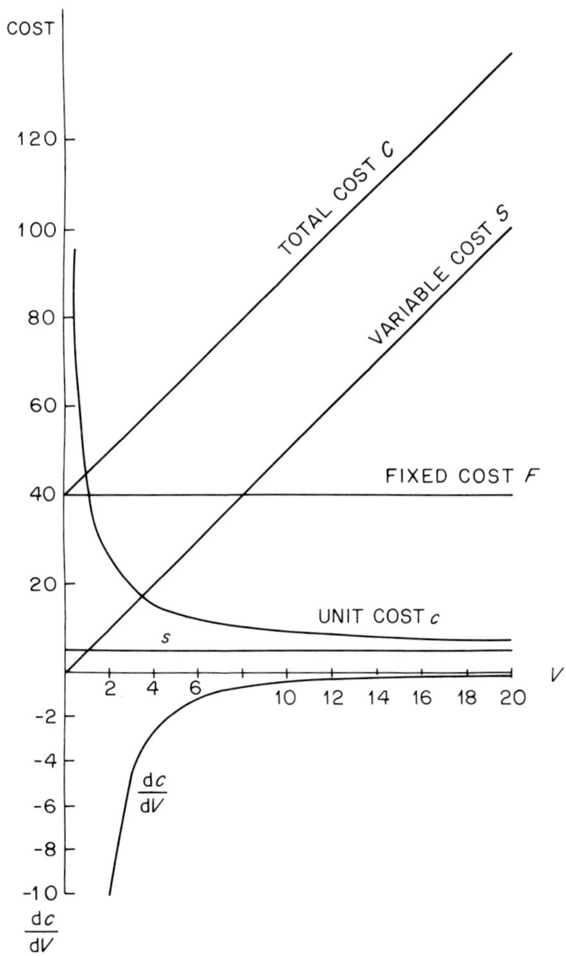

FIG. 5.11. Properties of the linear cost function (an example).

3. Divisional or product cost model

If the enterprise is organized on divisional or product lines, with each division or product constituting a separate cost or profit center, then the cost function is simply the sum of the cost centers

$$C = C_1 + C_2 + \ldots$$

Such a scheme implies that any central overhead costs (say, at the headquarters office) are allocated to the various business entities and are therefore included in C_1, C_2, etc. Alternatively, the central costs may be unallocated and treated as a separate item (say, F), so that

$$C = F + C_1 + C_2 + \ldots \tag{5.27}$$

with C_1, C_2 etc. being treated as a separate model as in Fig. 5.12.

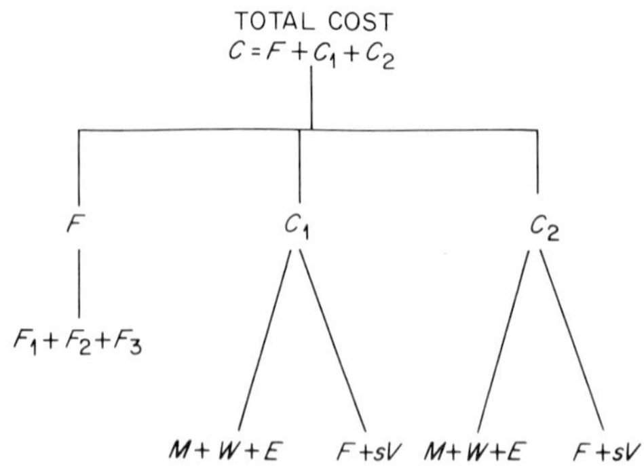

FIG. 5.12. A hierarchical cost structure.

The analysis in this volume is concerned with the first two models, which differ fundamentally from each other in their cost structure and purpose. As stated above, a judicious data system should make it possible for both models to coexist, as can be seen from the example in Table 5.1. The items in Table 5.1 are listed in three groups, which yield the sums M, W and E for Equation (5.20), and each item is categorized as fixed or variable (or divided into fixed and variable components), so that the total fixed cost F is determined as the sum of all the entries in its column, and similarly S is obtained, from which s is ascertained (by $s = S/V$) for Equation (5.22). The simple format in Table 5.1 can be replicated for a divisional/product cost model, and it is similarly possible to separate fixed and variable components in a functional cost model, when it is desirable to transform it to the format given in Equation (5.22).

It is, of course, possible to construct a hybrid model, again based on Table

TABLE 5.1
Data for two cost models.

Input	Fixed cost	Variable cost	Total
Raw materials		M_1	
Bought components		M_2	M
Materials for maintenance	M_3	M_4	
Fuel and energy	M_5	M_6	
Administration	W_1	W_2	
Production labor	W_3	W_4	W
Other labor costs	W_5	W_6	
Depreciation	E_1		
Interest charges	J_1	J_2	E
Rent	E_2		
Miscellaneous	E_3		
Total	F	S	

5.1, which follows the scheme shown in Fig. 5.13 and results in

$$C = F + M(V) + W(V) + E(V) \tag{5.28}$$

where all the fixed costs are aggregated under the column F in Table 5.1. $M(V)$, $W(V)$ and $E(V)$ denote materials, labor and other expenses, each expressed as a function of the output, so that the total variable cost consists of these elements

$$S = M(V) + W(V) + E(V) \tag{5.29}$$

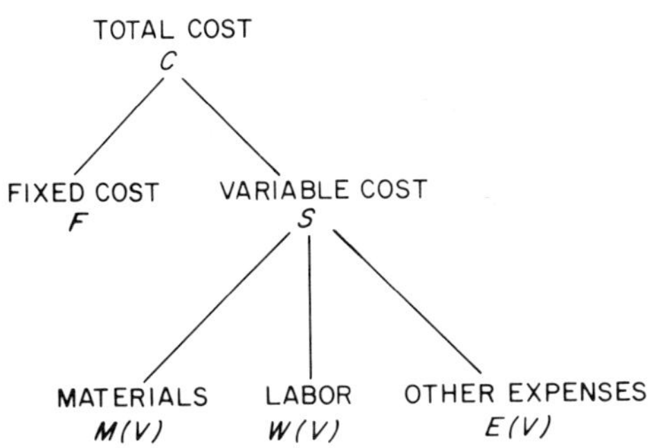

FIG. 5.13. A hybrid cost model.

Such a hybrid model combines the advantages of the factor inputs model and the fixed and variable cost model and may, therefore, be applicable in many circumstances.

A more detailed analysis of the two main cost models, with the aid of incremental calculus, follows in the next two chapters.

CONCLUSION

Total cost, and particularly unit cost, need to be closely controlled as a means of maintaining or improving the unit profit, the profit margin, the profit and the return on capital. The average unit cost is defined as

$$c = \frac{C}{V}$$

and the relative increment is given by

$$c^* = \frac{C^* - V^*}{1 + V^*}$$

so that an increase in unit cost is avoided when

$$c^* \leqslant 0 \quad \text{if } C^* \leqslant V^*$$

Economies of scale are said to exist when $c^* < 0$ for $V^* > 0$.

Many types of cost models can be constructed. Some allocate expenditure according to:

- organizational divisions
- geographical sites
- product groups
- manufacturing and service entities
- functions, such as production, marketing, distribution, personnel, etc.

This scheme of identifying cost centers is summarized in the cost function

$$C = F + C_1 + C_2 + \ldots$$

with F denoting the unallocated central cost. A further analysis of each cost center can then be undertaken, as shown in the hierarchical cost structure in Fig. 5.12. The most commonly used detailed models are:

the factor inputs model $C = M + W + E$ to incorporate specific resource inputs
the linear model $C = F + sV$ to highlight the effect of output

The latter is a special case of the general fixed and variable cost model

$$C = F + S$$

for which the marginal total unit cost is

$$\frac{\mathrm{d}c}{\mathrm{d}V} = \frac{1}{V}\left(\frac{\mathrm{d}S}{\mathrm{d}V} - c\right)$$

so that as volume increases the unit cost does not rise when

$$\frac{\mathrm{d}c}{\mathrm{d}V} \leqslant 0 \qquad \text{if } \frac{\mathrm{d}S}{\mathrm{d}V} \leqslant c$$

A general hybrid model, which combines the factor inputs model with the fixed and variable cost model, is given by (see Fig. 5.13)

$$\begin{aligned} C &= F + S \\ &= F + M(V) + W(V) + E(V) \end{aligned}$$

where the last three terms are each a function of the output V.

REFERENCES

[1] Brigham, E. F. and Pappas, J. L. (1972). "Managerial Economics". The Dryden Press.

[2] Cosmetatos, G. P. and Eilon, S. (1981). Analysis of unit cost, a modelling approach. In "Production Management Systems" (P. Falster and A. Rolstadås, Eds.). pp. 167–178. North-Holland Publishing Co.

[3] Eilon, S. and Cosmetatos, G. P. (1977). A profitability model for tactical planning. *Omega* **5**, 673–688.

EXERCISES

1. Consider a fixed fleet of aircraft of an airline company. Suggest what cost data would be useful to construct a cost model.
2. If the total cost function takes the form

$$C = F + a_1 V + a_2 V^2$$

   where $F = 40$, $a_1 = 1$ and $a_2 = 0.5$
   (a) Find whether the unit cost increases or decreases at the margin.
   (b) If the unit cost attains a minimum value, find it.
   (c) Plot the function of: total cost, unit cost, marginal cost, marginal unit cost.
3. For the linear total cost function, where $F = 40$ and $s = 5$, plot the following:

   The total cost C.
   The variable cost S.

The unit cost c.

The marginal unit cost $\dfrac{dc}{dV}$.

4. For the following data

$V =$	1	2	3	4	5	6	7	8	9	10
$C =$	30	40	49	57	64	72	81	90	100	110

where the fixed cost is 20, calculate:

The average unit cost.
The average variable cost.
The marginal cost.

5. A company has doubled its volume from 20 000 units to 40 000 per week while the cost has increased by 25% to 200 000. The selling price of the product is 6 per unit. Find the break-even point and the profit before and after the increase in output.
6. The average unit variable cost is known to follow the function

$$c_S = 10 + 2V^{-0.5}$$

while the average unit fixed cost is

$$c_F = 16\,000\,V^{-1}$$

If the selling price is 18 per unit, find:

The break-even point.
The profit.
The profit at break-even.
The marginal unit cost.

7. The total cost function takes the form

$$C = 100\,000 + 10V - V^2 + 0.5\,V^3$$

Determine whether economies of scale apply, the optimum value of V (if there is one), and the point at which the marginal unit cost begins to rise.
8. A hybrid cost model is constructed

$$C = F + M(V) + W(V) + E(V)$$

where F = fixed cost
$M(V) = a_1 V$
$W(V) = a_2 V + a_3 V^{1/2}$
$E(V) = a_4 C$

(a) Find the marginal cost and the marginal unit cost, and determine whether the unit cost function has a minimum.

(b) If $F = 2000$
$a_1 = 10$
$a_2 = 15$
$a_3 = -50$
$a_4 = 0.1$
calculate the following as a function of the output:

The unit cost.
The unit materials cost.
The unit labor cost.
The unit cost of other expenses.
The marginal unit variable cost.

Explain why the unit cost function is so flat and why $\mathrm{d}s/\mathrm{d}V > 0$ for $V > 1000$.

9. A plant is operating at output V, the ratio between V and the break-even point V_{bep} being k. The current price is double the marginal cost s and a linear cost function is assumed. Determine the relative incremental change for profit when changes take place in the values of ϕ, V and V_{bep}. Find the outcome for the following special cases and tabulate Z^* for various values of k (for $0.8 \leqslant k \leqslant 1.5$):
(a) $p^* = 0$, $V^* = 20\%$, $V^*_{\text{bep}} = 10\%$, $s^* = 10\%$
(b) $p^* = 0$, $V^* = V^*_{\text{bep}}$, $s^* = 0$

6
The Factor Inputs Cost Model

Themes in this chapter

Definition of the factor inputs cost model
Derivation of the relative increment of total cost and unit cost
Conditions leading the unit cost to decline
Decomposition of the wage cost into basic and overtime pay
Special cases when materials cost is proportional to output
The effect of basic hours remaining constant
Is it possible for the unit labor cost to decline while the average pay increases?
The effect of labor productivity

THE BASIC MODEL

The general hybrid cost model, given at the end of the last chapter, is of the form

$$C = F + M(V) + W(V) + E(V) \tag{6.1}$$

The factor inputs cost model is derived as a special case when $F = 0$, or—to be more precise—when the fixed cost elements are not separated from all the other cost items in the manner described in Table 5.1, but remain part of the factor inputs, resulting in the total cost function

$$C = M + W + E \tag{6.2}$$

and the unit cost, as stated in Chapter 5, is

$$c = \frac{C}{V} = c_M + c_W + c_E \tag{6.3}$$

where
$$\left.\begin{aligned} c_M &= \frac{M}{V} = \text{unit materials cost} \\ c_W &= \frac{W}{V} = \text{unit labor cost} \\ c_E &= \frac{E}{V} = \text{unit cost attributed to other expenses} \end{aligned}\right\} \tag{6.4}$$

In this formulation each of the inputs M, W and E may consist of fixed and variable components, and if need be the function (6.1) can be expanded to express them explicitly, though for the most part this chapter will not attempt to do so.

THE RELATIVE UNIT COST INCREMENT

From Equation (6.1) and the addition rule it follows that

$$C^* = f_M M^* + f_W W^* + f_E E^* \tag{6.5}$$

$$\left.\begin{aligned} \text{where } f_M &= \frac{M}{C} = \text{materials cost proportion} = \frac{M}{V}\frac{V}{C} = \frac{C_M}{c} \\ &= \text{materials unit cost proportion} \\ f_W &= \frac{W}{C} = \frac{C_W}{c} = \text{labor cost (or labor unit cost) proportion} \\ f_E &= \frac{E}{C} = \frac{C_E}{c} = \text{other expenses cost (or unit cost) proportion} \end{aligned}\right\} \tag{6.6}$$

and the sum of these cost proportions

$$f_M + f_W + f_E = 1 \tag{6.7}$$

Similarly, the effect of changes on the unit cost is derived from Equations (6.2) and (6.3), so that (as shown in [1])

$$c^* = \frac{C^* - V^*}{1 + V^*} = \frac{1}{1 + V^*}(f_M M^* + f_W W^* + f_E E^* - V^*) \tag{6.8}$$

and when the absolute change in volume is small ($V^* \ll 1$), the approximation

$$c^* \simeq C^* - V^* = f_M M^* + f_W W^* + f_E E^* - V^* \tag{6.9}$$

is adequate. From (6.8) (see also (5.5)) the conditions for the unit cost to change are

$$c^* \gtrless 0 \qquad \text{if } C^* = f_M M^* + f_W W^* + f_E E^* \gtrless V \tag{6.10}$$

The following conclusions and special cases are worth noting.

1. The case $F^* = 0$

If the starting point of the analysis is the general hybrid model, given at the beginning of the chapter, then

$$C^* = f F^* + f_M M^* + f_W W^* + f_E E^* \tag{6.11}$$

where $f = F/C$ = cost proportion of the fixed cost element. In the special case where cost changes are attributed to a change in volume, it follows from the definition of fixed costs that

$$F^* = 0$$

in which case the last equation for C^* is reduced to that in (6.5), and the condition in (6.10) similarly applies, except that the sum of cost proportions in (6.7) has to be modified to

$$f_M + f_W + f_E < 1$$

2. Changes equal to changes in volume

If each of the three factor inputs increases proportionately as an increase in volume (we assume here for convenience that $V^* > 0$), namely if all the cost inputs in the total cost function (6.2) are variable costs involving

$$M^*, W^*, E^* = V^*$$

then from (6.10) the condition for the unit cost to decline becomes

$$c^* < 0 \qquad \text{if } f_M\, M^* + f_W\, W^* + f_E\, E^* < V^* \tag{6.12}$$

Now, since it is assumed that $V^* > 0$, the last condition is reduced to

$$f_M + f_W + f_E < 1$$

but this condition is incompatible with the statement in (6.7), which generally holds when $F = 0$ and is also derived from Equation (6.8) for the case when the unit cost remains unchanged ($c^* = 0$). For the unit cost to decline, it is therefore necessary for at least one of the factor inputs to increase by less than the relative increase in volume, in order to satisfy (6.12). For example, a sufficient condition is

$$M^*, E^* = V^*; \qquad W^* < V^*$$

in other words, materials and other expenses rise in proportion to volume but the increase in labor cost lags behind. It is, of course, even possible for one or two factor inputs to rise by a higher percentage than the volume, provided this is sufficiently compensated for by a lower rise in the remaining input(s).

Consider the following example in Table 6.1, where the volume increase from period 1 to period 2 amounted to 10%:
Here,

$$f_M = 0.3, \qquad f_W = 0.4, \qquad f_E = 0.3$$

$$M^* = 10.8\,\%, W^* = 6.0\,\%,\ E^* = 12.0\,\%,$$

and

$$C^* = 9.3\,\% < V^* = 10\,\%$$

TABLE 6.1
An example.

	Materials M	Labor W	Other expenses E	Total C
Period 1	480	640	480	1600
Period 2	532	678.4	537.6	1748
Change (%)	10.8	6.0	12.0	9.2

The composition of C^* is found by

$$\begin{aligned} C^* &= f_M M^* + f_W W^* + f_E E^* \\ &= 0.3 \times 0.108 + 0.4 \times 0.06 + 0.3 \times 0.12 = 0.092 \text{ or } 9.2\% \end{aligned}$$

which satisfies the condition (6.12). In this case two factor-input increases exceed that of V^*, but the relatively low increase in labor cost, coupled with its relatively high cost proportion, causes the unit cost to decline, c^* being (by (6.8))

$$c^* = \frac{1}{1.10}(0.092 - 0.10) = -0.007$$

i.e. a decline of 0.7 %. Notice, incidentally, that the approximation (6.9), namely $c^* \simeq C^* - V^*$ yields here the very close result of -0.8%.

3. Materials cost proportionate to volume

It is often the case that while W and E do not vary proportionately with the volume V, *materials costs do tend to change proportionately with the volume*, so that $M^* = V^*$. The condition (6.12) for c to decline then becomes

$$f_W W^* + f_E E^* < (1 - f_M) V^*$$

but as

$$1 - f_M = f_W + f_E \qquad \text{(from equation (6.7))}$$

then

$$c^* < 0 \qquad \text{if } W^* - V^* < -\frac{f_E}{f_W}(E^* - V^*) \tag{6.13}$$

This expression with an equality sign is shown graphically in Fig. 6.1, where $W^* - V^*$ is plotted against $E^* - V^*$ for various values of f_E/f_W. The condition in (6.13) is satisfied below the line that corresponds to a given value of f_E/f_W.

Also, the value of f_E/f_W is plotted in Fig. 6.2 for five values of f_E (from 0.1 to

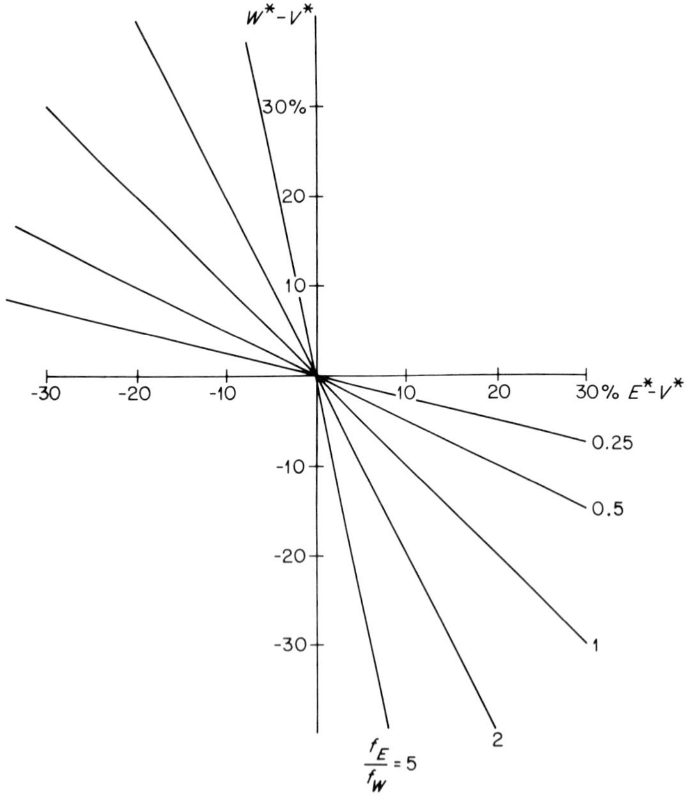

FIG. 6.1. The condition in (6.12).

0.5). The curves terminate at the points where $f_M = 0$ (i.e. whenever $f_W + f_E = 1$). For example, the top curve for $f_E = 0.5$ must terminate at point A, since beyond it f_W is greater than 0.5, and that would be contrary to the requirement that $f_M + f_W + f_E = 1$. It follows that at point A the value of f_M must be zero, as it must be at the termination points B, C and D. A line drawn through these points corresponds to this value of f_M.

Similarly, other iso-f_M lines (i.e. lines on which the value of f_M is constant) can be drawn for $f_M = 0.1, 0.2$, etc., as shown by the broken lines in Fig. 6.2. Thus, as we approach the origin, the materials input increases while the factor inputs for labor and other expenses decrease. A capital-intensive plant with, say, $f_E > 0.3$ and a relatively low labor content, say, $f_W < 0.2$, belongs to the upper left shaded region in Fig. 6.2, where the f_E/f_W curves are rather steep, so that (for given values of the starred variables) as f_W declines and f_E increases, the condition in (6.13) is increasingly difficult to meet.

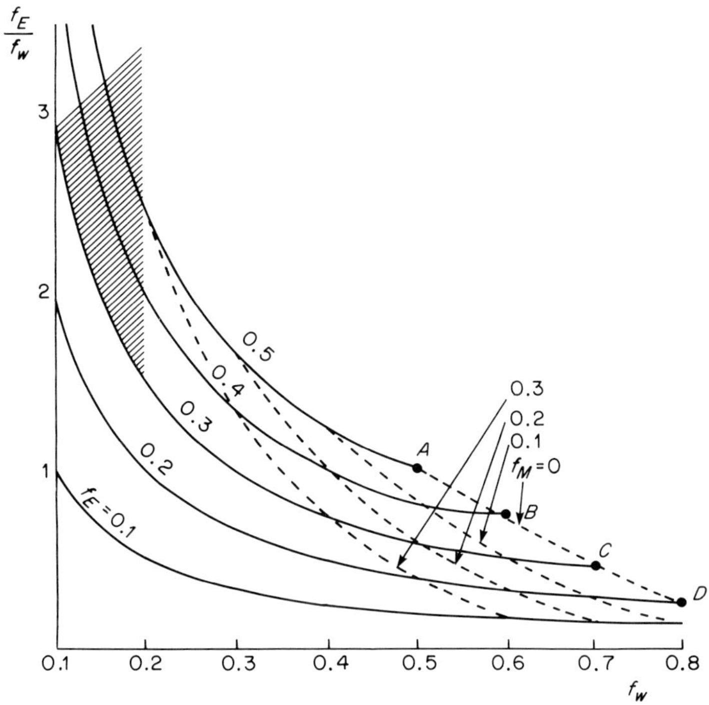

FIG. 6.2. The value of f_E/f_W.

Take, for example, the case where

$$f_W = 0.2 \text{ and } f_E = 0.4$$

so that $f_E/f_W = 2$ and the condition (6.13) becomes

$$W^* < -2E^* + 3V^*$$

this is shown in Fig. 6.3. If $V^* = 0$ and $E^* = 5\%$ then W will have to be reduced by at least 10% in this case. This requirement is mitigated (as is also clear from Fig. 6.1) by a volume increase ($V^* > 0$), and the larger this increase for a given value of E^*, the less pressure there is to reduce W. Thus, if $E^* = 5\%$ is coupled with $V^* = 2\%$, then W needs to be reduced by at least 4%, and if $V^* = 10\%$, then the wage cost W may even rise by up to 20% and yet the unit cost will decline. For the sake of comparison the line $E^* = 0$ is also shown in Fig. 6.3, demonstrating how the pressure on W^* is thereby further reduced. For any given value of E^* the domain below the line corresponds to a reduction in unit cost ($c^* < 0$).

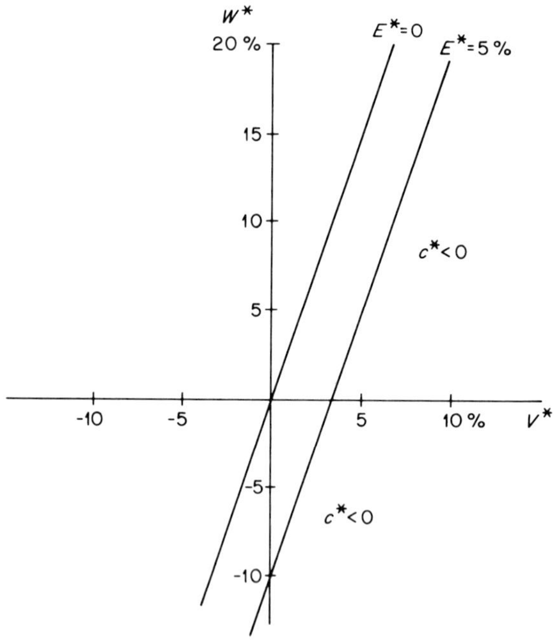

FIG. 6.3. The effect of V^* on the condition (6.12) for $f_E/f_W = 2$.

4. $M^* = V^* = 0$

For the special case where *material costs and volume remain unchanged*, i.e. $M^* = V^* = 0$, condition (6.13) becomes

$$W^* < -\frac{f_E}{f_W} E^* \tag{6.14}$$

This relationship is shown graphically in Fig. 6.1 (where $V^* = 0$ is substituted in the two axes). Also, if E increases ($E^* > 0$), for example owing to further capital investment, then the condition can be written as

$$-\frac{W^*}{E^*} > \frac{f_E}{f_W}$$

where the right hand size is given by the curves in Fig. 6.2.

5. $E^* = 0$; $M^* = V^*$

Another interesting special case is when *the other costs E do not change while materials costs are proportional to output*. Here

$$E^* = 0$$

$$M^* = V^*$$

which, substituted into (6.13), results in the condition

$$c^* < 0 \qquad \text{if } W^* < \left(1 + \frac{f_E}{f_W}\right) V^* \tag{6.15}$$

This relationship is shown in Fig. 6.4, where again $c^* < 0$ applies below the line corresponding to a given f_E/f_W value. Notice that as

$$1 + \frac{f_E}{f_W} > 1$$

a sufficient condition for (6.15) obtains when

$$W^* \leqslant V^*$$

This result is only to be expected, since if E remains constant while M and W increase proportionately with the output, then the unit input costs c_M and c_W

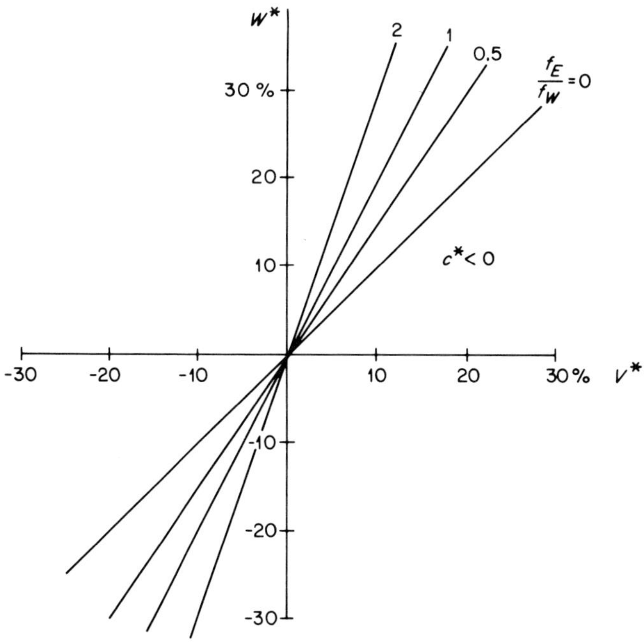

Fig. 6.4. The condition in (6.15) for $E^* = 0$, $M^* = V^*$.

remain constant and c_E falls as V increases, resulting in a decline in the total unit cost. Similarly, if under these circumstances the output falls, then unit cost will have to increase as a result.

DECOMPOSITION OF THE WAGE COST

The wage cost may be decomposed (as proposed in [1]) into

$$W = W_b + W_o$$

where W_b = wages for basic hours = $N\, w_b\, T_b$
W_o = wages for overtime hours = $N\, w_o\, T_o$
N = average number of employees
T_b = number of basic hours per employee
w_b = basic hourly rate
T_o = average number of overtime hours per employee
w_o = overtime rate per hour

Thus, the total average number of hours worked per employee is $T = T_b + T_o$.

In some cases it may well be appropriate to identify further constituents of W, for example

$$W = W_b + W_o + W_s + W_a$$

where W_b = $N\, w_b\, T_b$ for basic hours, as before
W_o = $N\, w_o T_o$ for overtime
W_s = $N\, w_s T_s$ for shift working, w_s being the hourly rate and T_s the number of shift hours worked per employee in addition to the basic and overtime hours
W_a = any ancillary benefits accruing to the employees, such as holidays, subsidized food, discounts, sick pay, various amenities, bonuses, etc. The total ancillary costs may be a function of the number of employees, in which case $W_a = w_a N$, where w_a is the average benefit per employee

Clearly, the degree of detail to which the total wage bill is disaggregated must depend on the circumstances and the purpose of the analysis. For the sake of simplicity, we shall proceed here with the function $W = W_b + W_o$, bearing in mind that when a more detailed analysis is justified, it can be developed on similar lines.

It is important to distinguish here between the following wage rates:

w_b = basic hourly rate, i.e. the hourly rate paid for basic hours
w_o = overtime hourly rate
w_H = average rate per hour, taking account of all hours worked (basic and overtime)
w = average pay per employee (consisting of basic and overtime pay) = W/N

As stated above,

$$W_b = N\, w_b\, T_b$$

$$W_o = N\, w_o\, T_o$$

The total number of hours worked by the workforce (basic time and overtime), denoted by H, is

$$H = N\,T = N\,(T_b + T_o)$$

Hence, the total wage cost is

$$\begin{aligned} W &= W_b + W_o \\ &= w\,N \\ &= w_H\,H \\ &= N\,(w_b\,T_b + w_o\,T_o) \end{aligned} \tag{6.16}$$

the average pay per employee is

$$w = \frac{W}{N} = w_b\,T_b + w_o\,T_o \tag{6.17}$$

and the average hourly rate is given by

$$\begin{aligned} w_H &= \frac{W}{H} = \frac{wN}{H} = \frac{w}{T} \\ &= \frac{1}{NT}\,(w_b\,T_b + w_o\,T_o) \end{aligned} \tag{6.18}$$

The relative change in wages then becomes

$$W^* = k_b\, W_b^* + k_o\, W_o^*$$

where $k_b = \dfrac{W_b}{W}$ = proportion of wages paid for basic hours = $\dfrac{w_b\,T_b}{w}$

$k_o = \dfrac{W_o}{W}$ = proportion of wages paid for overtime = $\dfrac{w_o\,T_o}{w}$

and

$$k_b + k_o = 1$$

Using the multiplication rule,

$$\begin{aligned} W_b^* &= (N\, w_b\, T_b)^* \\ &= N^* + w_b^* + T_b^* + N^*\, w_b^* + w_b^*\, T_b^* + T_b^*\, N^* + N^*\, w_b^*\, T_b^* \end{aligned}$$

and a similar expression can be obtained for W_o^*. For relatively small changes in the number of working hours and in the wage rates, we may use the approximations

$$W_b^* \simeq N^* + w_b^* + T_o^*$$

$$W_o^* \simeq N^* + w_o^* + T_o^*$$

so that (since $k_b + k_o = 1$)

$$W^* \simeq N^* + k_b\,(w_b^* + T_b^*) + k_o\,(w_o^* + T_o^*) \tag{6.19}$$

Special cases

(1) *If there is no change in working hours per employee and no change in the wage rate*, then

$$T_b^* = T_o^* = 0$$

and

$$w_b^* = w_o^* = 0$$

hence

$$w^* = 0 \quad \text{and} \quad W^* = N^*$$

i.e. a change in total wages is caused solely by a change in the number of employees, so that the total wage cost can decline only if the size of the workforce is reduced.

(2) *If basic hours and overtime hours change by the same percentage and if the percentage change is the same for basic rate and for overtime rate, then the average hourly rate changes by the same percentage as the basic rate.* Intuitively, this conclusion must apply, since the relative "mix" of basic hours and overtime hours remains the same and so is the relative "mix" of the corresponding hourly rates. In mathematical terms, given that

$$w_b^* = w_o^*$$

and

$$T_b^* = T_o^*$$

it follows that as $T = T_b + T_o$ then

$$\begin{aligned} T^* &= \frac{T_b}{T}T_b^* + \frac{T_o}{T}T_o^* \\ &= \left(\frac{T_b}{T} + \frac{T_o}{T}\right)T_b^* \\ &= T_b^* \end{aligned}$$

We know that the average pay per employee is, as in Equation (6.17),

$$w = w_b T_b + w_o T_o$$

hence, bearing in mind the definitions of k_b and k_o, we have

$$w^* = k_b (w_b T_b)^* + k_o (w_o T_o)^* \tag{6.20}$$

Here $w_b^* = w_o^*$ and $T_b^* = T_o^*$ and it follows that

$$w^* = (w_b T_b)^*$$

and since $w = w_H T$, we have

$$w^* = (w_H T)^*$$

hence $$(w_b T_b)^* = (w_H T)^*$$

and for $T_b^* = T^*$ this leads to the result stated above:

$$w_b^* = w_o^* = w_H^*$$

Furthermore, if there is no change in the hours per employee, i.e. if

$$T_b^* = T_o^* = T^* = 0$$

then

$$w^* = w_b^* = w_o^* = w_H^*$$

(3) *Another special case* is where $w_b^* + T_b^* = w_o^* + T_o^*$, and since $k_b + k_o = 1$ we have

$$W^* \simeq N^* + w_b^* + T_b^*$$

In conjunction with condition (6.14) for unit cost to decline it is clear that if $M^* = V^* = 0$, and if there is increased capital investment (i.e. $E^* > 0$), then W^* must become negative, which means it is necessary (but not a sufficient condition to ensure a reduction in unit cost) for $N^* + w_b^* + T_b^* < 0$. As hourly wage rates rarely decline, the total wage bill can obviously be reduced only by a fall in the number of employees and/or the number of paid hours worked.

(4) *A commonly found case is where the number of basic hours does not change* ($T_b^* = 0$) and any negotiations about hourly rates result in the same percentage change for regular time and for overtime. To signify that the two rates change by the same percentage, we introduce the notation w_{bo}^* for the relative change, i.e.

$$w_b^* = w_o^* = w_{bo}^*$$

Substituting the conditions prevailing in this case into Equation (6.20) yields

$$\begin{aligned} w^* &= w_{bo}^* + k_o T_o^* (1 + w_{bo}^*) \\ &\simeq w_{bo}^* + k_o T_o^* \end{aligned} \tag{6.21}$$

and the relative change in the total wage bill is

$$\begin{aligned} W^* &= N^* + w^*(1+N^*) \\ &= N^* + \left[w_{bo}^* + k_o T_o^* \, (1+w_{bo}^*)\right] (1+N^*) \\ &\simeq N^* + w_{bo}^* + k_o T_o^* \end{aligned} \tag{6.22}$$

All these approximations apply when the second order starred variables are ignored. Since wage rates tend to increase rather than decrease, i.e.

$$w_{bo}^* \geqslant 0$$

it follows that they contribute to an increase in both w and W.

A conflict of interests

Equations (6.21) and (6.22) reveal a fundamental conflict of interest between employees and management. The employees are anxious to increase their average pay, namely w, and from Equation (6.21) it is clear that they can achieve that objective primarily by an increase in the basic and overtime rates, and also by an increase in the amount of overtime worked. But these components also affect the total wage bill, as shown in Equation (6.22), in which the relative change in the number of employees also figures prominently. If output does not change, then management is concerned to reduce W if possible, and it is clear that to achieve both objectives, namely

$$w^* > 0$$

and

$$W^* < 0$$

it is necessary to reduce the number of employees ($N^* < 0$). This conflict of interest was referred to in Chapter 4 in connection with the multiplication rule for $W^* = (wN)^*$, where the interplay between the three starred variables W^*, w^* and N^* was demonstrated in Fig. 4.5.

In reality, the total wage bill is of less concern to management than the unit wage cost c_W. It is highly likely, for example, that when output rises, the wage cost will rise as well, and such an outcome, judged on its own, may well be warranted. Similarly, when the wage cost declines, as a consequence of a fall in output, the result may not necessarily be sufficiently reassuring. Thus, attention is naturally focused not on W but on c_W, since it is c_W that directly affects the unit cost and hence the unit profit. Management, therefore, aims to control the rate at which c_W is allowed to increase, and preferably it wishes to arrive at circumstances that cause c_W to decline.

Since

$$c_W = \frac{W}{V}$$

it follows that

$$c_W{}^* = \frac{W^* - V^*}{1 + V^*}$$

so that the general condition for the unit labor cost to decline is

$$c_W{}^* < 0 \qquad \text{if } W^* < V^* \tag{6.23}$$

THE EFFECT OF LABOR PRODUCTIVITY

Can management's desire for $c_W^* < 0$ (or at least $c_W^* \leqslant 0$) be compatible with the employees' aspiration for $w^* > 0$? The answer to this question lies with the *contribution of the productivity of labor* and whether it can be raised by an adequate amount (see, for example, [2]). Since account is taken here of the number of hours worked, it is appropriate to adopt the definition for the productivity of labor as (see Chapter 2)

$$\pi_H = \frac{V}{H} = \frac{V}{NT} \tag{6.24}$$

Starting from the identity

$$\frac{W}{V}\frac{V}{NT} = \frac{W}{N}\frac{1}{T}$$

and substituting

$$\frac{W}{V} = c_W$$

$$\frac{V}{NT} = \pi_H$$

$$\frac{W}{N} = w$$

we have

$$c_W \pi_H = \frac{w}{T} = w_H \tag{6.25}$$

where w_H = the average pay per hour. Thus, Equation (6.25) summarizes the following simple relationship (which is essentially the same as Equation (2.28) in Chapter 2 and follows directly from the definitions adopted):

> *The unit labor cost multiplied by labor productivity equals the average hourly rate of pay*

Clearly, for any given hourly rate, the labor content of the unit cost can only be reduced by higher productivity; similarly, higher productivity allows a higher hourly rate to be paid. From this equation we get

$$c_W^* + \pi_H^* + c_W^* \pi_H^* = w_H^* = \left(\frac{w}{T}\right)^* = \frac{w^* - T^*}{1 + T^*} \tag{6.26}$$

This relationship is shown in Fig. 6.5, which highlights the competition between c_W^* and w_H^*. One can only improve at the expense of the other when labor productivity is static ($\pi_H^* = 0$). The only way for both to "win" (shown by the shaded area) is when productivity increases ($\pi_H^* > 0$); the conflict between c_W and w_H becomes particularly acute when productivity falls. For relatively small changes, the following approximation applies

$$w^* \simeq c_W^* + \pi_H^* + T^* \tag{6.27}$$

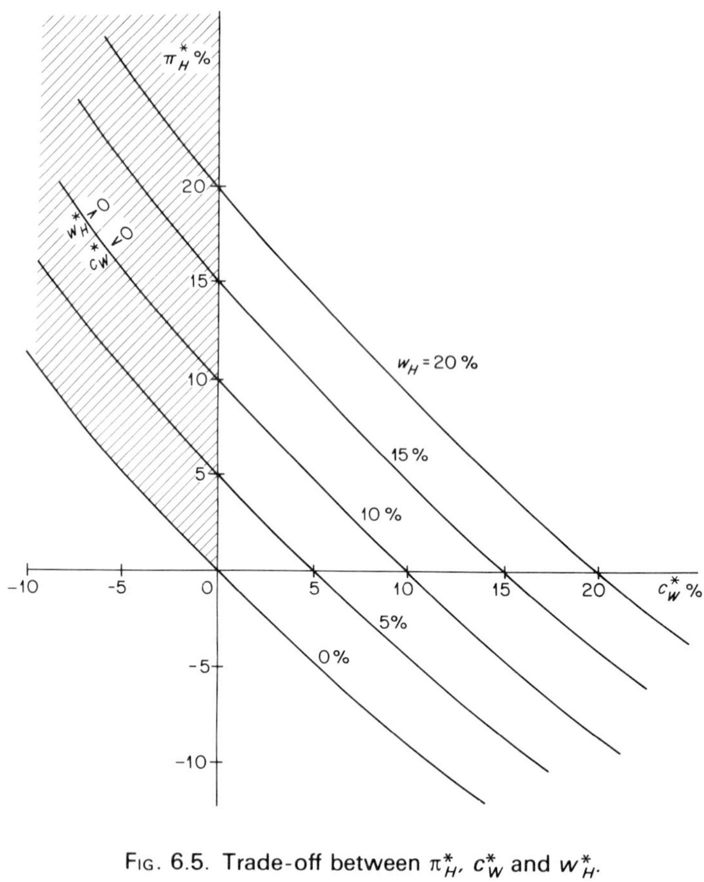

FIG. 6.5. Trade-off between π_H^*, c_W^* and w_H^*.

Measuring changes in labor productivity

This relationship can, incidentally, be used to measure changes in the level of labor productivity. For example, Table 6.2 summarizes data for two periods for the total wage bill, the number of employees, the number of hours worked and the labor unit cost. From this Table we find that the remuneration per employee was $w = W/N = 200$ in the base period, rising to 205 in period 2, i.e. $w^* = 2.5\%$. Also, we find that

$$c_W^* = \frac{7.864 - 8.200}{8.200} 100 = -4.1\%$$

Table 6.2
Operational data for two periods.

	Wage bill W	Employees N	Hours worked H	Labor unit cost c_W
Period 1	20 000	100	5000	8.200
Period 2	22 550	110	5808	7.864

The average hours per employee in period 1 was $\frac{5000}{100} = 50$ and in period 2 it rose to $\frac{5808}{110} = 52.8$, hence

$$T^* = \frac{52.8 - 50.0}{50.0} 100 = 5.6\%$$

The estimated change in productivity from period 1 to period 2 is obtained from the last approximation for w^*, namely

$$\pi_H^* \simeq w^* - c_W^* - T^* = 2.5 + 4.1 - 5.6 = 1.0\%$$

i.e. an increase of 1 %. A more accurate value is obtained from Equation (6.26), yielding

$$\pi_H^* = \frac{w^* - T^*}{(1 + T^*)(1 + c_W^*)} - \frac{c_W^*}{1 + c_W^*}$$
$$= \left(\frac{0.025 - 0.056}{1.056 \times 0.959} + \frac{0.041}{0.959}\right) 100 = 1.2\%$$

The discrepancy between the two values reflects the effect of the second order terms, which are ignored in the approximation.

Condition for $w^* \geqslant 0$

Equation (6.26) can now be used to determine the conditions for $w^* \geqslant 0$ and $c_W^* \leqslant 0$ respectively. The equation can be rewritten as

$$w^* = (c_W^* + \pi_H^* + c_W^* \pi_H^*)(1 + T^*) + T^* \tag{6.28}$$

or

$$c_W^* = \frac{w^* - T^*}{(1 + T^*)(1 + \pi_H^*)} - \frac{\pi_H^*}{1 + \pi_H^*} \tag{6.29}$$

Hence, the objective of the employees to achieve $w^* > 0$ can be met if the right-hand side of (6.28) is positive, so that for given values of c_W^* and T^* there is a

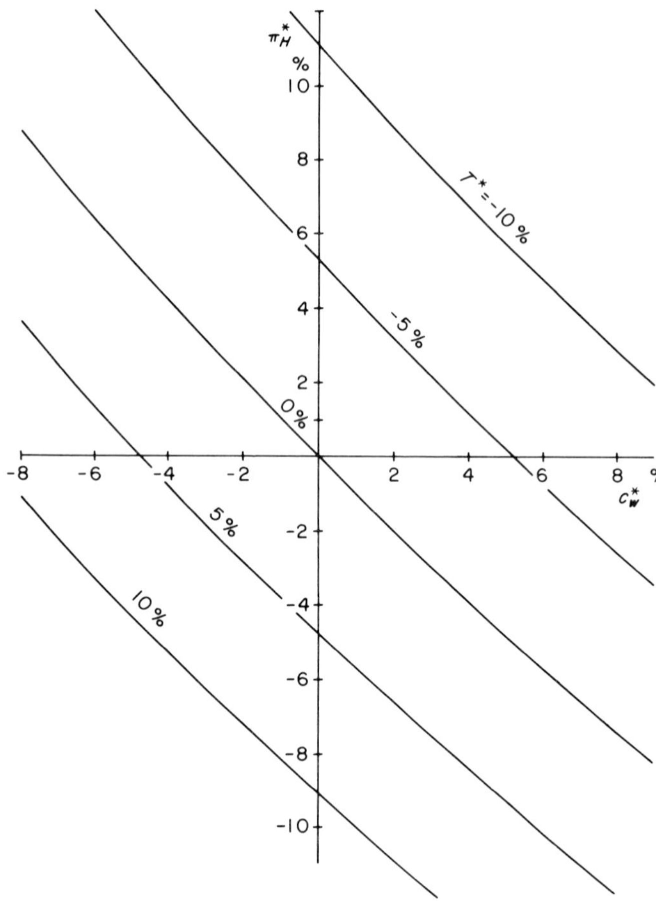

FIG. 6.6. Lower bound for π_H^* to achieve $w^* \geqslant 0$.

bound which the productivity of labor π_H must exceed for w to increase. If this bound is not achieved, then the specified value of c_W^* cannot be attained with the given value of T^*; alternatively, $w^* > 0$ will not be feasible. The expression (6.28) thus highlights the conflict between the relevant variables. This lower bound for π_H^* is given in Table 6.3 for various values of c_W^* and T^* and is also shown in Fig. 6.6.

Examples

Consider the following examples:

(1) Unit labor cost should not change ($c_W^* = 0$) and total number of hours per employee should be decreased by 10% ($T^* = -10\%$). Table 6.3 shows that these conditions, as well as $w^* = 0$, can be met if labor productivity increases by 11.1%.
(2) As above, except that it is intended to reduce the unit labor cost by 8%. This can be achieved if labor productivity increases by 20.8%.
(3) No change in hours is sought ($T^* = 0$) and the unit labor cost is allowed to increase by 6% ($c_W^* = 6\%$). Here the bound is $\pi_H^* = -5.7\%$, i.e. labor productivity can even fall and yet $w^* > 0$ is achievable (because the drop in productivity is absorbed by the increase in the cost element c_W).

Conditions for $c_w^* \leqslant 0$

Similarly, we can explore conditions to ensure that c_W does not rise and the lower bound for the productivity of labor π_H is tabulated in Table 6.4 and is also shown in Fig. 6.7. It is clear from the Table that a reduction in hours is possible either by a compensatory increase in productivity, or a decline in average pay per employee, or both. For example, a reduction of 10% in hours worked per employee, coupled with no change in productivity, can be achieved through a 10% reduction in the rate of pay. An increase in pay by 4% and a reduction of hours by 5% demands a 9.5% increase in productivity, otherwise the unit labor cost will increase.

Summary of conditions affecting c_w, w and π_H

There are, of course, many combinations of changes that can take place. Some are listed in Table 6.5 with their consequences for c_W, w and π_H. These results are derived directly from the foregoing analysis. One special case to note is where there is little change in the number of hours T (i.e. it is assumed that

TABLE 6.3

Lower bound for π_H^* to achieve $w^* \geqslant 0$ (figures in %).

c_W^* (%)	−10	−8	−6	−4	−2	0	2	4	6	8	10
$T^* = -10\%$	23.5	20.8	18.2	15.7	13.4	11.1	8.9	6.8	4.8	2.9	1.0
−5	17.0	14.4	12.0	9.6	7.4	5.3	3.2	1.2	−0.7	−2.5	−4.3
0	11.1	8.7	6.4	4.2	2.0	0	−2.0	−3.8	−5.7	−7.4	−9.1
5	5.8	3.5	1.3	−0.8	−2.8	−4.8	−6.6	−8.4	−10.2	−11.8	−13.4
10	1.0	−1.2	−3.3	−5.3	−7.2	−9.1	−10.9	−12.6	−14.2	−15.8	−17.4

TABLE 6.4

Lower bound for π_H^* to achieve $c_W^* \leqslant 0$ (figures in %).

w^* (%)	−10	−8	−6	−4	−2	0	2	4	6	8	10
$T^* = -10\%$	0	2.2	4.4	6.7	8.9	11.1	13.3	15.6	17.8	20.0	22.2
−5	−5.3	−3.2	−1.1	1.1	3.2	5.3	7.4	9.5	11.6	13.7	15.8
0	−10.0	−8.0	−6.0	−4.0	−2.0	0	2.0	4.0	6.0	8.0	10.0
5	−14.3	−12.4	−10.5	−8.6	−6.7	−4.8	−2.9	−1.0	1.0	2.9	4.8
10	−18.2	−16.4	−14.6	−12.7	−10.9	−9.1	−7.3	−5.5	−3.6	−1.8	0

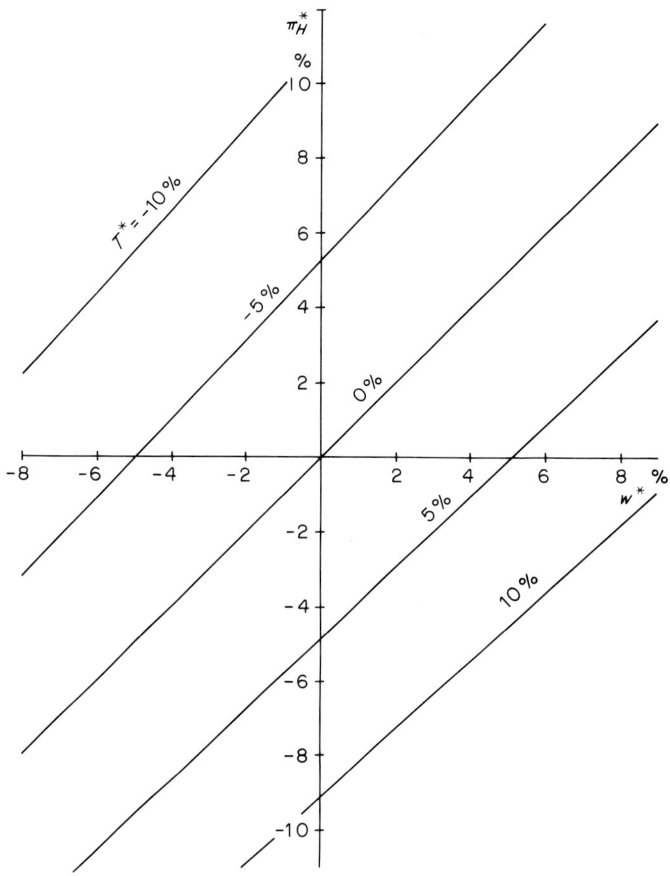

FIG. 6.7. Lower bound for π_H^* to achieve $c_W^* \leqslant 0$.

TABLE 6.5
Effect on c_W, w and π_H.

Conditions	Consequences for c_W^*	w^*	π_H^*	From Equation
$w_b^*, w_o^*, T_b^*, T_o^* \geqslant 0$		$\geqslant 0$		(6.20)
$W^* \leqslant V^*$	$\leqslant 0$			(6.4)
$V^* \geqslant H^*$			$\geqslant 0$	(6.24)
$c_W^*, \pi_H^*, T^* \geqslant 0$		$\geqslant 0$		(6.28)
$w^* \geqslant T^*$; $c_W^* \leqslant 0$			$\geqslant 0$	(6.26)
$W^* \geqslant N^*$		$\geqslant 0$		(6.17)
$w^* \leqslant T^*$; $\pi_H^* \geqslant 0$	$\leqslant 0$			(6.29)

$T^* = 0$), so that Equation (6.26) becomes

$$w^* = c_W^* + \pi_H^* + c_W^* \pi_H^* = (c_W \pi_H)^* \qquad \text{for } T^* = 0 \tag{6.30}$$

This equation perhaps best demonstrates the trade-off between unit labor cost and labor productivity, namely that for any given rise in pay per employee, productivity must rise by an adequate amount to avoid an increase in unit labor cost. This relationship is shown in Fig. 6.5, except that now $T^* = 0$ and hence $w_H^* = w^*$.

Of particular interest is *the case where an increase in productivity is linked with the same increase in overall pay*, so that (as discussed in [3])

$$w^* = \pi_H^*$$

In applying this case to Equation (6.29), where $T^* = 0$, we find that

$$c_W^*(1 + \pi_H^*) = 0$$

so that irrespective of how labor productivity changes, the result is $c_W^* = 0$, i.e. the whole benefit of increased productivity goes into pay and *the unit labor cost remains unaffected*. On the other hand, in the more general case where $T^* \neq 0$, the substitution of $w^* = \pi_H^*$ into Equation (6.29) yields

$$c_W^* = -\frac{T^*}{1 + T^*} \qquad \text{for } w^* = \pi_H^* \tag{6.31}$$

This function is tabulated in Table 6.6, from which it follows that c_W and T move in opposite directions (as shown in Fig. 6.8): when T decreases then c_W increases, and vice versa. The reason for this interesting result is that an increase in the number of hours T per employee helps to reduce the requirement for labor, so that the total wages tend to increase less rapidly than the increase in volume, thereby causing the component cost c_W to decline. The special case $T^* = 0$ in Equation (6.31) results in $c_W^* = 0$, as cited above.

The domain where $w^* > 0$ and $c_W < 0$

It is obviously desirable to explore the conditions that enhance the objective of the employees for higher remuneration per employee ($w^* > 0$) and that of management to reduce unit labor costs ($c_W^* < 0$). We derive first from (6.29) the condition for the latter, namely

$$w^* < \pi_H^*(1 + T^*) + T^*$$

Substituting

$$\pi_H^*(1 + T^*) + T^* = (\pi_H T)^*$$

$$= \left(\frac{V}{H}T\right)^* = \left(\frac{V}{N}\right)^* = \pi_N^*$$

TABLE 6.6

c_W^* for given values of T^* (when $w^* = \pi_H^*$).

T^* (%)	−20	−18	−16	−14	−12	−10	−8	−6	−4	−2	0
c_W^* (%)	25.0	22.0	19.0	16.3	13.6	11.1	8.7	6.4	4.2	2.0	0
T^*	0	2	4	6	8	10	12	14	16	18	20
c_W^*	0	−2.0	−3.8	−5.7	−7.4	−9.1	−10.7	−12.3	−13.8	−15.3	−16.7

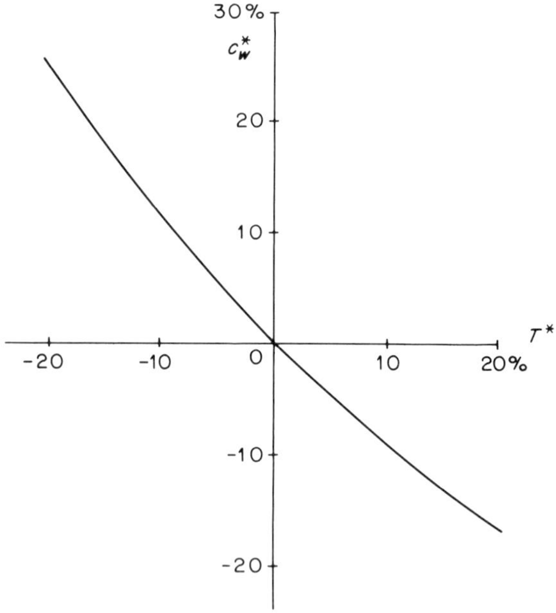

FIG. 6.8. c_W^* as a function of T^* for $w^* = \pi_H^*$.

we find that the condition is reduced to

$$w^* < \pi_N^*$$

i.e. the unit labor cost can decline if the rise in remuneration per employee is exceeded by the rise of productivity per employee. Hence the dual objectives for $w^* > 0$ and $c_W^* < 0$ can be achieved:

$$w^* > 0 \quad \text{and} \quad c_W^* < 0 \text{ when } \pi_N^* > 0 \tag{6.32}$$

Note that the attainment of $w^* > 0$ does not necessarily imply a rise in the average hourly rate w_H. The domain where both w_H increases while c_W declines is shown by the shaded area in Fig. 6.5.

Effect of labor productivity on the total unit cost

Turning to the total unit cost, consider the case where material costs are proportional to the output (i.e. $M^* = V^*$), so that the condition for unit cost to decline is given in Equation (6.13). Let us examine the case where $w_b^* = w_o^* = w_H^* = w^*$ and $T_b^* = T_o^* = T^* = 0$, from which it follows that

$\pi_N^* = \pi_H^*$ (the change in productivity per employee is the same as the change in productivity per hour). From

$$W^* = w^* + (1 + w^*)\,N^*$$

and

$$N^* = \frac{V^* - \pi_N^*}{1 + \pi_N^*}$$

we get

$$W^* = w^* + \frac{1 + w^*}{1 + \pi_N^*}\left(V^* - \pi_N^*\right) \tag{6.33}$$

and we may draw the following conclusions (we assume here that $w^* \geqslant 0$):

(1) if $w^* > \pi_N^*$ then the increase in the wage bill exceeds the increase in the pay rate ($W^* > w^*$) provided $V^* > \pi_N^*$, but will lag behind (i.e. $W^* < w^*$) if $V^* < \pi_N^*$.

(2) For the special case where an increase in productivity is rewarded by the same proportionate increase in pay rate, i.e. $w^* = \pi_N^*$, we have

$$W^* = V^*$$

and if $E^* = 0$ then the condition in (6.13) is met when the volume increases, so that the unit cost declines. However, if $E^* - V^* > 0$, the condition is not met and the unit cost will increase.

(3) It is further interesting to note that when the relative changes are comparatively small, then (6.33) becomes

$$W^* \simeq w^* + V^* - \pi_N^* \tag{6.34}$$

which shows clearly how increases in w and V increase the wage bill while an increase in labor productivity causes the wage bill to decline. Thus,

$$W^* - V^* \simeq w^* - \pi_N^*$$

and substituting this into Equation (6.13), then

$$w^* - \pi_N^* < -\frac{f_E}{f_W}(E^* - V^*) \tag{6.35}$$

This is the condition for the unit cost to decline. Hence, if $E^* - V^* > 0$ then a necessary (but not sufficient) condition for $c^* < 0$ is $\pi_N^* > w^*$, i.e. if the increase in E (owing to investment and capital charges) is proportionately higher than the relative increase in volume, then the relative increase in labor productivity must exceed the relative increase in the wage rate by an adequate amount, in order to achieve a reduction in unit cost. The condition (6.35) is shown in Fig. 6.1, except that the vertical axis should read $w^* - \pi_N^*$ instead of $W^* - V^*$.

CONCLUSION

The factor inputs cost model is given by

$$C = M + W + E$$

and the unit cost is

$$c = \frac{C}{V}$$

The relative unit cost increment is

$$c^* = \frac{1}{1 + V^*}(f_M M^* + f_W W^* + f_E E^* - V^*)$$

from which the following conditions need to be met in order to ensure that the unit cost declines:

(1) The general case is summarized in the condition (6.12).
(2) When $M^* = V^*$ then (6.13) and Fig. 6.1 apply.
(3) When $M^* = V^* = 0$, condition (6.13) is reduced to (6.14).
(4) If $M^* = V^*$ and $E^* = 0$ then (6.15) applies.
(5) If $M^* = E^* = V^*$ then the unit cost declines provided $W^* < V^*$.

Each of the factor inputs M, W and E can be decomposed for further detailed analysis. A decomposition of W allows the following variables and their respective effects to be accounted for:

hourly wage rates for basic time and for overtime
average hourly rate
average pay per employee
proportions of basic pay and overtime pay
number of basic hours and overtime hours
number of employees
productivity of labor per hour and per employee

The effect on the total wage bill is given in Equation (6.16) and on the pay per employee in (6.17) and (6.20), while the case for no change in basic hours is summarized in (6.21) and (6.22).

It is of particular interest to explore the circumstances under which it is possible for the labor unit cost to be contained ($c_W^* \leqslant 0$) while employee pay is allowed to increase ($w^* > 0$). Lower bounds of labor productivity for this purpose are given in Tables 6.3 and 6.4, following the maxim that *the unit labor cost multiplied by labor productivity equals the average hourly rate of pay*, i.e.

$$c_W \pi_H = w_H$$

The relationship between the relative increments of these variables, as well as the effect of changes in the pay per employee and the number of hours worked per employee, is stated in Equation (6.26). The domain for both $c_W^* < 0$ and $w^* > 0$ to hold is provided when the productivity per employee rises ($\pi_N^* > 0$).

In the special case where an increase in labor productivity is rewarded by an equal proportional increase in pay ($w^* = \pi_H^*$), and when the total hours worked per employee is constant ($T^* = 0$), the effect on the unit labor cost is that it remains unchanged ($c_W^* = 0$).

A decomposition of the materials cost element and analysis of the effect of productivity of materials can be developed on similar lines to the method adopted for the decomposition of the wage cost.

The potential conflict between aspirations to increase the level of remuneration per employee on the one hand and to reduce unit cost on the other, is further discussed in Chapter 13.

REFERENCES

[1] Cosmetatos, G. P. and Eilon, S. (1981). Analysis of unit cost, a modelling approach. In "Production Management Systems" (P. Falster and A. Rolstadås, Eds). pp. 167–178. North-Holland Publishing Co.

[2] Easterfield, T. E. (1965). Productivity—target or conceptual tool? *Operational Res. Q.* **16**, 177–187.

[3] Eilon, S. (1982). Effects of labour productivity and wage rate on costs. *Omega* **10**, 703–704.

EXERCISES

1. Data for monthly operations are summarized in the following Table:

	Volume (units)	Materials costs	Labor costs	Total costs
Current	10 000	20 000	22 000	50 000
Envisaged	10 500	21 200		

As the Table suggests, it is envisaged that volume should rise to 10 500 units per month and the estimated materials cost will then rise by 6 %. Other expenses are likely to rise by a percentage either (a) as that of the volume V or (b) as 1.2 times the percentage rise in total cost. Find in both cases the maximum level of labor costs that will not cause an increase in the unit costs.

2. (a) If other expenses E are constant and materials costs change in proportion to the labor costs, state the conditions that would lead to $c^* < 0$.
   (b) Is $W^* < V^*$ a sufficient condition in this case? Illustrate your answer with a numerical example.
3. In the input factor cost model, where $M^* = V^*$, is it sufficient for one or both of the following conditions to ensure that unit cost declines?
   (a) $W^* < V^*$, $E^* = 0$.
   (b) $E^* < V^*$.
4. (a) If basic hours and overtime do not change and if $w_o^* = w_b^*$ prove that $w_H^* = w_o^* = w_b^*$.
   (b) If $T_b^* = T_o^*$ and $w_b^* = w_o^*$, find the value of w_H^*.
   (c) If $T_b^* = 0, T_o^* \neq 0$ and $w_b^* = w_o^*$, show that $w_H^* \neq w_o^*$.
5. Prove that if $N^* = T_b^* = T_o^* = 0$ and if $w_b^* = w_o^* = w_H^*$, then $w_H^* = w^*$.
6. Explore how good the approximation in Equation (6.27) is.
7. Complete the following Table, where the starred variables refer to a comparison between one period and the previous period.

Period	W	N	T	H	V	w	c_W	π_H	T^*	w^*	c_W^*	π_H^*
1	20 000	100	5000		4000							
2	18 000	90	5000		3900							
3	21 000			506 000						−5.3%	2.7%	

8. (a) Find an expression for c_W^* for the case $T^* = 0$. Verify your conclusion by the example in the previous exercise when $T_1 = T_2 = 5000$.
   (b) Check the result for c_W^* for period 3 by use of Equation (6.29).
   (c) Test the approximation (6.27) for periods 2 and 3.
9. What is the relationship between c_W^* and w^* when $T^* = -\pi_H^*$?
10. When the total number of hours worked per employee and the remuneration per employee are unchanged, then the percentage reduction in the unit labor cost is equal to the percentage increase in labor productivity. True or false?
11. If the same increase in hourly rate is contemplated both for basic hours and for overtime hours, and if there is to be no change in the number of basic hours and in the number of employees, find the condition that will result in

$$w_H^* > w_b^*$$

namely, that the relative increase in the average hourly rate will exceed that of the relative increase in the basic hourly rate.

7
The Fixed and Variable Linear Cost Model

Themes in this chapter

Simple forms of unit cost models that incorporate:
fixed and variable costs
interest charges on borrowings
expenditure on promotion and advertising
Networks of relationships to show how a change in one variable triggers off a series of other changes
The relative incremental unit cost
Effect of changes in the working capital
Effect of increasing advertising expenditure to stimulate demand
Capital expenditure to reduce marginal costs (including increased mechanization and automation)

THE UNIT COST FUNCTION

In its simplest form the total cost model is given in Chapter 5 as

$$C = sV + F \tag{7.1}$$

so that the unit cost (as used in [1–4]) is

$$c = s + \frac{F}{V} \tag{7.2}$$

The interest charges to finance loans may be included either as part of marginal cost or the fixed cost, as indicated in Table 5.1, since some of the loans and bank overdrafts are intended for the financing of working capital requirements, while others are associated with capital expenditure on a medium or long-term basis. The high rates of interest experienced in recent years have had a significant effect on total costs; it may, therefore, be highly desirable to identify this cost component explicitly as a separate term in the cost function, and it is convenient to do so in the form

$$c = s + \frac{F}{V} + \frac{J}{V} \tag{7.3}$$

where $J = jB =$ the interest paid on the borrowings (loans and overdrafts) B at an average rate of interest j. Thus, J/V is the unit interest cost and it is excluded from the marginal cost s and the unit fixed cost element F/V. It is possible, of course, to segregate B into separate loans B_1, B_2, etc., each borrowed at its own rate j_1, j_2, etc., namely

$$B = B_1 + B_2 + \ldots$$
$$J = j_1 B_1 + j_2 B_2 + \ldots$$

with the average rate being $j = B/J$.

There are occasions where a further decomposition of the cost function is desirable, for example when a change in advertising expenditure is contemplated. Rather than have this cost item subsumed in the fixed cost F it may then be denoted as a separate term, i.e.

$$c = s + \frac{F}{V} + \frac{J}{V} + \frac{D}{V} \tag{7.4}$$

where $D =$ advertising and promotion expenditure. Other cost items, such as the cost of distribution, central computing and administration facilities, fuel and energy, etc., can be accommodated in a similar fashion, if singling them out is thought to be desirable. Thus, the methodology of analyzing the unit cost model by the use of incremental calculus is flexible and can be adapted to the needs of any given particular situation.

THE RELATIVE UNIT COST INCREMENT

Pursuing the unit cost model in Equation (7.3), we get (with the aid of incremental calculus)

$$c^* = f_1 s^* + f_2 \left(\frac{F}{V}\right)^* + f_3 \left(\frac{J}{V}\right)^*$$

$$= f_1 s^* + \frac{1}{1 + V^*}\left[f_2 F^* + f_3 J^* - (f_2 + f_3) V^*\right] \tag{7.5}$$

where the cost proportions are

$$\left.\begin{aligned} f_1 &= \frac{s}{c} \\ f_2 &= \frac{F}{cV} \\ f_3 &= \frac{J}{cV} \end{aligned}\right\} \tag{7.6}$$

so that

$$f_1 + f_2 + f_3 = 1 \tag{7.7}$$

Special cases

Three special cases are worthy of note:

(1) If $f_3 = 0$, i.e. the interest charges J are included in the fixed cost F and the unit cost function is as in Equation (7.2), then

$$c^* = f_1 s^* + \frac{f_2}{1 + V^*}(F^* - V^*) \tag{7.8}$$

(2) If, in addition, the fixed cost component remains unchanged, namely $F^* = 0$, then

$$c^* = f_1 s^* - \frac{f_2 V^*}{1 + V^*} \tag{7.9}$$

(3) Furthermore, if working practices and the cost of materials remain unchanged, so that the marginal cost s is constant and is unaffected by a change in volume (this is the basic assumption in (7.2)), then $s^* = 0$ and

$$c^* = -\frac{f_2 V^*}{1 + V^*} \tag{7.10}$$

Results for c^* of Equation (7.10) are tabulated in Table 7.1 and some are shown in Fig. 7.1, from which it is clear that the effect on the unit cost is not symmetrical with respect to a change in volume. For example, for $f_2 = 0.5$ a 20% increase in volume is expected to reduce the unit cost by 8.3%, whereas a decline of 20% in volume will increase c by 12.5%. The reason for this asymmetry stems from the denominator $1 + V^*$ in Equation (7.10) and it becomes more pronounced as the absolute value of V^* increases, so that for $f_2 = 0.5$ an increase of 50% in volume leads to $c^* = -16.7\%$, whereas a decrease of 50% in volume results in $c^* = 50.0\%$, thereby reflecting the effect of having to carry the fixed cost by a drastically reduced physical volume.

TABLE 7.1

The effect of a change in volume on unit cost (c^* in % for values of V^* and f_2, see Equation (7.10)).

V^* (%) =	−25	−20	−15	−10	−5	0	5	10	15	20	25
f_2 = 0.1	3.3	2.5	1.8	1.1	0.5	0	−0.5	−0.9	−1.3	−1.6	−2.0
0.2	6.7	5.0	3.5	2.2	1.1	0	−1.0	−1.8	−2.6	−3.3	−4.0
0.3	10.0	7.5	5.3	3.3	1.6	0	−1.4	−2.7	−3.9	−5.0	−6.0
0.4	13.3	10.0	7.1	4.4	2.1	0	−1.9	−3.6	−5.2	−6.7	−8.0
0.5	16.7	12.5	8.8	5.6	2.6	0	−2.4	−4.5	−6.5	−8.3	−10.0
0.6	20.0	15.0	10.6	6.7	3.2	0	−2.9	−5.5	−7.8	−10.0	−12.0
0.7	23.3	17.5	12.4	7.8	3.7	0	−3.3	−6.4	−9.1	−11.7	−14.0
0.8	26.7	20.0	14.1	8.9	4.2	0	−3.8	−7.3	−10.4	−13.3	−16.0
0.9	30.0	22.5	15.9	10.0	4.7	0	−4.3	−8.2	−11.7	−15.0	−18.0
1.0	33.3	25.0	17.6	11.1	5.3	0	−4.8	−9.1	−13.0	−16.7	−20.0

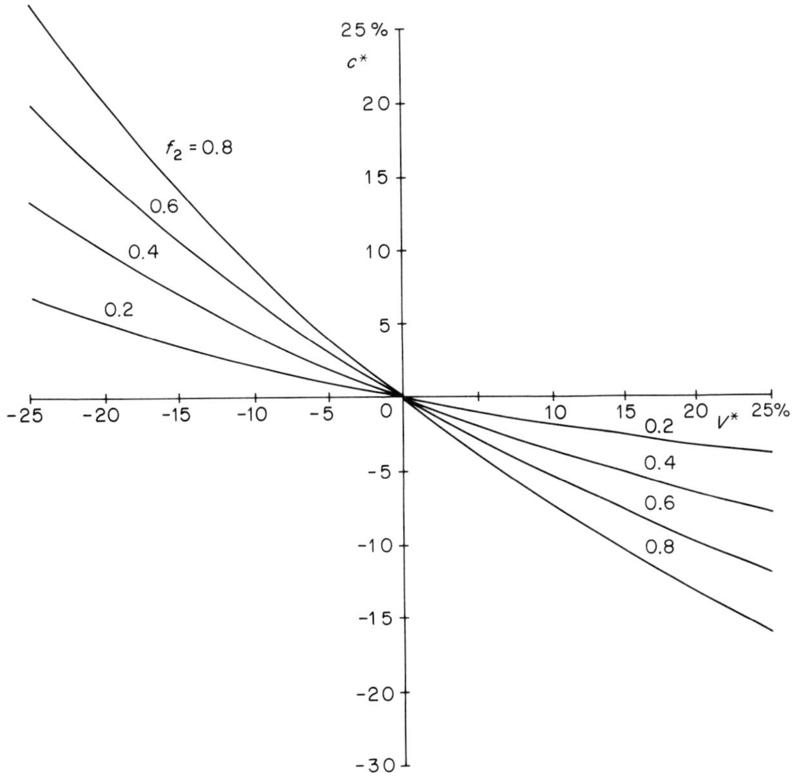

FIG. 7.1. Effect of change in volume on unit cost (for $s^* = F^* = 0$).

This result highlights first the general argument propounded for "economies of scale", which are aimed at reducing unit cost by building ever larger manufacturing facilities. Such an argument cannot be sustained if the larger plants cause the fixed cost element and/or the marginal cost to increase to an extent that they offset the otherwise expected benefit. Secondly, the effect of a substantial decline in volume (either because of competitive pressures or because of a general economic depression) is obviously so serious, that management then needs to resort to a drastic reduction in overhead fixed costs in a desperate attempt to contain the resultant increase in unit costs.

THE WORKING CAPITAL [1]

In Equation (7.3), the interest charge J is explicitly stated. This charge may be affected in two ways:

(1) a change in the average interest rate j owing to external factors;
(2) a change in the level of borrowings B owing to a change in the level of activity and/or costs incurred in the plant.

These effects are shown graphically in Fig. 7.2:

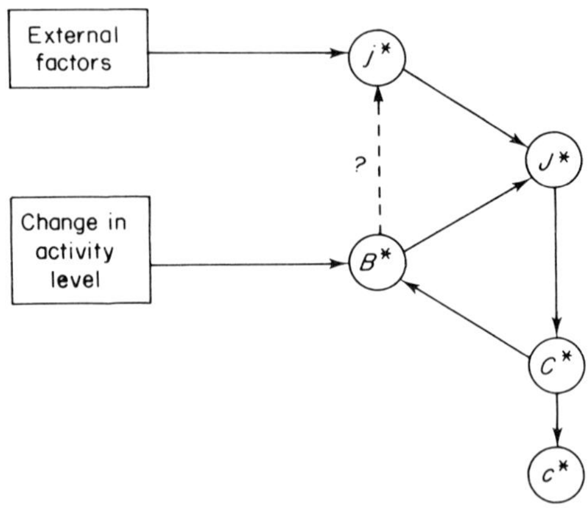

FIG. 7.2. Effect of interest rate and borrowings on costs.

- market rates cause a change j^*
- a change in the level of activity causes B^*
- both cause J^*
- the change in interest charges affects costs, hence C^*
- a change in cost may in turn affect B^*, etc.
- C^* results in c^*

Since the interest charge $J = jB$ it follows that

$$J^* = j^* + B^* + j^* B^* \tag{7.11}$$

If a change in interest rate occurs on its own, while $B^* = 0$, then $J^* = j^*$, i.e. J^* is caused solely by a change in market interest rates; if the level of borrowing changes, the rate j may or may not be affected as a result, depending on the level of borrowings already outstanding and the change in the level of risk involved (as perceived by the banks), hence the broken arc and the question mark in Fig. 7.2. We shall assume here that j is not affected by a change in B.

The borrowings B under consideration here are for the purpose of financing the working capital I_W, which is assumed to consist also of a sum K (being cash and equity finance), so that

$$I_W = K + B \tag{7.12}$$

and if it is further assumed that K is not subject to change ($K^* = 0$), then

$$I_W^* = \frac{B}{I_W} B^* = \lambda B^*$$

or

$$B^* = I_W^*/\lambda \tag{7.13}$$

i.e. the change in borrowings is caused by a change in working capital requirements, and the magnitude of the change in borrowings depends on $\lambda = (B/I_W)$, which is the proportion of the working capital financed by the borrowings. In the special case where the whole working capital is financed in this way, then $K = 0$ and $\lambda = 1$.

Changes in working capital

Economists, accountants and financial controllers do not postulate a single factor which causes a change in working capital. Since its primary purpose is to finance current operations, which largely depend on costs, one view is that a change in working capital is caused by a change in total cost, or

$$I_W^* = h_1 C^* \tag{7.14}$$

Another view is that working capital is needed to finance stocks of finished goods and debtors, so that the change in working capital is associated with a change in the total revenue

$$I_W^* = h_2 R^* \tag{7.15}$$

where h_1 and h_2 are parameters (or elasticities) that financial controllers can estimate, based on their past experience. Empirical evidence does not seem to universally support one hypothesis or the other, and it has been suggested [1] that a more general assumption may be made by incorporating both into

$$I_W^* = h_1 C^* + h_2 R^* \tag{7.16}$$

to allow the effect of both a change in revenue and a change in total cost to play their parts, and in this way the relationships postulated in (7.14) and (7.15) become special cases of (7.16).

The direct effect of any changes in the basic ingredients in the model (7.5), such as a change in s, F and j, or in any combination of them, causes c to change by c_0^* (see Fig. 7.3). This change causes a change in the working capital requirement, thereby triggering off the following sequence of changes:

- a change in the borrowings, by Equation (7.13)
- a change in the interest charge, by Equation (7.11)
- a change in the unit cost, by Equation (7.5)

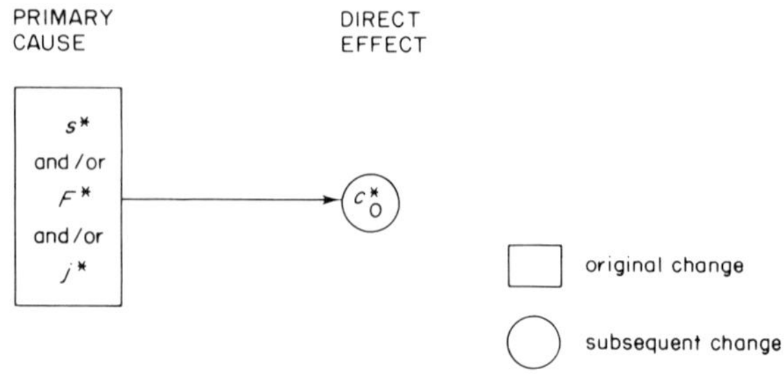

FIG. 7.3. The direct effect of s^* and/or F^* and/or j^* [3].

- a change in the working capital, by Equation (7.16)
- a change in the borrowings, by Equation (7.13)

and so on. This loop of secondary effects is shown in Fig. 7.4 and eventually the unit cost converges to its ultimate value, the total relative change amounting to c^*.

Note that the assumption throughout this book is that the same value V applies both to the production output and to the physical sales volume, in other words, the stock level of finished goods between the production system and sales is assumed to remain constant. In reality, however, this stock is subject to fluctuations, so that in a given time period the level of sales may either exceed the production volume (the excess being taken from stock), or may fall below the production rate (in which case the excess of production is added to stock). If such stock fluctuations occur, it may be necessary to make appropriate adjustments in the definition of V and also to account specifically for the effect of movements in the stock level on changes in the working capital.

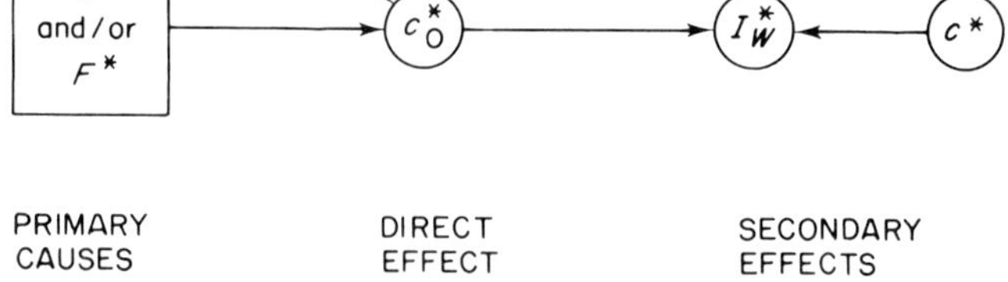

FIG. 7.4. The effect of s^* and/or F^* and/or j^* (adapted from [1]).

EFFECT OF WORKING CAPITAL CHANGES ON THE UNIT COST (for $p^* = 0$)

The case $V^* = 0$

The result for c^*, following the primary causes in Fig. 7.4, may be derived as follows:

(1) If it is assumed that $V^* = 0$, then the primary effect of the changes in s and/or F and/or j leads to

$$c_0^* = f_1 s^* + f_2 F^* + f_3 j^* \tag{7.17}$$

from Equation (7.5)

(2) The effect on the working capital is found from (7.16), where we assume $p^* = 0$ and $V^* = 0$, hence $R^* = 0$ and $C^* = c_0^*$, leading to

$$I_W^* = h_1 c_0^*$$

(3) Thus, the effect on the borrowings, by Equation (7.13), is

$$B^* = \frac{h_1}{\lambda} c_0^*$$

(4) The primary cause j^* has a direct effect on c_0^*, as already accounted for in Equation (7.17). In addition, it affects J^* by its interaction with B^*. Thus, in calculating J^* in this cycle, we have the additional effect of $B^* + B^* j^*$ or

$$J^* = (1 + j^*) B^*$$

Substituting B^* from the previous step, the additional relative interest charge is

$$J^* = (1 + j^*) \frac{h_1}{\lambda} c_0^*$$

(5) The next step is to find the new value of c^*. The primary causes s^* and/or F^* and/or j^* have all been accounted for in c_0^*, so that we need only note the effect of J^*, established in the previous step, i.e. from Equation (7.5), where s^* and F^* are disregarded, we have the added effect of the previous step

$$\begin{aligned} c_1^* &= f_3 J^* \\ &= (1 + j^*) \frac{f_3 h_1}{\lambda} c_0^* \end{aligned}$$

At the end of the first cycle, the cumulative effect of $c_0^* + c_1^*$ is

$$c_0^* (1 + u_1)$$

where

$$u_1 = (1 + j^*)\frac{f_3 h_1}{\lambda} \tag{7.18}$$

(6) Proceeding with the second cycle, the process is similar to that undergone in the first cycle, except that the starting point now is $c_0^*(1+u)$ instead of c_0^*. Hence,

similar to step 2 above: $I_W^* = h_1 c_0^*(1+u_1)$

following steps 3 and 4: $B^* = \dfrac{h_1}{\lambda} c_0^*(1+u_1)$

$$J^* = (1+j^*)\frac{h_1}{\lambda} c_0^*(1+u_1)$$

completing the cycle with: $c_2^* = f_3 J^*$

$$= \frac{f_3(1+j^*)h_1}{\lambda} c_0^*(1+u_1)$$

substituting u_1: $c_2^* = c_0^*(u_1 + u_1^2)$

At the end of the second cycle, the cumulative effect of $c_0^* + c_2^*$ is

$$c_0^*(1 + u_1 + u_1^2)$$

and so on. The ultimate value of c^* is, therefore,

$$c^* = c_0^*(1 + u_1 + u_1^2 + u_1^3 + \ldots)$$

$$= \frac{c_0^*}{1-u_1} \qquad \text{for } u_1 < 1 \tag{7.19}$$

The convergence of c^* in Equation (7.19) depends on the condition

$$u_1 = (1+j^*)\frac{f_3 h_1}{\lambda} < 1$$

or

$$h_1 < \frac{\lambda}{f_3(1+j^*)}$$

The meaning of the right-hand side is clarified when we substitute $\lambda = B/I_W$ and $f_3 = J/C$, yielding

$$h_1 < \frac{BC}{I_W J(1+j^*)} = \frac{C}{I_W j(1+j^*)}$$

where the numerator is the total cost and the denominator is the interest charge that would apply if the whole working capital I_W were to be borrowed at the new rate $j(1+j^*)$. A sufficient condition for h_1 is obtained when $f_3(1+j^*) \leqslant 1$, in which case

$$h_1 < \lambda$$

is sufficient for $u_1 < 1$ to be satisfied and for the convergence in Equation (7.19) to apply. Note that when the interest charge is not subject to change then

$$j^* = 0$$

and since $f_3 < 1$ it again follows that

$$h_1 < \lambda$$

is a sufficient condition to ensure $u_1 < 1$.

A simpler way of deriving c^* is found as follows (as in [1]): from Equation (7.5) for $V^* = 0$

$$c^* = f_1 s^* + f_2 F^* + f_3 J^*$$

Now

$$J^* = j^* + B^*(1+j^*)$$

and

$$B^* = \frac{I_W^*}{\lambda} = \frac{h_1 c^*}{\lambda}$$

hence

$$J^* = j^* + (1+j^*)\frac{h_1}{\lambda}c^*$$

substituting into c^*:

$$c^* = f_1 s^* + f_2 F^* + f_3 j^* + (1+j^*)\frac{f_3 h_1}{\lambda}c^*$$

Denoting

$$c_0^* = f_1 s^* + f_2 F^* + f_3 j^*$$

and substituting u_1 we get

$$c^* = c_0^* + u_1 c^*$$

hence

$$c^* = \frac{c_0^*}{1-u_1}$$

Example

If the cost proportion $f_3 = 0.2$, the coefficient $h_1 = 0.25$ and the borrowings proportion $\lambda = 0.1$ then

$$u_1 = (1+j^*)\frac{0.2 \times 0.25}{0.1} = 0.5(1+j^*)$$

We see that although λ is relatively small (only one tenth of the working capital is borrowed) the condition $u_1 < 1$ holds even for substantially high values of j^*. Suppose the interest rate increases by 20%, then

$$j^* = 0.2$$

and

$$u_1 = 0.6$$

so that

$$c^* = \frac{c_0^*}{0.4} = 2.5\, c_0^*$$

On the other hand, if in this case the whole of the working capital is financed by borrowings, then

$$\lambda = 1$$

and

$$u_1 = 0.06$$

resulting in

$$c^* = \frac{c_0^*}{0.94} = 1.06\, c_0^*$$

In the former example the result c^* is significantly higher than the initial value c_0^*, whereas in the latter case it is relatively close to c_0^*.

The case $V^* \neq 0$, $p^* = 0$

In the more general case, where $V^* \neq 0$, the network of relationships is shown in Fig. 7.5, where in addition to the direct effects of s^* and/or F^* and/or j^*, there is a direct effect of V^*, resulting in c_0^* given by

$$c_0^* = f_1 s^* + \frac{1}{1+V^*}[f_2 F^* + f_3 j^* - (f_2 + f_3) V^*] \tag{7.20}$$

The final result is

$$c^* = f_1 s^* + \frac{1}{1+V^*}[f_2 F^* + f_3 J^* - (f_2 + f_3) V^*]$$

where, as before,

$$J^* = j^* + B^*(1 + j^*) = j^* + \frac{I_W^*}{\lambda}(1 + j^*)$$

so that

$$c^* = c_0^* + \frac{1 + j^*}{1 + V^*} \frac{f_3}{\lambda} I_W^*$$

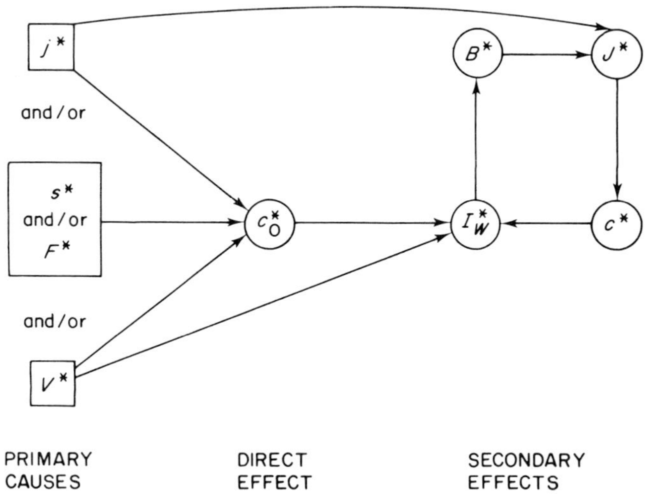

FIG. 7.5. The additional effect of V^*.

But now, by Equation (7.16),

$$I_W^* = h_1 C^* + h_2 R^*$$

If there is no change in unit price ($p^* = 0$), then $R^* = V^*$, hence

$$\begin{aligned} I_W^* &= h_1(c^* + V^* + c^* V^*) + h_2 V^* \\ &= (h_1 + h_2) V^* + h_1 c^*(1 + V^*) \end{aligned} \tag{7.21}$$

and the effect on the interest charge is obtained by substituting into J^*

$$J^* = j^* + \frac{1+j^*}{\lambda} [(h_1 + h_2) V^* + h_1 c^*(1 + V^*)]$$

Substituting J^* into c^* we get

$$c^* = c_0^* + \frac{(1+j^*) f_3}{\lambda} \left[(h_1 + h_2) \frac{V^*}{1+V^*} + h_1 c^* \right]$$

Denoting

$$\left.\begin{aligned} u_1 &= \frac{(1+j^*) f_3 h_1}{\lambda} \\ u_2 &= \frac{(1+j^*) f_3 h_2}{\lambda} \\ v &= (u_1 + u_2) \frac{V^*}{1+V^*} \end{aligned}\right\} \tag{7.22}$$

and the expression for c^* is reduced to

$$c^* = c_0^* + v + u_1 c^*$$

hence

$$c^* = \frac{c_0^* + v}{1 - u_1} \tag{7.23}$$

The result in Equation (7.19) is a special case of (7.23) obtained when $v = 0$, which occurs when $V^* = 0$. We thus note that in (7.23):

c_0^* is the direct effect of s^* and/or F^* and/or j^* and/or V^* as in Fig. 7.5 and given in Equation (7.20);

v is the added direct effect of V^* through I_W^* as in Fig. 7.5;

$\dfrac{1}{1-u_1}$ is the factor to apply to $c_0^* + v$ to represent the secondary effects;

c^* is the final outcome of the relative change in c due to s^* and/or F^* and/or j^* and/or V^*.

The convergence factor $1/(1-u_1)$ is tabulated in Table 7.2 for various values of h_2 and f_3/λ, and for $j^* = 0$, 0.1 and 0.2 respectively. When $u_1 \simeq 0$, which occurs when $h_1 \simeq 0$, then $1/(1-u_1) \simeq 1$ so that instead of (7.23) the approximation

$$c^* \simeq c_0^* + v \qquad (\text{for } u_1 \ll 1)$$

applies. However, as shown in Table 7.2, the convergence factor increases rapidly in value as h_1 increases, particularly when f_3/λ approaches 1.

Note that the results given here refer to the case where there is no change in unit price. The more general case, where $V^* \neq 0$ and $p^* \neq 0$, is discussed in Chapter 11.

Example

Consider the following case: to cater for an expected increase in volume by 15%, a modification in the plant is planned, whereby fixed costs will increase by 20% but marginal costs will remain unchanged. Currently, half the working capital is financed by borrowings, the working capital requirement coefficients are $h_1 = h_2 = 0.2$ and one fifth of the total cost is due to interest charges, the other two cost proportions being equal. The rate of interest is expected to rise from 10% to 12%.

Hence,

$s^* = 0$	$F^* = 20\%$	$j^* = 20\%$	$V^* = 15\%$
$h_1 = h_2 = 0.2$	$\lambda = 0.5$	$f_1 = f_2 = 0.4$	$f_3 = 0.2$

TABLE 7.2

The convergence factor $\frac{1}{1-u_1}$ for $j^* = 0$, 0.1 and 0.2.

$h_1 =$	0	0.1	0.2	0.3	0.4	0.5	0.6	0.7	0.8	0.9	1.0
$j^* = 0$											
$f_3/\lambda = 0.1$	1.000	1.010	1.020	1.031	1.042	1.053	1.064	1.075	1.087	1.099	1.111
0.2	1.000	1.020	1.042	1.064	1.087	1.111	1.136	1.163	1.190	1.220	1.250
0.4	1.000	1.042	1.087	1.136	1.190	1.250	1.316	1.389	1.471	1.563	1.667
0.6	1.000	1.064	1.136	1.220	1.316	1.429	1.563	1.724	1.923	2.174	2.500
0.8	1.000	1.087	1.190	1.316	1.471	1.667	1.923	2.273	2.778	3.571	5.000
1.0	1.000	1.111	1.250	1.429	1.667	2.000	2.500	3.333	5.000	10.000	–
$j^* = 0.1$											
$f_3/\lambda = 0.1$	1.000	1.011	1.022	1.034	1.046	1.058	1.071	1.083	1.096	1.110	1.124
0.2	1.000	1.022	1.046	1.071	1.096	1.124	1.152	1.182	1.214	1.247	1.282
0.4	1.000	1.046	1.096	1.152	1.214	1.282	1.359	1.445	1.543	1.656	1.786
0.6	1.000	1.071	1.152	1.247	1.359	1.493	1.656	1.859	2.119	2.463	2.941
0.8	1.000	1.096	1.214	1.359	1.543	1.786	2.119	2.604	3.378	4.808	8.333
1.0	1.000	1.124	1.282	1.493	1.786	2.222	2.941	4.348	8.333	100	—
$j^* = 0.2$											
$f_3/\lambda = 0.1$	1.000	1.012	1.025	1.037	1.050	1.064	1.078	1.092	1.106	1.121	1.136
0.2	1.000	1.025	1.050	1.078	1.106	1.136	1.168	1.202	1.238	1.276	1.316
0.4	1.000	1.050	1.106	1.168	1.238	1.316	1.404	1.506	1.623	1.761	1.923
0.6	1.000	1.078	1.168	1.276	1.404	1.563	1.761	2.016	2.358	2.841	3.571
0.8	1.000	1.106	1.238	1.404	1.623	1.923	2.358	3.049	4.310	7.353	25.000
1.0	1.000	1.136	1.316	1.563	1.923	2.500	3.571	6.250	25.000	–	–

From Equation (7.20),

$$c_0^* = \frac{1}{1.15}(0.4 \times 0.2 + 0.2 \times 0.2 - 0.6 \times 0.15)$$

$$= 0.026 \text{ or } 2.6\%$$

The value of u_1 is

$$u_1 = (1 + j^*)\frac{f_3 h_1}{\lambda} = 1.2\frac{0.2 \times 0.2}{0.5} = 0.096$$

and the convergence factor is

$$\frac{1}{1 - u_1} = 1.106$$

The added effect v on unit cost (here $h_1 = h_2$) is

$$v = 2u_1 \frac{V^*}{1 + V^*} = 0.025$$

i.e. almost the same as the value of c_0^*. Hence, the overall change in unit cost is

$$c^* = \frac{c_0^* + v}{1 - u_1} = (0.026 + 0.025)1.106 = 0.056 \text{ or } 5.6\%$$

Clearly, in this case an approximation $c^* \simeq c_0^*$ would grossly underestimate the effect of the envisaged changes, whereas the approximation $c^* \simeq c_0^* + v$ would yield 5.1%, which is only 10% lower than the expected value.

THE EFFECT OF ADVERTISING ON UNIT COST

Identifying promotion and advertising expenditure as a separate cost item (as explored in [2]) leads to the cost function in Equation (7.4), from which we get

$$c^* = f_1 s^* + \frac{1}{1 + V^*}[f_2 F^* + f_3 J^* + f_4 D^* - (f_2 + f_3 + f_4)V^*] \qquad (7.24)$$

where f_1, f_2 and f_3 are defined as before and $f_4 = D/C$ so that $f_1 + f_2 + f_3 + f_4 = 1$. Since the purpose of advertising is to stimulate further sales, a change in the level of expenditure on this item has two effects, as shown in Fig. 7.6: first, it affects the unit cost by the term $f_4 D^*$ in the last equation, and secondly it causes a change in demand in addition to whatever intrinsic change V_0^* the market experiences. It is, therefore, reasonable to postulate that the total change in the market is

$$V^* = V_0^* + V_D^*$$

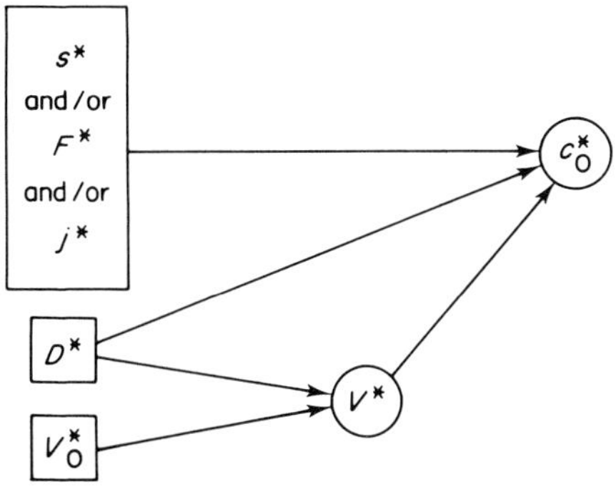

FIG. 7.6. The additional effect of D^*.

where V_D^* is due to the change in advertising expenditure and is given by

$$V_D^* = \sigma D^*$$

and σ may be described as the *promotion elasticity of demand* [2]. The concept of elasticity of demand is well known in economics and is discussed further in Chapter 8. Suffice it to say here that σ represents either the ratio found in practice between the percentage change in sales due to a known percentage change in advertising expenditure, or the expectation of the decision maker regarding the value of this ratio. Thus, $\sigma = 1$ simply means that if D increases by 10% then the volume increases by 10%.

The direct effect on unit cost takes the form of c_0^*, as shown in Fig. 7.6, which is similar to Equation (7.20), except that now the direct effect of D^* needs to be accounted for by Equation (7.24), i.e.

$$c_0^* = f_1 s^* + \frac{1}{1+V^*}(f_2 F^* + f_3 j^* + f_4 D^* - f V^*) \tag{7.25}$$

where

$$f = f_2 + f_3 + f_4 = 1 - f_1$$

The effects of j^* and V^* directly on J^* and I_W^* respectively, and all the secondary effects are shown in Fig. 7.5, and can be handled exactly as before. Thus, the result in (7.23) applies here, except that c_0^* is now given by the last equation.

An interesting special case is one where the fixed and marginal costs do not change and where the interest rate remains static, i.e. $s^* = F^* = j^* = 0$, reducing Equation (7.25) to

$$c_0^* = \frac{1}{1+V^*}(f_4 D^* - f V^*)$$

If, in addition, there is no change in demand other than that affected by a change in advertising expenditure, then $V^* = V_D^*$ or

$$V^* = \sigma D^* \tag{7.26}$$

and if γ is defined as

$$\gamma = \frac{f_4}{\sigma} - f \tag{7.27}$$

then the direct effect on the unit cost becomes

$$c_0^* = \frac{\gamma V^*}{1 + V^*} = \frac{\gamma \sigma D^*}{1 + \sigma D^*}$$

where the value of γ depends on the coefficients f, f_4 and σ. It is interesting to note that if γ and D^* are both positive or negative then c_0^* is positive, but if either γ or D^* is negative then c_0^* is negative. The value of γ is linear with f and f_4 and a simple tabulation is given in Table 7.3 for three alternative values of σ. A further conclusion is that $c_0^* = 0$ when $\gamma = 0$ or when

$$\sigma = \frac{f_4}{f}$$

and since

$$f = f_2 + f_3 + f_4 \geqslant f_4$$

TABLE 7.3
The value of γ.

$f_1 =$	0	0.2	0.4	0.6	0.8
$f_4 = 0$	−1.0	−0.8	−0.6	−0.4	−0.2
$\sigma = 0.5$					
$f_4 = 0.1$	−0.8	−0.6	−0.4	−0.2	0
0.2	−0.6	−0.4	−0.2	0	0.2
0.3	−0.4	−0.2	0	0.2	—
$\sigma = 1.0$					
$f_4 = 0.1$	−0.9	−0.7	−0.5	−0.3	−0.1
0.2	−0.8	−0.6	−0.4	−0.2	0
0.3	−0.7	−0.5	−0.3	−0.1	—
$\sigma = 2$					
$f_4 = 0.1$	−0.95	−0.75	−0.55	−0.35	−0.15
0.2	−0.9	−0.7	−0.5	−0.3	−0.1
0.3	−0.85	−0.65	−0.45	−0.25	−0.05

it follows that the denominator exceeds the numerator (assuming that the other cost proportions f_2 and f_3 are not zero), so that a necessary (but certainly not a sufficient) condition for $c_0^* = 0$ is that $\sigma \leqslant 1$. Similarly, for $c_0^* < 0$ (and assuming $D^* > 0$ and $\sigma > 0$) the condition is $\gamma < 0$ or

$$\sigma > \frac{f_4}{f}$$

so that $\sigma > f_4$ becomes a sufficient condition for such an outcome. In the event that the value of v in Equation (7.23) is negligible, these conclusions for c_0^* will apply to the total change c^* as well. However, it should be noted that as it is assumed here that D^* and hence V^* are positive, it follows from its definition in Equation (7.22) that $v > 0$, so that while c_0^* is negative the sum $c_0^* + v$ may become positive and the resultant unit cost c will then increase.

EFFECT OF MECHANIZATION

The simple cost function in Equation (7.2) can be used to explore the effect of capital expenditure on the unit cost. Usually such expenditure is undertaken for one or several of the following reasons [3]:

(1) A need to extend the production facilities to meet expanding demand
(2) A need to replace old plant
(3) A desire to increase the level of mechanization and automation and to use new technology in order to contain or reduce the ever increasing labor costs.

The effect of any of these causes is to increase the level of the fixed cost element in the cost function, but there is often an explicit intention in such investment decisions, namely to reduce the marginal cost element. This idea is based on a postulated trade-off between fixed costs and marginal costs, as shown in Fig. 7.7, so that while capital investment is expected to increase F, it is also expected to reduce s. The question is whether the end result would be a reduction in the total unit cost or not.

Our starting point [4] is Equation (7.8), namely

$$c^* = f_1 s^* + \frac{f_2}{1 + V^*}(F^* - V^*)$$

where F is assumed to include the interest charges J. Assuming two primary causes of change in F and V respectively, the network of relationships is shown in Fig. 7.8: an increase in the fixed investment I_F^* leads to F^* which causes s^*, and these combined causes of change affect the unit cost.

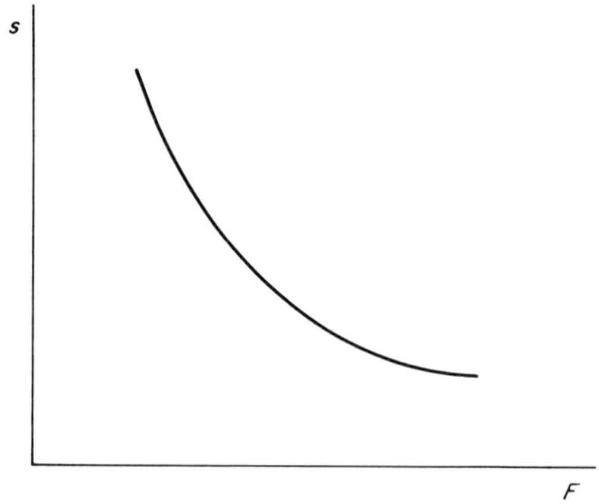

FIG. 7.7. Trade-off between fixed and marginal costs.

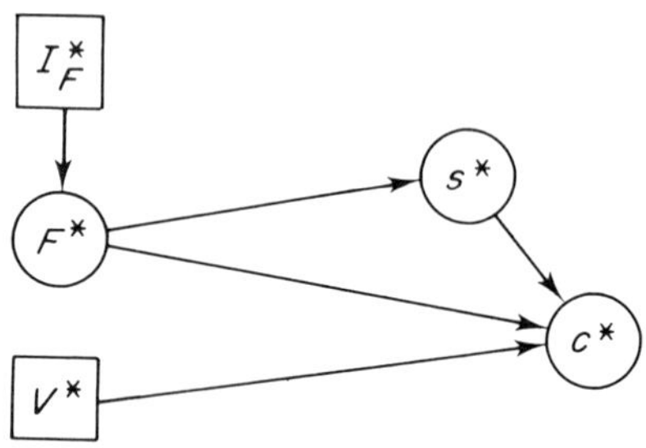

FIG. 7.8. Effect of investment and volume on unit cost.

It is assumed that a functional relationship between F^* and s^* can be established from past experience, published data relating to the trade-off in Fig. 7.7, and the expectations of management in evaluating the investment decision in question to the effect that an increase δF will be coupled with a decline δs. This relationship may be expressed as

$$s^* = -\mu F^* \tag{7.28}$$

where μ is described as the *mechanization elasticity*. This term is chosen for convenience, though it should be stressed that the analysis is not confined to investment aimed at increased mechanization, but to any investment that has the effect of increasing F and reducing s simultaneously. It is assumed that $F^* > 0$ and $s^* < 0$, i.e. that μ is positive. Substituting (7.28) into (7.8), we have

$$c^* = -f_2 F^* \left(\frac{f_1}{f_2} \mu - \frac{1 - V^*/F^*}{1 + V^*} \right) \tag{7.29}$$

so that the unit cost will decline ($c^* < 0$) if

$$\mu > \mu_0 = \frac{f_2}{f_1} \frac{1 - V^*/F^*}{1 + V^*} \tag{7.30}$$

The limiting value of the right-hand side of this expression is denoted by μ_0 and this is where $c^* = 0$, i.e. there is no change in the unit cost. For the special case $V^* = 0$, the value of μ_0 ($= f_2/f_1$) is shown in Fig. 7.9, so that the domain above the graph corresponds to a reduction in unit cost and the domain below to an increase. A highly labor-intensive plant is characterized by a relatively high value of f_1 (if the marginal cost s includes a significant proportion of direct labor), corresponding to a low value of μ_0. It appears, therefore, that for any given value of μ it is more likely for the unit cost to be reduced by investment in a labor-intensive plant. When f_1 is low, however, a substantially high value of μ is needed for the outcome to be above the graph of μ_0.

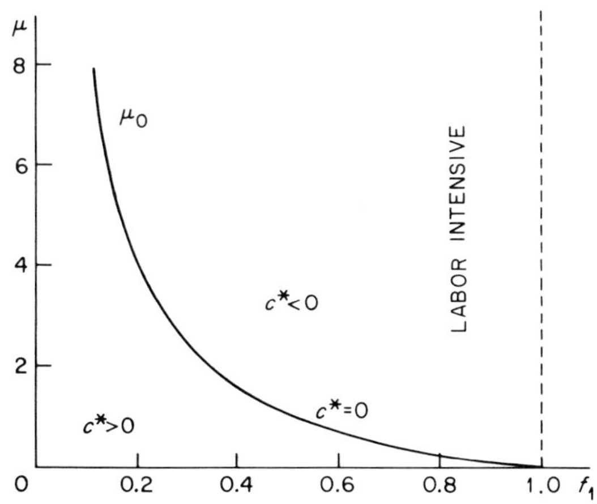

FIG. 7.9. Effect of μ and f_1 on c^* (for $V^* = 0$).

A more general case is obtained when the change in the unit cost function in Equation (7.5) is taken as a starting point, i.e. when $f_3 \neq 0$. If it is assumed that the interest rate does not change (i.e. $j^* = 0$), so that

$$J^* = B^* = \frac{I_W^*}{\lambda}$$

where I_W^* is given (for $p^* = 0$) in Equation (7.21), then the direct and secondary effects of changes in F and V follow the network of relationships in Fig. 7.10: an incremental investment, represented by I_F^*, causes an increase in the fixed cost,

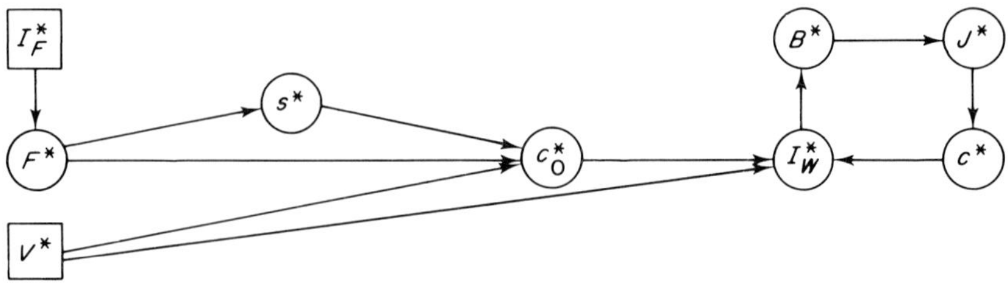

FIG. 7.10. Direct and secondary effects of investment and volume on unit cost.

shown by F^*, which causes s^*. Both F^* and s^* affect the unit cost, as indicated by c_0^*, and a change in volume also contributes to c_0^*. This part of the network is similar to that shown in Fig. 7.8. Following the direct effect c_0^* there are secondary effects, the pattern being as in Fig. 7.4. From Equation (7.5) and after substitution of (7.28) we get

$$c^* = \frac{-f_1 F^*}{1+V^*}\left[(1+V^*)\mu - \frac{f_2}{f_1} - \frac{f_3}{f_1}\frac{B^*}{F^*} + \frac{f_2+f_3}{f_1}\frac{V^*}{F^*}\right] \tag{7.31}$$

so that the unit cost declines ($c^* < 0$) when

$$\mu > \mu_0 = \frac{1}{1+V^*}\left(\frac{f_2}{f_1} + \frac{f_3}{f_1}\frac{B^*}{F^*} - \frac{f_2+f_3}{f_1}\frac{V^*}{F^*}\right) \tag{7.32}$$

and clearly Equations (7.29) and (7.30) are special cases of (7.31) and (7.32) respectively. When the change V^* is relatively small ($V^* \ll 1$) then all the four equations may be simplified by ignoring $1+V^*$.

CONCLUSION

A general fixed and variable linear cost model is given by

$$C = sV + F + J$$

and the unit cost by Equation (7.3), namely

$$c = s + \frac{F}{V} + \frac{J}{V}$$

the relative unit cost increment being as stated in Equation (7.5), i.e.

$$c^* = f_1 s^* + \frac{1}{1+V^*}(f_2 F^* + f_3 J^* - f V^*)$$

where $f = 1 - f_1$.

The effect of a change in output on its own (when $s^* = F^* = f_3 = 0$) is found by

$$c^* = -\frac{f_2 V^*}{1+V^*}$$

as tabulated in Table 7.1 and shown in Fig. 7.1, from which it is clear that (a) the effect is proportional to the value of $f_2 = F/C$ and (b) a decrease in volume by a given percentage is more significant in its effect on unit cost than an increase in volume by the same amount.

Various primary changes,such as changes in the fixed cost, the variable cost and the rate of interest, have a direct effect on the unit cost. The resultant change in the unit cost—coupled with a change in volume, if it takes place—affect the working capital requirements, which in turn affect the interest charges and thereby the unit cost. These networks of relationships are shown in Figs 7.4 and 7.5 and the results (for $p^* = 0$) are given in Equations (7.20) and (7.23). The case $p^* \neq 0$ and the effect of changes in the working capital are further considered in Chapter 11. The effect of additional expenditure on advertising to stimulate further demand can also be accounted for, as in Fig. 7.6 and Equations (7.23) and (7.25), stemming from Equation (7.4).

Capital expenditure, particularly in the introduction of new technology and modernization of plant, often aims at reducing the marginal cost per unit at the expense of increasing the fixed cost component, the intention being a reduction in the overall unit cost. The condition for achieving this objective is given in Equations (7.30) and (7.32).

REFERENCES

[1] Eilon, S. and Cosmetatos, G. P. (1977). A profitability model for tactical planning. *Omega* **5**, 673–688.

[2] Eilon, S. and Cosmetatos, G. P. (1979). Effect of price and allied changes on unit cost and profit. *Appl. Econ.* **11**, 117–131.

[3] Eilon, S. (1981). Some effects of changes in fixed and variable costs on unit cost and profit. *In* "Proceedings of the VI International Conference on Production Research, Novi Sad", pp 723–727.

[4] Eilon, S. and Cosmetatos, G. P. (1982). The effect of investment on unit cost and profit. *Omega* **10**, 35–42.

EXERCISES

1. If the unit cost function is

$$c = s + \frac{F}{V}$$

and if an increase in volume causes an increase in the marginal cost while the fixed costs are unaffected, find the relationship between s^* and V^* that will ensure that the unit cost does not increase.

If $f_1 = 0.6$ and $V^* \geqslant 10\%$ causes s to increase by 10% (irrespective of the actual value of V^*), at what level of increase in volume will the unit cost cease to rise? Tabulate values of c^* for $10\% \leqslant V^* \leqslant 30\%$.

2. An increase in volume necessitates an increase in fixed costs without requiring a change in marginal costs. Determine the condition for the unit cost to decline.

If the relationship between F^* and V^* (values in %) is

$$F^* = 10 + 0.6\,V^*$$

tabulate c^* for $0 \leqslant V^* \leqslant 40\%$ when $f_2 = 0.5$.

3. The following data are given: the cost proportion for variable costs, fixed costs and interest charges are 0.5, 0.4 and 0.1. If $s^* = j^* = 10\%$, and if $h_1 = h_2 = 0.2$ while $\lambda = 1$, plot (or tabulate) the change in the unit cost with a change in volume (assume $F^* = 0$).
4. If $j^* = 0$ and $f_3/\lambda = 1$, show (with the aid of graphs or tables) how u_1, u_2 and v depend on h_1, h_2 and V^*.
5. The effect of advertising is assumed to be as follows:

$D^*(\%) =$	0	5	10	15	20	25	30
$V^*(\%) =$	0	5	9	12	13	14	15

If the cost proportions are 0.4, 0.4, 0.1, 0.1 and if $h_1 = h_2 = 0$, $s^* = F^* = j^* = 0$, find γ and the effect of advertising on the unit cost.

6. If an increase in fixed costs by 10% is expected to reduce marginal costs by 10%, and if no change in the rate of interest is expected, plot the resultant relative change in the unit cost when the volume is expected to change from -30% to $+30\%$. Other data: $f_1 = 0.5$, $f_2 = 0.4$, $f_3 = 0.1$; the whole of the working capital is financed by a bank loan; $h_1 = 0.15$, $h_2 = 0.25$.
7. Show that if

$$\frac{f_3}{f_2} = \frac{F^* - V^*}{V^* - J^*}$$

then

$$c^* = f_1 s^*$$

8. For the cost function

$$C = S + F$$

any change S^* and/or F^* causes the total cost to change by C^* and hence the cost proportions f_1 and f_2 undergo changes. Prove that

$$f_1 f_1^* + f_2 f_2^* = 0$$

9. A plant was operating at $f_2 = 0.3$ in the base year. In three successive years the fixed costs remained the same, but volume dropped by 10% in each year compared with the previous year. Determine the effect on the unit cost and on f_2 for each of the years in question and for the final year compared with the base year.
10. An investigation of the mechanization elasticity in a particular plant revealed that a reasonable approximation is

$$\mu = \frac{1}{1 + F^*}$$

If $f_1 = 0.6$ and $f_2 = 0.4$ find the optimal increase in fixed costs to minimize unit cost

(a) if $V^* = 0$
(b) if $V^* \neq 0$

In the case of $V^* \neq 0$ show what effect capital expenditure (in the shape of $F^* > 0$) has on reducing the unit cost (compared with $F^* = 0$).
11. If C_0^* and C^* are the initial and eventual relative incremental changes respectively in the total cost function, find their values for the circumstances depicted in Fig. 7.5.

8
Price and Volume

Themes in this chapter

Relationship between price and volume
Demand and supply curves
Pricing methods
Definitions of price elasticities of demand:
point elasticity
arc elasticity
relative change elasticity
Demand curves for constant point elasticity
The linear demand function
Expressions for relative incremental volume
Relative incremental revenue

THE MEANING OF PRICE

Price and value

As intimated in previous chapters, the essential ingredients that determine profit are unit price, volume and costs. In Chapters 5–7 our attention was centered on the cost function, and to complete the picture we have to turn to consider revenue. As revenue is the product of unit price and volume, the implications of price and its relationship with volume need to be explored.

The theory of price, its purpose in the economy, its effects, and how it is determined, have long been a subject of great fascination to economists. As suggested by Dorfman

> "the price system is a system of economic organization in which each individual, including in that term legal individuals such as corporations, decides for himself what contribution he will make to the economy with the understanding that he can sell that contribution at a price acceptable to him and the buyer, and that he can obtain the goods and services contributed by other individuals only at prices acceptable to them" [3, p. 3]

In short, price is the amount of money paid in exchange for goods or services. In a free economy, where a price system is not imposed by a central authority on traders, the price represents an accommodation between supply and demand, namely a payoff acceptable to both seller and buyer. The seller wishes to sell his goods or services for as high a price as possible, while the buyer wishes to pay as little as possible. If the seller tries to charge too much, he will not find a buyer; if the buyer is not prepared to pay enough, he will not be able to acquire what he wants. An agreement between them about price signifies that each is satisfied with his prospective payoff and is willing for the transaction to proceed.

Price is not synonymous with value, but is indicative of it. An individual is willing to pay more for an article if he attaches a high value to it, hence the expression "getting good value for money". In this sense, price represents the lower bound of value as perceived by the buyer, if he is willing to buy. By being prepared to exchange a sum of money x for an article, he reasons to himself that the article is worth at least x (thus, the difference between the asking price and what he would be prepared to pay is what is called "the consumer surplus"). And similarly, the seller may argue that the price x represents his perception of the upper bound of value of the article, in believing that the worth of the article to him does not exceed x.

Naturally, value is not absolute, and different individuals are willing to pay different prices for the same object. The differences in perceptions of value are partly due to differences in individual tastes and backgrounds, to changes in fashion, to social pressures, and to what people already possess. Individual A desperately wants a swimming pool, while B would rather have a luxurious automobile. The differences in their preferences represent their respective expectations of enjoyment or utility that they will derive from such acquisitions. And when A gets his swimming pool, it is unlikely that he will desperately wish to have another one; instead, a new desirable object will attract his attention.

Scarcity

A crucial element in the determination of price is that of *scarcity*. In a hypothetical world of boundless resources, price would be meaningless. Anything you wanted would be available, and you could always afford it. But in the real world resources are limited, and so are the incomes of individuals. *The more scarce an object is, for which there is demand, the more valuable it becomes to some sections of the market, and the higher the price which the object will fetch.* This is why suppliers are most anxious to regulate the amounts available on the market, in order to ensure that a reasonable price level is maintained. Oversupply by suppliers in competition with each other invariably causes the price to drop, as often occurs in the world commodity markets, for example when good weather yields bumper crops of wheat or cocoa. When there is a risk of oversupply, there is an incentive for suppliers to get together in an attempt to control the total amount available by allocating quotas, which the individual suppliers are required to adhere to. Occasionally, suppliers are anxious to increase their revenue and breach these quotas, as has happened from time to time with the oil producing countries and suppliers of other commodities, and then price levels tend to decline. In the absence of arrangements to regulate the total supply, there is a temptation for suppliers to agree on minimum prices, but such actions are strongly resisted in many countries by anti-monopolies and anti-price-fixing legislation.

On the supply side, therefore, price is affected by the total amount available and by competition. The other side of the equation is that of demand, which is governed by scarcity, namely the limited funds available to customers for spending. Thus, there is competition not only between suppliers of the same goods or services, but between different goods as well. His limited resources compel the customer to rank the various goods and services on offer; he will regard some as essential, while others will fall into the category of discretionary expenditure. The more money he is willing and able to spend, the more he will be prepared to pay for a given article; but the more expensive the article, the less likely it is to retain a high position in the ranking order.

Equilibrium between supply and demand

The conventional economic model postulates that in a free market, i.e. one where competitors are free to decide on the amounts they produce and on their prices, without interference or collusion, the resultant price may be viewed as a point of equilibrium between the demand and supply curves, an example of which is shown in Fig. 8.1. Demand is assumed to decline with price, while the willingness of the supplier to increase volume is stimulated when the price is high, so that the supply curve is assumed to increase with price. Figure 8.1 suggests that:

- At price p_2 the corresponding demand is V'_2 but the available supply is V''_2
- At price p_3 the demand rises to V'_3 but at that price only V''_3 is available
- At p_1 the quantities in demand and supply are in balance.

It is argued, therefore, that when the price exceeds p_1 the excess of goods forces the supplier to reduce the price, and when it is below p_1 the scarcity of goods forces the customer to pay more, so that in both cases there is a movement towards the point of equilibrium at p_1.

How can this point of equilibrium be changed? When the supply and/or demand curves are shifted, a new equilibrium is established, as shown in

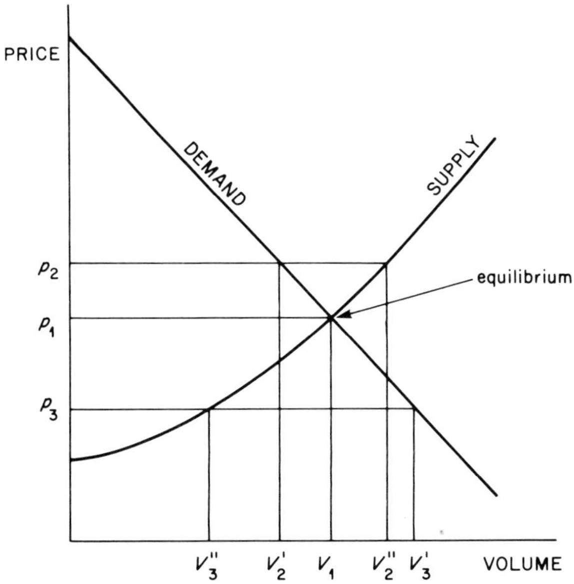

FIG. 8.1. Supply and demand curves.

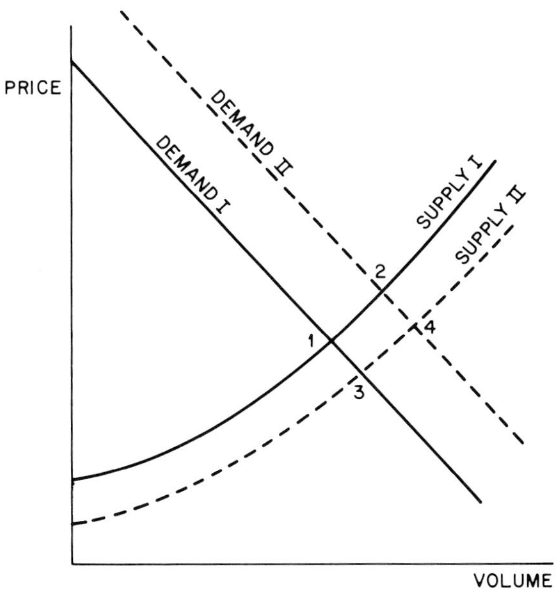

FIG. 8.2. Changes in the supply and demand curves.

Fig. 8.2. If the original demand curve I moves to position II, then the market is willing to consume a larger volume for the same price. The reasons for such a positive shift may be due to an intrinsic realignment of customers' preferences, the effect of advertising, or increased spending power. The net result is for the equilibrium to move from point 1 to 2, corresponding to a higher volume and a higher price. A shift in the supply curve from I to II may come about with improved production methods, so that the supplier is able and willing to produce more at the same price. In that case, the equilibrium moves from point 1 to 3. When both the supply and demand curves shift, the resultant equilibrium is at point 4. If either shift is positive (i.e. to the right), then by definition the volume increases, but notice that a positive shift in demand causes the price to rise, while a positive shift in supply causes it to fall. If the two shift simultaneously then volume increases, but the outcome for price depends on the amounts of shift involved. Similarly, the overall effect on volume of a positive shift in one curve and a negative shift in the other, depends on the amounts by which the shifts occur.

Pricing methods

The model in Fig. 8.1 assumes that for any given price there is a corresponding level of output that suppliers are willing to make available. But why should this level depend on price? Why would a supplier be unwilling to supply any quantity provided a minimum price is obtained? In Fig. 8.3 this proposition is depicted by a horizontal supply line, representing the price level p_1 which the supplier is prepared to accept irrespective of volume. As before, the supply line crosses the demand line at V_1, and the area of the shaded rectangle is $p_1 V_1$, namely the revenue.

Pricing policies by companies vary enormously, but the following methods are probably predominant:

(1) *A cost-plus philosophy*, which starts with costing the manufacturing activities required to meet given product specifications. The result of this costing exercise yields the function of unit cost with respect to volume, so that for any postulated level of demand it is possible to state what the expected total cost, average unit cost and marginal cost would be. Price is then determined by adding an appropriate amount to the unit cost (hence the term "cost-plus") as a profit element. Methods for determining this additional element may take the form of:

- (a) a fixed unit profit z irrespective of volume;
- (b) a unit profit z which follows a step function, namely that beyond a specified volume a lower z may be tolerated;
- (c) a fixed profit margin α irrespective of volume;

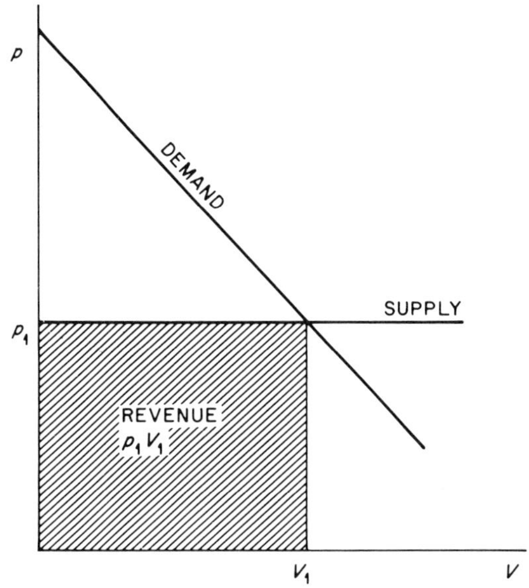

FIG. 8.3. A horizontal supply line.

(d) a profit margin α which follows a step function, similar in concept to (b);
(e) a multiplier which is applied to a characteristic element of the cost (for example, price = materials cost × a multiplier, or price = expected man-hours × a multiplier, or price = cost of labor × a multiplier).

(2) *A market-oriented philosophy*, which involves charging what the market will bear in conjunction with an expected volume. Here the starting point is the customer. How much will he be prepared to pay, bearing in mind the choices available to him and prevailing prices? Starting with the price, the supplier can then decide whether he can produce the product at an acceptable profit and take action to control his costs accordingly. Pricing methods then take the form of:

(a) fixing a price which maintains a constant ratio to average market price (for example, management may aim to keep the price at 10% above the average price charged in the market for a prescribed range of products of a similar or comparable type);
(b) following a major competitor (for example, undercutting him by 5%);
(c) becoming a price leader, to enhance the quality image of the product and to exploit the reputation of the supplier in the market place;

(d) experimenting with different price levels for different sectors of the market and for essentially identical products under different brand names.

The supply curve

Whatever pricing philosophy is employed, the supply curve cannot be divorced for long from the costs incurred, and several examples are shown in Fig. 8.4. Curve 1 is linear and would be appropriate in the hypothetical case where the unit cost c increases linearly with volume and where management wishes to maintain a constant unit profit z. It then follows that the unit price

$$p = c + z$$

also increases linearly with volume. In that case, it should be noted that the profit margin α must decline with volume, since

$$\alpha = \frac{p-c}{p} = \frac{z}{p} \tag{8.1}$$

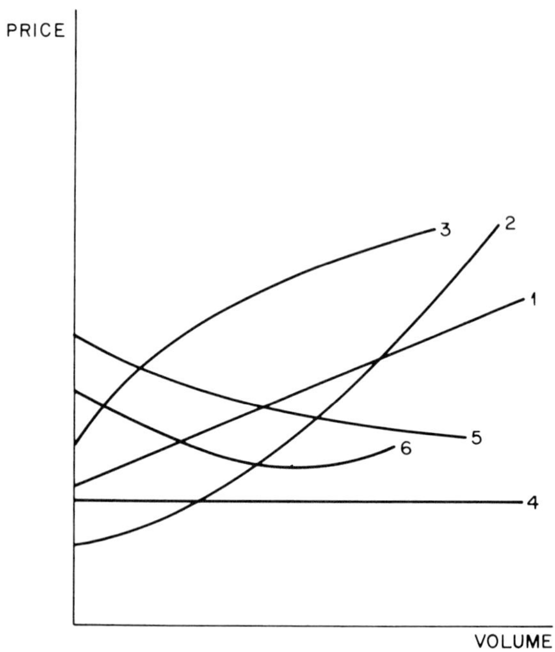

FIG. 8.4. Some types of supply curves.

When management aims at keeping α constant, then the unit price is determined by

$$p = \frac{c}{1-\alpha} \tag{8.2}$$

and if α is kept constant, while c is a linear function of V, then p is linear as well.

Curve 2, in Fig. 8.4, is the one shown in Fig. 8.1 and suggests that there are diseconomies of scale in manufacture, so that as volume increases, price must rise at a faster rate to justify continued supply. Alternatively, long-run marginal cost may rise when relatively inefficient new firms enter the industry. Curve 3 implies some advantage in producing larger volumes, so that the rate of increase in the required price eventually falls below that in curves 1 and 2. Curve 4 is a horizontal supply curve, already shown in Fig. 8.3, corresponding to the case of zero or negligible fixed costs, so that the cost per unit is constant and equal to the marginal unit cost, and if z or α is constant, it follows from the discussion above that the price is independent of the volume. Curve 5 corresponds to the case where the fixed and variable cost model applies, as in Equation (5.23). Since the unit cost c then declines as the volume increases, the supplier is prepared to accept a lower price, and this is a common enough phenomenon when economies of scale are evident. Curve 6 applies when a minimum unit cost is obtained at a certain level of output, beyond which the unit cost begins to rise.

Any of these supply curves can be used in conjunction with the model in Fig. 8.1. Notice, however, that when both the demand and supply curves decline with volume, the point of intersection is not as clearly determined as in Fig. 8.1 Furthermore, even the slightest shift in one of the curves can cause a significant displacement of the point of equilibrium, as shown in Fig. 8.5.

It is possible, in certain circumstances, to have more than one intersection between the supply and demand curves. Figure 8.6 shows an example where the supply curve is above the demand curve to start with. This means that up to output V_1 the supplier's expectation of an equitable price exceeds what the customer is willing to pay. Economies of scale allow costs to decline, so that the supplier is willing to reduce his price provided volume increases. Beyond V_1 the customer is prepared to pay more than the minimum level set by the supplier. As volume increases, diseconomies of scale begin to assert themselves in Fig. 8.6 and beyond V_2 the supplier can no longer afford to match the expectations of the customer, who continues to press for a lower price. The range from V_1 to V_2, therefore, represents the feasible domain within which the supplier can operate, without having to resort to intensive advertising (to shift the demand curve), or to improve production methods and use new technology (to shift the supply curve).

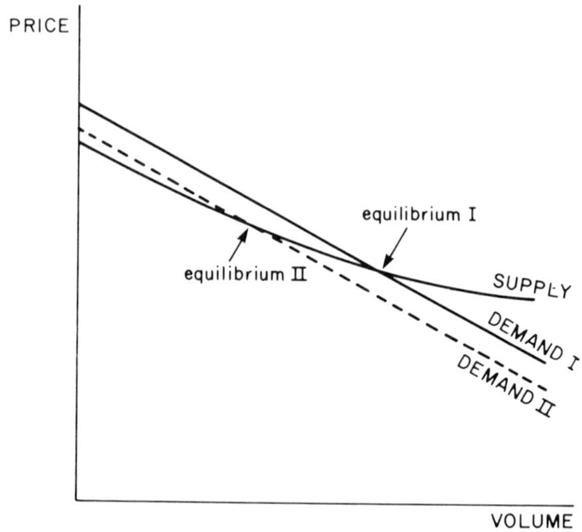

FIG. 8.5. Declining supply and demand curves.

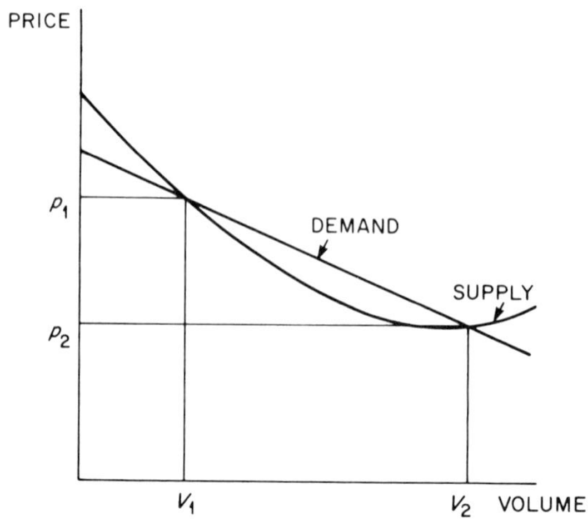

FIG. 8.6. Two equilibrium points.

The demand curve

In many text books, the demand curve in the model shown in Fig. 8.1 is depicted as a declining linear function of the volume. This assumption implies that when a certain volume is reached, the customer expects to get it free, as

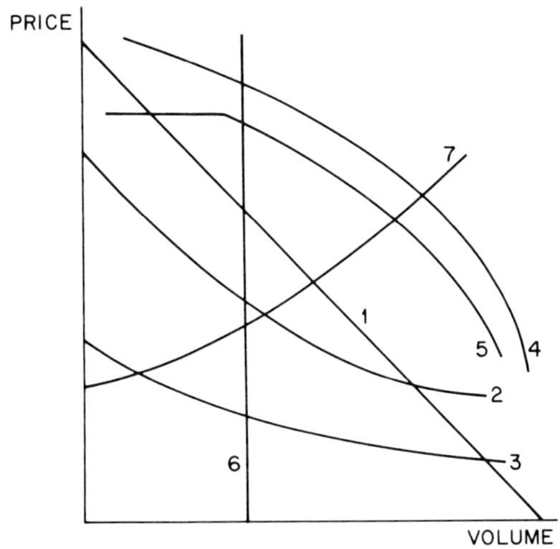

FIG. 8.7. Demand curves.

suggested by the intersection of curve 1 with the horizontal axis in Fig. 8.7. In reality such a relationship rarely exists, and the linear form is used merely for convenience. Many other demand curves are possible, as shown in Fig. 8.7:

Curve 2—demand increases at an increasing rate as price falls.
Curve 3—similar to 2, but demand is very sensitive even to a small change in price.
Curve 4—here the sensitivity operates the other way, with volume increase being difficult to achieve even when price is substantially reduced (possibly due to market saturation).
Curve 5—similar to 4, except that when a certain high price is reached, demand collapses (shown by the horizontal portion of the curve).
Curve 6—demand is constant irrespective of the price.
Curve 7—here demand increases with price, and although such a relationship is rare, it can occur when the supplier manages to convey the impression that a high price has an intrinsic attraction, which succeeds in stimulating demand (it has been suggested, for example, that the catch phrase "diamonds are a girl's best friend" has for many years helped to push up the price of diamonds with demand constantly on the increase); strictly, what this curve means is a shift in the perception of the customer regarding the value of the product.

Curve 7 is uncommon, and the general hypothesis for demand curves is that the sales volume rises when price is reduced and vice versa. The curves 1–5 in

Fig. 8.7 conform with this notion, but it must be pointed out that each of these curves is a static representation of all pairs of p and V that pertain to a given product or service in a given market, so that for any given price there is a corresponding volume, irrespective of the prevailing previous price. In reality, the effect on volume may well depend on whether the price increases or decreases. Figure 8.8 shows an example where the initial price p_1 is reduced to p_2, causing the volume to increase from V_1 to V_2, but when the price is then restored to the level p_1, the new volume is $V_3 \neq V_1$.

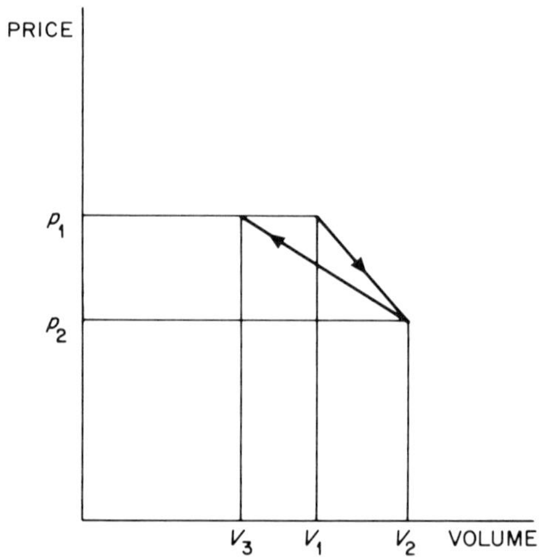

FIG. 8.8. The possible effect of two price changes.

In the example shown in Fig. 8.8 the resistance to a price increase is so marked that the final outcome is $V_3 < V_1$. Even when the initial and final states are the same, as in p_1 and V_1 in Fig. 8.9, a series of price reductions follows one path to p_2 with the corresponding volume V_2, whereas a subsequent series of price increases may follow a different path. In this case, the market reacts quickly to a price reduction, but its reaction to a price increase is more sluggish, so that for any given price between p_1 and p_2 the corresponding volume is greater on the upward path than on the downward path. This phenomenon is known as *the hysteresis effect* (a term borrowed from the field of magnetism) which, in the case of Fig. 8.9, can be exploited by the supplier to his advantage by judicious manipulations of the price level.

Another point to note is that all the demand curves discussed, hitherto, ignore the effect of time. A shift in the demand curve was mentioned earlier, for

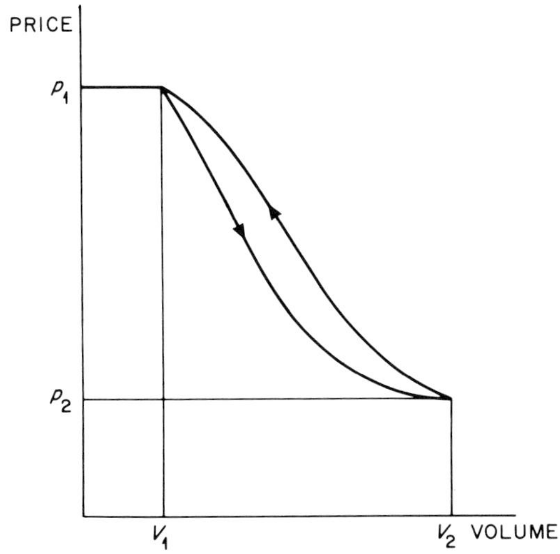

FIG. 8.9. The hysteresis effect.

example as a consequence of an advertising campaign. The effect of such a campaign may not, in fact, be lasting, and Fig. 8.10 shows an example of how demand is stimulated by advertising to a new level, which is held for a while, and then the effect wears off, causing demand to drift downwards. In this example the final demand is higher than the original level, signifying a residual

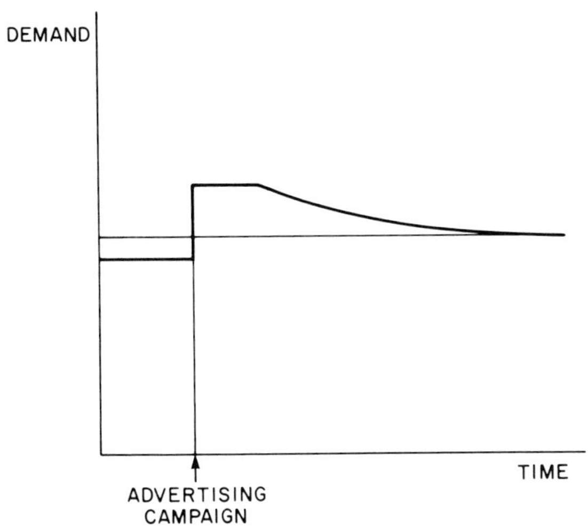

FIG. 8.10. Effect of an advertising campaign on demand.

positive advantage, but circumstances may well arise (for example, in response to advertising by competitors) where the residual effect is negative.

The rationale of a demand curve in economic theory

The demand curves 1–5 depicted in Fig. 8.7 are embedded in the notion of rational market behavior, which postulates that

- the lower the price for a given product or service, the greater is the demand for it ("demand" is defined here as the volume that customers are willing to buy);
- the scarcer a product or a service is, the more the customer is willing to pay to obtain it.

These hallmarks of rationality do not always accord with actual customer response. As Dorfman puts it [3, pp. 11–12]:

> "At bottom, an economy consists of people, and people are more complicated than we care to reason about. To keep things manageable, economists populate their theories with *homo economus*, a very simple species of mankind who know a great deal, reason accurately, behave predictably, and are driven by uncomplicated motivations, quite unlike you and me. *Homo economus* always finds out when the price of bread rises, and promptly consumes less of it. He knows the prices of everything that he uses or could use."

In reality, however, people are not cast in this simple mold. For a start, information is not fully and instantly available, so that many customers will take a while to learn about price reductions by competitors, comparative merits of competing products, even the existence of other products. Secondly, some customers are inherently conservative in their behavior and will continue to stick with their suppliers, long after competitive alternatives are brought to their attention, while others are experimental by nature and will try new suppliers at the slightest excuse. These factors, coupled with the fact that customers differ in their preferences and systems of values, mean that they do not respond uniformly to changes in the market place. Economic analyses of market behavior, therefore, tend to be based on *expectations of the balance of responses of the customer population* (rather than predictions of how individual customers will react), and this is what demand curves attempt to depict. In practice, these demand curves are based on historical statistical data, on market research, on the views of managers with knowledge and experience of the market scene, or on deliberate experiments with price changes to ascertain market response.

PRICE ELASTICITY OF DEMAND

As we have seen, the purpose of a demand curve is to provide information about the response of market demand to different price levels. If the characteristics of a particular curve are known, for example of one of the curves 1–5 in Fig. 8.7 or of the curve in Fig. 8.9, then managers can proceed to formulate a pricing policy for their products, and in particular to consider whether they should change prevailing price levels.

Because evaluation of performance is so inexorably wedded to making comparisons (as discussed at length in previous chapters) between two time periods, or between the state of affairs before and after given events, it is convenient to describe market behavior in terms of *demand response to a price change*, a common measure of which is given by *the price elasticity of demand.* The general definition of price elasticity is

$$\text{elasticity} = \frac{\text{proportional change in volume}}{\text{proportional change in price}}$$

This definition, however, is not precise enough, as it does not specify how the proportional changes in price and volume are to be measured. The commonly found definitions in the economic literature are those of *point elasticity* and *arc elasticity.* The third definition, that of *relative change elasticity*, is more convenient for use in models associated with analysis of performance and managerial decisions.

POINT ELASTICITY

Definition

The notion of point elasticity stems from the idea of *marginal revenue.* Since revenue is

$$R = pV$$

the marginal revenue is given by the derivative with respect to volume, i.e.

$$\frac{\mathrm{d}R}{\mathrm{d}V} = p + V\frac{\mathrm{d}p}{\mathrm{d}V}$$

so that the marginal revenue does not change when

$$\frac{\mathrm{d}R}{\mathrm{d}V} = 0$$

or when

$$-\frac{\frac{\mathrm{d}V}{V}}{\frac{\mathrm{d}p}{p}} = 1$$

The point elasticity, denoted as ε_p, is then defined as

$$\varepsilon_p = -\frac{\mathrm{d}V}{V}\bigg/\frac{\mathrm{d}p}{p} \tag{8.3}$$

or

$$\frac{\mathrm{d}V}{V} = -\varepsilon_p\frac{\mathrm{d}p}{p} \tag{8.4}$$

The minus sign ensures that when either $\mathrm{d}p$ or $\mathrm{d}V$ is negative, then the elasticity measure is positive. Substituting the definition of ε_p into the expression for the marginal revenue, we get

$$\frac{\mathrm{d}R}{\mathrm{d}V} = p + V\frac{\mathrm{d}p}{\mathrm{d}V} = p\left(1 - \frac{1}{\varepsilon_p}\right) \tag{8.5}$$

Similarly, the derivative of R with respect to p is

$$\begin{aligned}\frac{\mathrm{d}R}{\mathrm{d}p} &= V + p\frac{\mathrm{d}V}{\mathrm{d}p}\\ &= V(1-\varepsilon_p)\end{aligned} \tag{8.6}$$

Both $\mathrm{d}R/\mathrm{d}V$ and $\mathrm{d}R/\mathrm{d}p$ are expressions of marginal revenue, the former with respect to the variable V and the latter with respect to the variable p. When the term *marginal revenue* is used without specifying the variable with respect to which the revenue function is differentiated, it is generally understood to refer to $\mathrm{d}R/\mathrm{d}V$ (namely to a marginal change in revenue related to a marginal change in volume). From Equations (8.5) and (8.6) we note the well-known results (see, for example, [4]) summarized below:

Condition	Marginal revenue $\mathrm{d}R/\mathrm{d}V$	Marginal revenue $\mathrm{d}R/\mathrm{d}p$	Revenue when p increases
$\varepsilon_p = 1$	0	0	unchanged
$\varepsilon_p > 1$	positive	negative	falls
$\varepsilon_p < 1$	negative	positive	rises

Determining the value of point elasticity

From Equations (8.5) and (8.6) it is clear that the point elasticity definition is governed by the slope of the demand curve, as shown in Fig. 8.11. At point A on the demand curve the tangent is BC and the slope at A is $-\mathrm{d}p/\mathrm{d}V$ (the minus sign is due to the fact that the demand curve is a decreasing function). The elasticity at point A is

$$\varepsilon_p\,(\text{at A}) = -\frac{\mathrm{d}V}{V}\bigg/\frac{\mathrm{d}p}{p} = -\frac{p}{V}\bigg/\frac{\mathrm{d}p}{\mathrm{d}V}$$

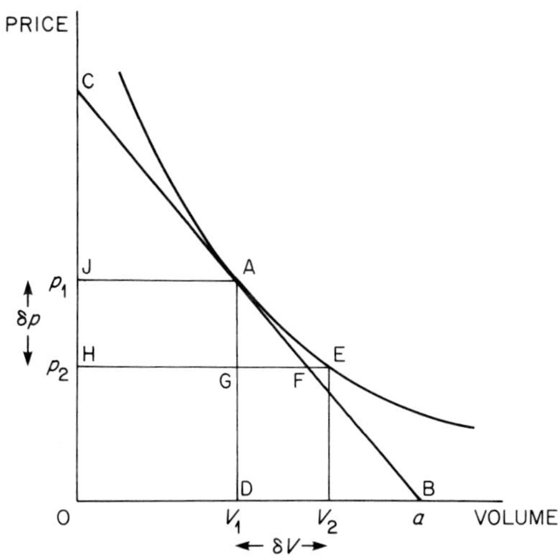

FIG. 8.11. Point elasticity of demand.

Substitute

$$p = p_1 = \text{AD}$$

$$V = V_1 = \text{OD}$$

$$-\frac{\mathrm{d}p}{\mathrm{d}V} = \frac{\text{AD}}{\text{BD}}$$

and from the similarity of triangles

$$\frac{\text{BD}}{\text{OD}} = \frac{\text{AB}}{\text{AC}}$$

to get

$$\left.\begin{aligned}\varepsilon_p\,(\text{at A}) &= \frac{\text{AD}}{\text{OD}}\frac{\text{BD}}{\text{AD}} = \frac{\text{BD}}{\text{OD}} = \frac{\text{AB}}{\text{AC}} \\ &= \frac{a - V_1}{V_1}\end{aligned}\right\} \tag{8.7}$$

where a is the intercept of the tangent AB on the volume axis.

Thus, irrespective of the shape of the demand curve, the elasticity $\varepsilon_p = 1$ if the point A bisects the tangent BC, so that whether the curve is concave, convex or linear, the elasticity just above the midpoint is $\varepsilon_p > 1$ and below it $\varepsilon_p < 1$ (as shown in Fig. 8.12).

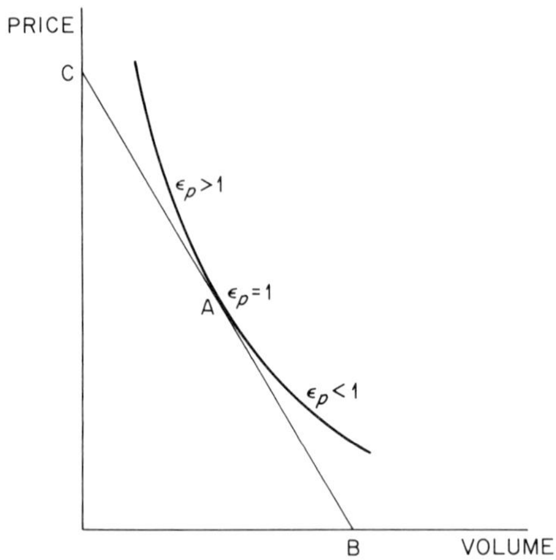

FIG. 8.12. Values of point elasticities (where AB = AC).

The applicability of the point elasticity of demand is confined to continuous demand functions where the derivative exists. Even then, the computation of the absolute change in demand is valid only for relatively small changes in price, as can be seen from the following example.

Suppose that in Fig. 8.11 a change is contemplated from price p_1 at point A to price p_2. From Equation (8.4), we can estimate the change in volume by

$$\mathrm{d}V = -\varepsilon_p \frac{V}{p}\,\mathrm{d}p$$

If instead of $\mathrm{d}V$ and $\mathrm{d}p$ we use the increments δp and δV then

$$\delta V = -\varepsilon_p \frac{\delta p}{p_1} V_1$$

Substituting ε_p from (8.4) and using the similarity of triangles in Fig. 8.11 we get

$$\delta V = -\frac{\text{AB}}{\text{AC}}\frac{(-\text{AG})}{\text{AD}}\text{OD}$$

$$= \frac{\text{AB}}{\text{AC}}\frac{\text{GF}}{\text{BD}}\text{OD}$$

but

$$\frac{\text{AB}}{\text{AC}} = \frac{\text{BD}}{\text{OD}}$$

therefore

$$\delta V = \text{GF}$$

However, as can be seen from Fig. 8.11, the actual increment in demand from V_1 to V_2 is GE and not GF. Hence, the point elasticity can be used to estimate the outcome either when the change from point A is quite small or when the slope changes very little, so that the expected point F and the actual point E are very close to each other.

Further comments on the ramifications of the point elasticity of demand are given below.

1. Constant ε_p

If the elasticity ε_p remains constant along the demand curve, then integration of Equation (8.4) yields

$$\ln V = -\varepsilon_p \ln p + \ln K$$

where K is a constant resulting from the integration; alternatively,

$$V = Kp^{-\varepsilon_p} \tag{8.8}$$

As an illustration of this family of demand functions, curves for $K = 100$ and for three values of the point elasticity ($\varepsilon_p = 0.5$; 1; 1.5) are given in Fig. 8.13. Starting at $p = 1$ and $V = 100$ (which is a common point to all three curves) the volume decreases as price increases, the effect being more marked for higher values of ε_p (the higher ε_p the closer the curve is to the origin point 0). The curves are concave, so that initially demand is sensitive to a change in price, whereas when the price is already high, a price change seems to have a lesser effect on demand. In fact, this impression is misleading, as can be seen from Table 8.1. Four cases are explored, the first two involving a price increase of 20% and the other two a price reduction of 20%. Cases 1 and 3 have a low initial price, while the other two cases have a high initial price. We find that a 20% price increase when $\varepsilon_p = 0.5$ causes volume to decline by 8.7%

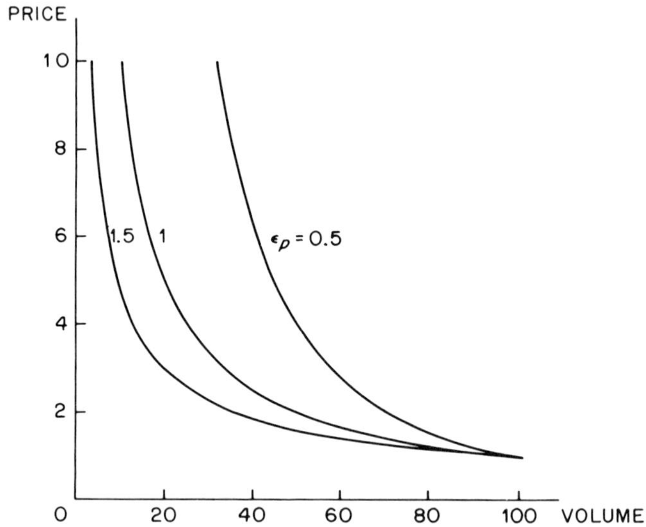

FIG. 8.13. Demand curves for constant ε_p.

TABLE 8.1
Effect of price change for constant ε_p ($V = 100\,p^{-\varepsilon_p}$).

Case	Initial price	Subsequent price	Price change %	Volume change (%) $\varepsilon_p = 0.5$	$\varepsilon_p = 1.0$	$\varepsilon_p = 1.5$
1	1.0	1.2	20	−8.7	−16.7	−23.9
2	8.0	9.6	20	−8.7	−16.7	−23.9
3	1.0	0.8	−20	11.8	25.0	39.8
4	8.0	6.4	−20	11.8	25.0	39.8

irrespective of the initial price, and when the price increases by 20% the volume increases by 11.8%, again independently of the initial price. Similar results obtain for other values of ε_p. We further note the asymmetry in the effect of a price change on volume: in all the cases in Table 8.1 a 20% price increase has a lesser absolute effect (in percentage terms) on the volume than a 20% price decrease.

2. Effect on the relative change in demand

The examples in Table 8.1 suggest that *when the point elasticity ε_p is constant, the relative change in demand depends solely on the relative change in price* (*and* *on* ε_p). This result is simple to derive from Equation (8.8). If the initial state is p and V and the subsequent state $p+\delta p$ and $V+\delta V$, then

$$\frac{V+\delta V}{V}=\left(\frac{p+\delta p}{p}\right)^{-\varepsilon_p}$$

But as $V+\delta V=V(1+V^*)$, and a similar expression applies to $p+\delta p$, it follows that

$$V^*=(1+p^*)^{-\varepsilon_p}-1 \tag{8.9}$$

Here $V^*=\delta V/V=$ the actual relative change in the volume (relative to the initial state V), following the definition of starred variables in previous chapters, whereas $\mathrm{d}V/V$ in Equation (8.4) is the relative change for an infinitesimal change in the volume, so that Equation (8.4) prescribes the effect of change at the margin, while (8.9) gives the effect of the *actual change* (relative to the initial state V). Some results of how p^* affects V^* in Equation (8.9) are tabulated in Table 8.2 and shown graphically in Fig. 8.14. As implied by Equation (8.9), for any given value of the point elasticity, these results do not depend on the initial levels of volume or price. If, for example, $\varepsilon_p=1$ then a price increase of 15% is expected to affect volume adversely by 13%, and this result applies to all initial price levels. The asymmetry in the effect of a given price change on volume, noted earlier, is again clear from Table 8.2 and Fig. 8.14.

3. Revenue for constant ε_p

The implications of the model in Equation (8.8) on revenue can be readily ascertained. Since

$$R=pV$$

it follows that for the demand function in (8.8)

$$R=Kp^{1-\varepsilon_p}=K^{1/\varepsilon_p}V^{(\varepsilon_p-1)/\varepsilon_p} \tag{8.10}$$

The marginal revenues with respect to volume or price are given in Equations (8.5) and (8.6), which are general results, irrespective of whether ε_p is constant or not. Needless to say, they apply here (for the case $\varepsilon_p=$ constant) as well, so that maximum revenue is obtained when $\varepsilon_p=1$, whereas for $\varepsilon_p>1$ the revenue increases with volume and for $\varepsilon_p<1$ it declines when volume increases. These conclusions are illustrated in the examples shown in Figs 8.15 and 8.16 for R. The value of $\mathrm{d}R/\mathrm{d}V$ is plotted in Fig. 8.17 for three values of ε_p (0.5; 1.0; 1.5) and $K=100$, and we note (as is also clear from Fig. 8.14) that the results are sensitive to ε_p.

TABLE 8.2

V^* for given values of p^* by Equation (8.9).

p^* (%)	−30	−25	−20	−15	−10	−5	0	5	10	15	20	25	30
V^* (%)													
$\varepsilon_p = 0.5$	19.5	15.5	11.8	8.5	5.4	2.6	0	−2.4	−4.7	−6.7	−8.7	−10.6	−12.3
1.0	42.9	33.3	25.0	17.6	11.1	5.3	0	−4.8	−9.1	−13.0	−16.7	−20.0	−23.1
1.5	70.7	54.0	39.8	27.6	17.1	8.0	0	−7.1	−13.3	−18.9	−23.9	−28.4	−32.5

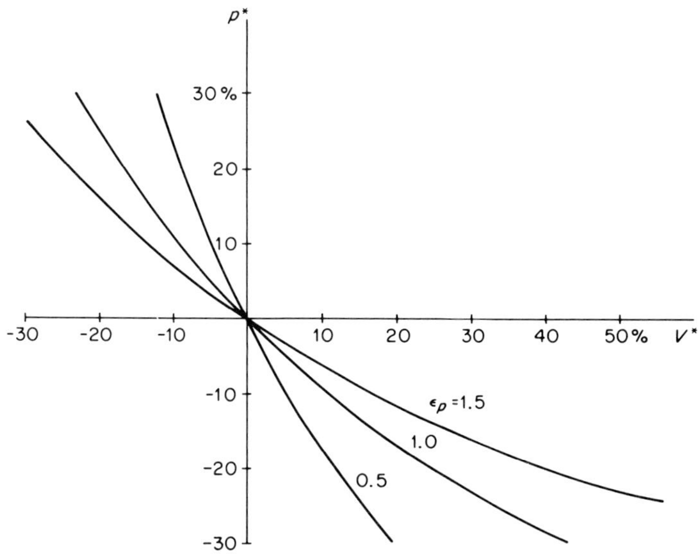

FIG. 8.14. Effect on relative demand for ε_p = constant.

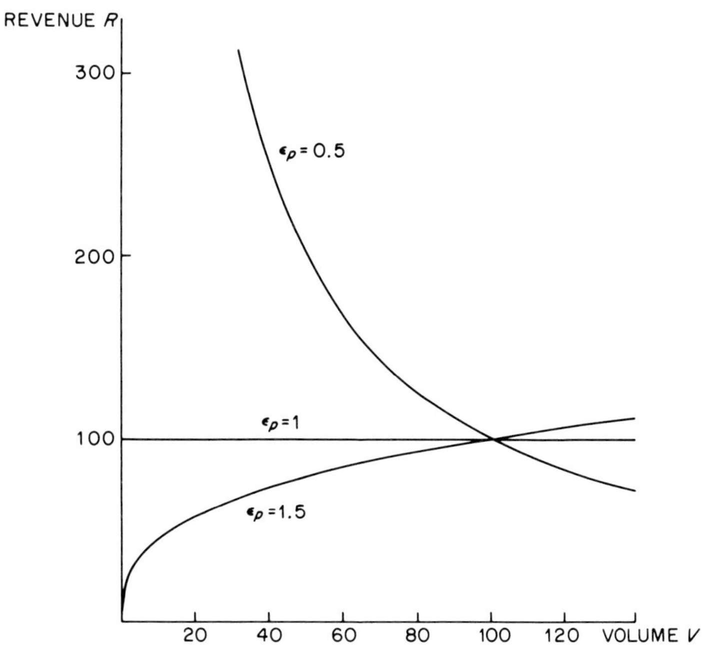

FIG. 8.15. The revenue function for $V = 100p^{-\varepsilon_p}$.

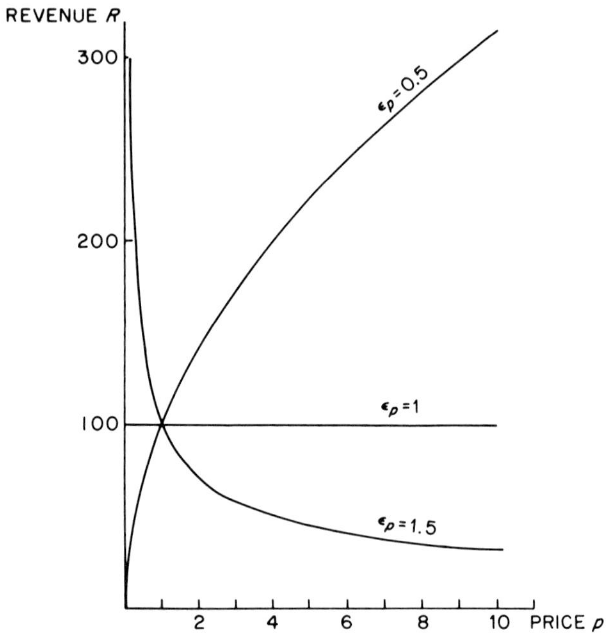

FIG. 8.16. The revenue function $R = 100p^{-\varepsilon_p}$.

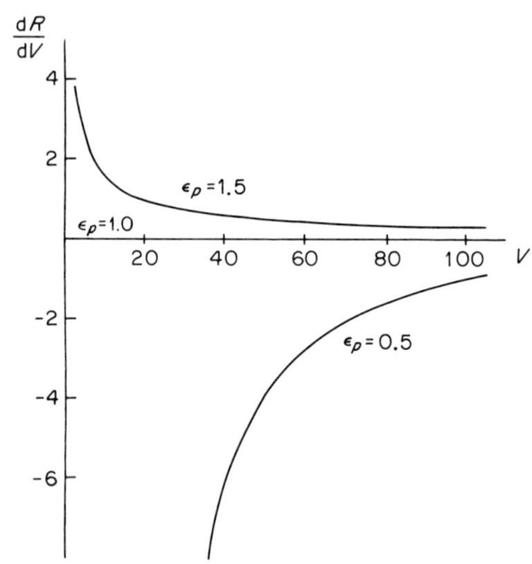

FIG. 8.17. Marginal revenue dR/dV for $V = 100p^{-\varepsilon_p}$.

4. Relative change in revenue for constant ε_p.

From earlier discussions we know that

$$R^* = V^* + p^* (1 + V^*)$$

Substituting (8.9) we have

$$\begin{aligned} R^* &= (1 + p^*)^{-\varepsilon_p} - 1 + p^* (1 + p^*)^{-\varepsilon_p} \\ &= (1 + p^*)^{1-\varepsilon_p} - 1 \end{aligned} \tag{8.11}$$

Hence, the relative change in revenue depends solely on p^* for any given value of the point elasticity ε_p. Again we find that for $\varepsilon_p = 1$ there is no change in revenue for any value of p^*, since in this case $R^* = 0$. To illustrate the implications of Equation (8.11), the example $V = 100\, p^{-\varepsilon_p}$ is explored for three values of ε_p and the results are shown in Table 8.3 and Fig. 8.18.

5. Fitting piece-wise demand functions

First impressions of the demand curves in Fig. 8.13 suggest that in many practical cases, where the curve is concave, it is possible to fit a function of the type given in Equation (8.8) with a constant ε_p. However, the implications of Equation (8.11) are obviously serious, since unless $\varepsilon_p = 1$ (at which point no movement along the demand curve can improve the revenue position) management must be tempted to increase or decrease prices relentlessly. The assumption that market response in the form of the elasticity ε_p will then remain unaffected (i.e. that ε_p will remain constant) is impossible to sustain. There must come a point when the price is so high that market resistance will stiffen by increasing ε_p and thereby cause a dramatic fall in demand, in excess of the fall expected for the original value of ε_p. We therefore have to conclude that fitting the function in (8.8) can apply only to a limited volume range, so that the fitting exercise would consist of several segments with ε_p = constant for each. An example of how this can be done is described below.

Suppose that the relationship between price and demand is to be established from empirical data, given by the scatter in Fig. 8.19, where the points are extracted from market data, with the exception of the two extreme points which are based on managerial estimates. A fitted curve clearly does not suggest that ε_p = constant in Fig. 8.19, where three curves following the function $V = Kp^{-\varepsilon_p}$ are shown for $K = 100$ and different values of ε_p. In Fig. 8.20 three curves are shown for this function with $\varepsilon_p = 0.6$ and different values of K, and again the demand curve does not fit any of them. To achieve a reasonable fit, the demand curve needs to be segmented. We notice from Fig. 8.19 that the function $V = 100\, p^{-0.8}$ is a good fit for $V > 30$. The remaining

TABLE 8.3

R^* for the example $V = 100\,p^{-\varepsilon_p}$.

p^* (%)	−30	−25	−20	−15	−10	−5	0	5	10	15	20	25	30
	R^* (%)												
$\varepsilon_p = 0.5$	−16.3	−13.4	−10.6	−7.8	−5.1	−2.5	0	2.5	4.9	7.2	9.5	11.8	14.0
1.0	0	0	0	0	0	0	0	0	0	0	0	0	0
1.5	19.5	15.5	11.8	8.5	5.4	2.6	0	−2.4	−4.7	−6.7	−8.7	−10.6	−12.3

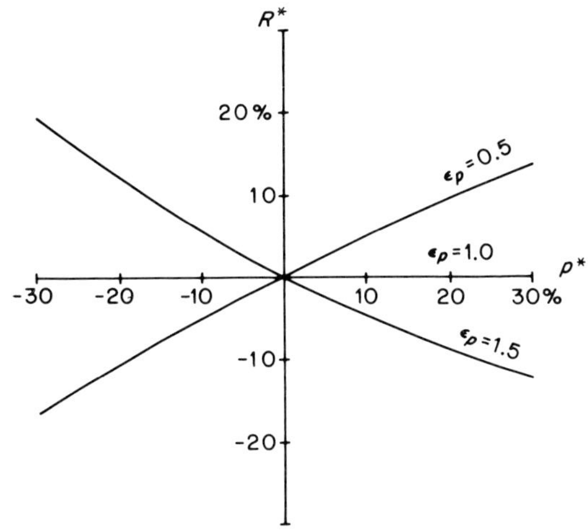

FIG. 8.18. Relative change in revenue for ε_p = constant.

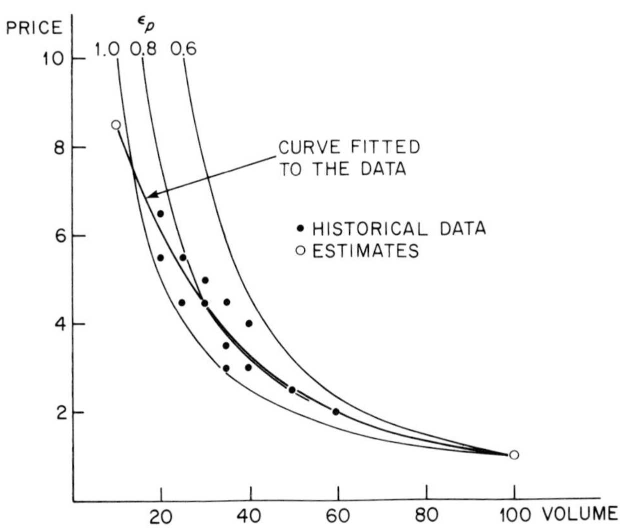

FIG. 8.19. A curve fitted to empirical data.

segment of the demand function is far less steep in the range $10 < V < 30$ than the three curves in Fig. 8.19, suggesting that a higher value for ε_p should be explored. After some trial and error, a reasonable fit is obtained with $K = 300$

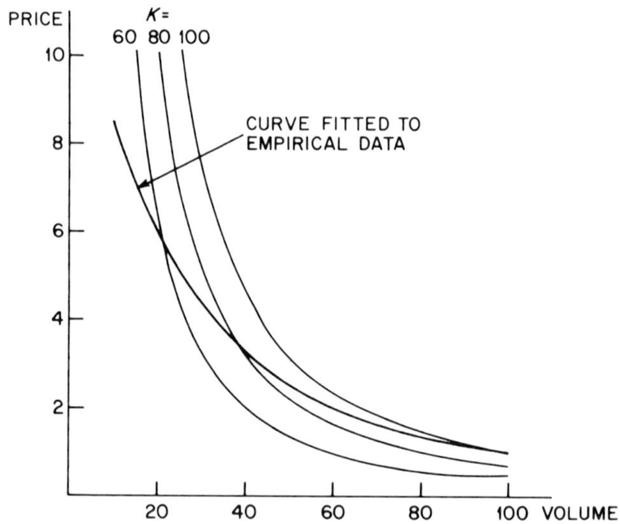

FIG. 8.20. The fitted curve in Fig. 8.19 and three other curves for which $\varepsilon_p = 0.6$.

and $\varepsilon_p = 1.5$. Thus, we can fit two segments to the demand curve:

$$V = 100\,p^{-0.8} \qquad \text{for } 30 < V < 110$$
$$V = 300\,p^{-1.5} \qquad \text{for } 10 < V < 30$$

The two segments, shown by the broken lines in Fig. 8.21, are close to the fitted demand curve and correspond well with the data, as can also be seen from Table 8.4, where values for the fitted segments are compared with values read from the demand curve fitted in Fig. 8.19. Bearing in mind the scatter of the original data (also shown in Fig. 8.21), this piece-wise fitting exercise is satisfactory. The example demonstrates that it is often possible for a concave-type segment to be represented by a function of the type $V = Kp^{-\varepsilon_p}$.

6. The linear demand curve

A linear relationship between demand and price, as shown in Fig. 8.1, is often cited in textbooks as a convenient demand curve, particularly when empirical information is limited and/or when extrapolation is used. As an example, consider the empirical data in Fig. 8.19. If the two extreme points are ignored (since, as stated, they are based on estimates), then the remaining 13 points cover the range $20 \leqslant V \leqslant 60$, and—as shown in Fig. 8.22—a reasonably good linear demand function can be fitted by regression, yielding

$$V = 67.04 - 7.90p$$

TABLE 8.4

Goodness of fit of the segments in Fig. 8.21 (figures in %).

$p =$	1	1.5	2	3	4	4.5	5	6	7	8	8.5
Values of V from fitted demand curve in Fig. 8.19 (also shown in Fig. 8.21)	100	73.5	60	43	33.5	30	26	20	15.5	12	10
Segment $V = 100\,p^{-0.8}$	100	72.3	57.4	41.5	33	30					
Segment $V = 300\,p^{-1.5}$						31.4	26.8	20.4	16.2	13.3	12.1

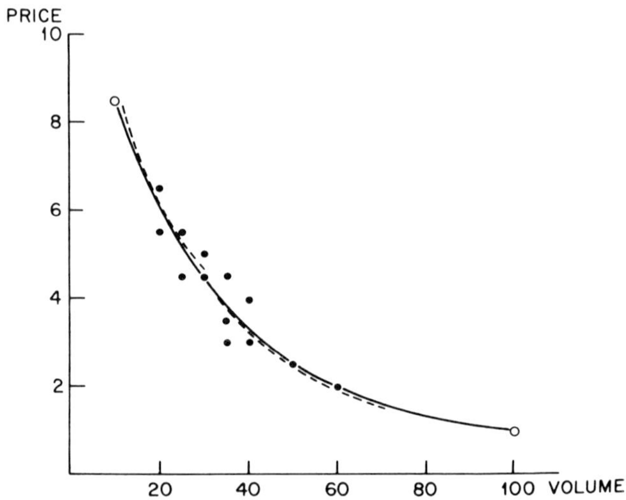

FIG. 8.21. Piece-wise curve fitting.

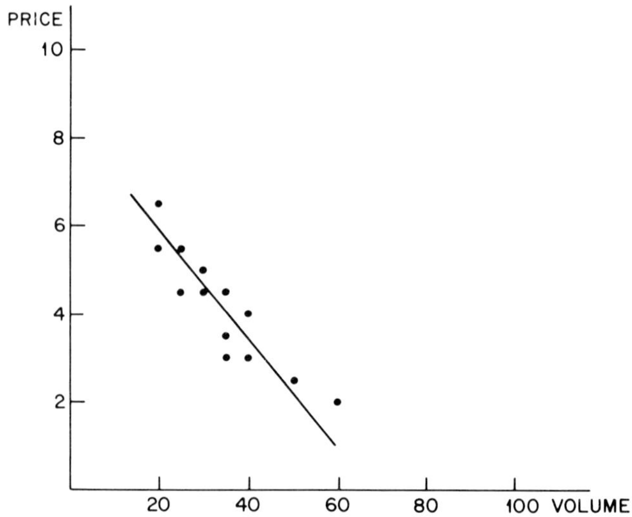

FIG. 8.22. A linear demand curve fitted to empirical data.

with a regression coefficient of -0.901, which signifies a reasonably good fit. In this example the range of values is 3:1 for volume and even more than that for price, so that if operations are conducted within this range, the fitted curve should be adequate to cover changes that are contemplated or likely to occur.

In practice, data are usually available within much more confined ranges, and a linear demand curve often offers as good a fit as can be obtained by other alternatives. If necessary, piece-wise linear segments can be fitted to satisfy managerial requirements for all practical purposes.

A straight line for the demand curve has the added advantage of simplicity in presentation: in Fig. 8.22, for example, the demand line implies that for every unit decrease in price, the sales volume is expected to expand by 7.9 units. If the given demand line is BC, as shown in Fig. 8.23, Equation (8.7) can be used to determine the value of ε_p at every point on this demand line so that $\varepsilon_p = \infty$ at point C, declining to $\varepsilon_p = 1$ at the midpoint A (where AB = AC) and to $\varepsilon_p = 0$ at B.

It should be noted that all linear demand curves that intersect the volume axis at B have the same point elasticity ε_p for a given value of V. For example, the point X′ on the line BC′ in Fig. 8.23 corresponds to the same volume V as point X on the line BC, and as

$$\frac{\mathrm{XB}}{\mathrm{XC}} = \frac{\mathrm{X'B}}{\mathrm{X'C'}}$$

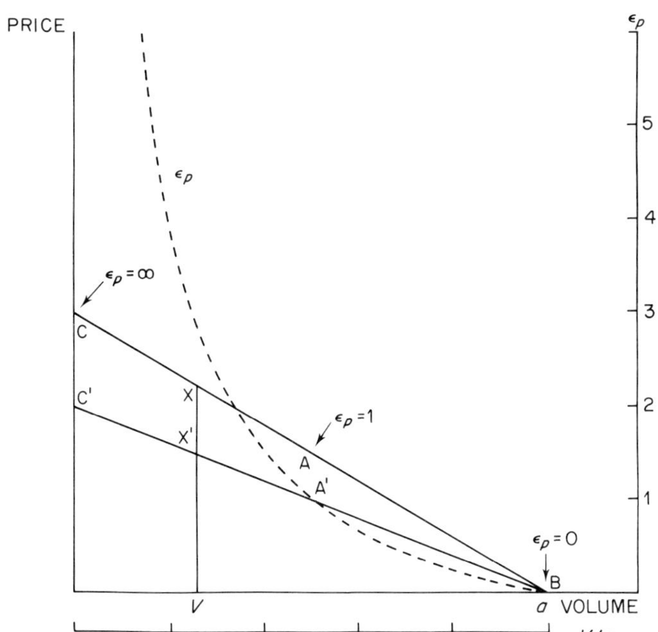

FIG. 8.23. Point elasticity for a linear demand curve.

the value of ε_p is the same in both cases, as a result of Equation (8.7). Hence, if the intercept of the demand line on the volume axis is a, then V/a is the proportion of this intercept which corresponds to the volume V. From Equation (8.7) it follows that at any given point

$$\varepsilon_p = \frac{a - V}{V} = \frac{1}{V/a} - 1 \tag{8.12}$$

so that the value of the point elasticity is solely determined by the proportion V/a as shown in Fig. 8.23 and Table 8.5.

TABLE 8.5
Point elasticity ε_p for values of V/a.

$V/a =$	0.1	0.2	0.3	0.4	0.5	0.6	0.7	0.8	0.9	1.0
$\varepsilon_p =$	9.00	4.00	2.33	1.50	1.00	0.67	0.43	0.25	0.11	0

7. Classification of the point elasticity

It is conventional (see, for example, [1, 2]) for the range of values of ε_p to be classified as follows, irrespective of the shape of the demand curve (unless specifically stated otherwise, it is assumed throughout that ε_p is non-negative):

(a) $\varepsilon_p = 0$ denotes *completely inelastic demand*, as shown by the vertical line in Fig. 8.24 (also by line 6 in Fig. 8.7). In this case $\mathrm{d}V/\mathrm{d}p = 0$, so that irrespective of the values of p and V, the value of ε_p—by its definition in (8.3)—is zero.
(b) $\varepsilon_p < 1$ is taken to represent *inelastic demand*, where revenue moves in the same direction as price, as seen in Equation (8.6).
(c) $\varepsilon_p = 1$ is defined as *unitary elasticity*, where revenue does not change with price.
(d) $\varepsilon_p > 1$ denotes *elastic demand*, where revenue and price move in opposite directions, again from Equation (8.6).
(e) $\varepsilon_p = \infty$ is regarded as *perfectly elastic demand*, given by the horizontal line in Fig. 8.24. Here $\mathrm{d}V/\mathrm{d}p = \infty$, so that the definition in (8.3) yields ∞ for ε_p.

The case of $\varepsilon_p < 0$ is not listed here, as negative elasticity implies that demand is stimulated by a price increase, and when that occurs we have a case of *super-inelastic demand.*

The term "inelastic demand" for $0 < \varepsilon_p < 1$, although widely used, is in some ways unfortunate, in that it may give the impression that the market does not respond to price changes, whereas a response does in fact exist, except that it is more muted than in the region $\varepsilon_p > 1$, the degree of responsiveness being

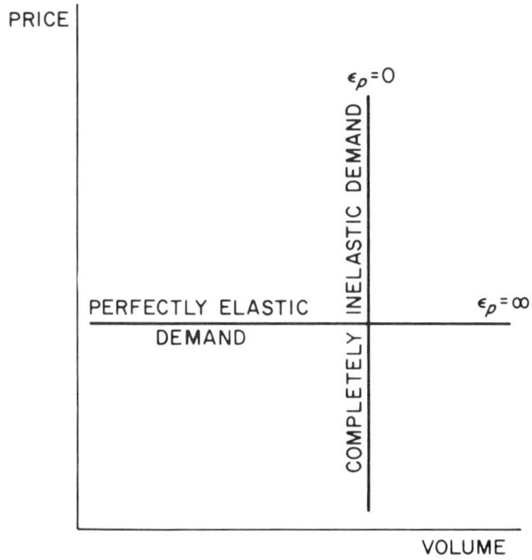

FIG. 8.24. Completely inelastic and perfectly elastic demands.

sensitive to the prevailing value of the elasticity ε_p. The terms "elastic" and "inelastic" are used to indicate the regions where revenue is expected to decrease or increase respectively when price rises. As Bach says [1, p. 585]

> "If demand is 'inelastic', note that a reduction in quantity of the commodity offered on the market would increase total revenue. Thus, a businessman with control over his selling price and output would be foolish to produce at a point where the demand for his commodity is inelastic, for then he could obtain more revenue by producing and selling less!."

8. Revenue for the linear demand curve

If the linear demand function is expressed as

$$V = a - bp \tag{8.13}$$

where a is the intercept on the V axis at $p = 0$ and $-b$ is the slope (both a and b being positive), then the relative change in volume (employing the subtraction rule) is

$$V^* = -\frac{bp}{V}p^*$$

Now, from the definition of ε_p in Equation (8.3)

$$\varepsilon_p = -\frac{\mathrm{d}V}{\mathrm{d}p}\frac{p}{V}$$

but here

$$\frac{\mathrm{d}V}{\mathrm{d}p} = -b$$

hence

$$\varepsilon_p = \frac{bp}{V}$$

Substituting this value into the result for V^* we get

$$V^* = -\varepsilon_p p^* \tag{8.14}$$

where ε_p is tabulated in Table 8.5 for various points on the demand curve and is also shown in Fig. 8.23.

The revenue for the linear demand curve is

$$R = pV = ap - bp^2 \tag{8.15}$$

The marginal revenue with respect to volume is

$$\frac{\mathrm{d}R}{\mathrm{d}V} = \frac{1}{b}(a - 2V)$$

$$= 2p - \frac{a}{b}$$

$$= \frac{a}{b}\frac{\varepsilon_p - 1}{\varepsilon_p + 1} \tag{8.16}$$

and the marginal revenue with respect to price

$$\frac{\mathrm{d}R}{\mathrm{d}p} = a - 2bp$$

$$= a\frac{1 - \varepsilon_p}{1 + \varepsilon_p}, \tag{8.17}$$

From the last two equations it is clear that maximum revenue is obtained when $\mathrm{d}R/\mathrm{d}V = 0$ or $\mathrm{d}R/\mathrm{d}p = 0$, i.e. at

$$p = \frac{a}{2b} \qquad \text{or } V = \frac{a}{2} \tag{8.18}$$

(this is a point of maximum, since the second derivative is negative), at which point the value of R is

$$R_{\max} = \frac{a^2}{2b} - \frac{a^2 b}{4b^2} = \frac{a^2}{4b} \tag{8.19}$$

and it is also obvious that at this point $\varepsilon_p = 1$. An example showing how R, dR/dp and dR/dV change with price is given in Fig. 8.25.

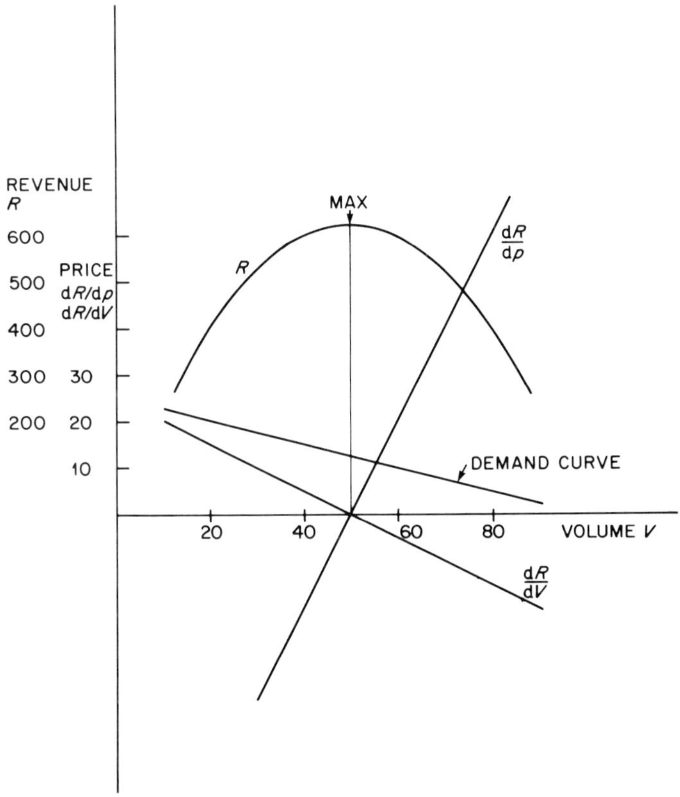

FIG. 8.25. Revenue R and its derivatives for a linear demand function.

9. Relative change in revenue

The relative revenue increment is

$$R^* = p^* + V^* + p^* V^*$$

and in the case of a linear demand curve, V^* can be substituted from Equation (8.14), hence

$$R^* = p^*(1 - \varepsilon_p - \varepsilon_p p^*) \tag{8.20}$$

and for relatively low values of p^* and not inordinately high value of ε_p (i.e.

when $\varepsilon_p p^* \ll 1$) the following approximation is adequate

$$R^* \simeq p^*(1 - \varepsilon_p) \tag{8.21}$$

It is clear from Equation (8.20) that when $\varepsilon_p = 1$ then

$$R^* = -(p^*)^2$$

namely, *revenue is expected to decline* (irrespective of whether p^* is positive or negative, but provided $p^* \neq 0$). The larger p^* is the more R^* deviates from the marginal revenue dR/dV.

ARC ELASTICITY

Definition

As indicated earlier, the concept of point elasticity of demand relates to marginal changes in price and volume, and this is specifically expressed in the definition of ε_p in Equation (8.3). The definition cannot apply when the demand curve does not constitute a continuous mathematical function, since the differential dV/dp does not then exist, and this is one of the obvious shortcomings of the concept of point elasticity for practical purposes. Consider also the following two examples:

(1) Two points on the demand curve are known, the first point corresponding to p_1 and V_1 and the second to p_2 and V_2, as shown in Fig. 8.26. What is the

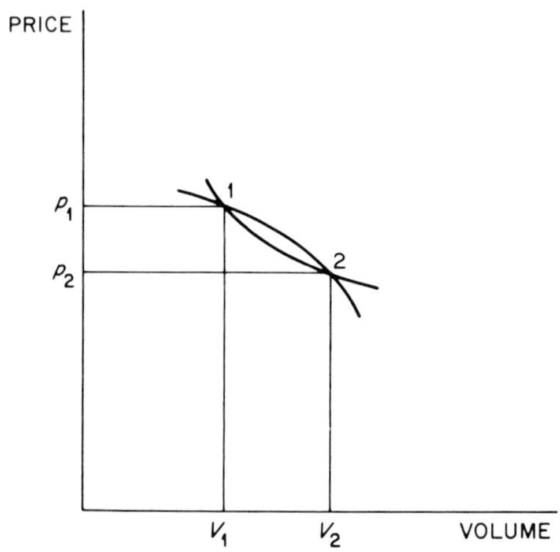

FIG. 8.26. Alternative curves for two given points.

point elasticity of demand at point 1? The answer depends on the tangent at that point, and that depends on the shape of the curve passing through the point. In the absence of any further information, it is clearly possible to draw an infinite number of curves through the two points (two examples are shown in Fig. 8.26), each curve having its own ε_p. The information is, therefore, not adequate for the point elasticity to be determined.

(2) Even when three points are known, as shown in Fig. 8.27, it is still possible to draw alternative demand curves. Furthermore, the curve may consist of two separate segments (e.g. linear arcs), in which case the slope at point 2, where they meet in Fig. 8.27, depends on whether we proceed towards point 1 or 3. This means that when the elasticity is plotted against price or volume, a discontinuity is found at point 2. Such a discontinuity may be perfectly credible, as market response may well depend on whether the price is increased or decreased. But the basic problem remains, namely that the two values of the point elasticity at point 2 depend on the shape of the segments at that location.

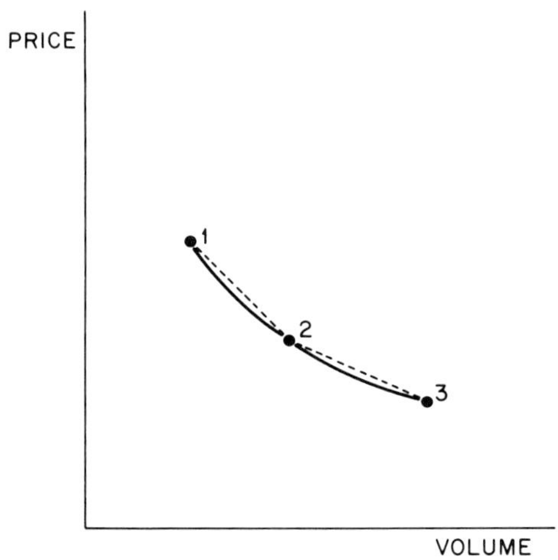

FIG. 8.27. Curves through three given points.

In practice, a complete definition of the demand curve is rarely available and management has to rely on scanty information expressed in discrete form, such as: if prices are increased by x_1 % then demand is expected to fall by y_1 % and if the increase is by x_2 %, the expected fall in demand is y_2 %. A simple definition of elasticity for the case in Fig. 8.26 relies on joining the two points by a straight line and finding an elasticity value for the midway point. This is called the *arc elasticity of demand*, denoted by ε_a and defined as

$$\varepsilon_a = -\frac{V_2 - V_1}{V_2 + V_1} \bigg/ \frac{p_2 - p_1}{p_2 + p_1} \tag{8.22}$$

Incorporating V^* and p^*

Now, we may rewrite the fraction

$$\frac{V_2 - V_1}{V_2 + V_1}$$

in Equation (8.22) as

$$\frac{V_2 - V_1}{V_2 + V_1} = \frac{(V_2 - V_1)/V_1}{(V_2 + V_1)/V_1}$$

Using the definitions for starred variables (and here $V_1 = V$),

$$\frac{V_2}{V_1} = 1 + V^*$$

hence

$$\frac{V_2 - V_1}{V_2 + V_1} = \frac{V^*}{2 + V^*}$$

A similar expression is obtained for the denominator of ε_a so that Equation (8.22) becomes

$$\varepsilon_a = -\frac{V^*}{p^*}\frac{2 + p^*}{2 + V^*} \tag{8.23}$$

from which the relative volume increment is

$$V^* = -\frac{2\varepsilon_a p^*}{2 + p^* + \varepsilon_a p^*} \tag{8.24}$$

and it can easily be shown that in the special case where $\varepsilon_a = 1$ the following results hold:

$$V^* = -\frac{p^*}{1 + p^*}$$

$$R^* = 0$$

Also

$$\left.\begin{aligned} R^* &> 0 \quad \text{if } \varepsilon_a < 1 \\ R^* &< 0 \quad \text{if } \varepsilon_a > 1 \end{aligned}\right\} \text{ when } p^* > 0$$

i.e. similar to the result derived for the point elasticity ε_p.

Example

As a simple example, consider the following data:

	point 1	point 2	
Price	5.0	4.5	hence $p^* = -10\%$
Volume	100	115	hence $V^* = 15\%$

The arc elasticity of demand is

$$\varepsilon_a = \frac{0.15}{0.10}\,\frac{1.90}{2.15} = 1.33$$

It is not possible to compare this value with ε_p at either point 1 or point 2, since we have no information on the demand curve and the slope at either point. On the assumption that the demand curve is linear, then the slope is constant along the line, and

$$\frac{\mathrm{d}V}{\mathrm{d}p} = \frac{V_2 - V_1}{p_2 - p_1} = \frac{115 - 100}{4.5 - 5.0} = -30.0$$

Using the definition (8.3)
at point 1

$$\varepsilon_p = -\frac{\mathrm{d}V}{\mathrm{d}p}\,\frac{p}{V} = 30 \times \frac{5}{100} = 1.50$$

at point 2

$$\varepsilon_p = 30 \times \frac{4.5}{115} = 1.17$$

The linear demand curve

If the demand curve between the two points in Fig. 8.26 is assumed to be linear of the form

$$V = a - bp$$

then

$$V_2 - V_1 = -b(p_2 - p_1)$$

and

$$\begin{aligned} V_2 + V_1 &= 2a - b(p_2 + p_1) \\ &= 2(a + bp_m) = 2V_m \end{aligned}$$

where $p_m = \frac{1}{2}(p_1 + p_2)$ is the average price of p_1 and p_2
$V_m = \frac{1}{2}(V_1 + V_m) = a + bp_m$ is the average value of V_1 and V_2

Substituting these values into the definition of ε_a we get

$$\varepsilon_a = \frac{bp_m}{V_m} = \frac{a - V_m}{V_m} \tag{8.25}$$

and it is interesting to compare this arc elasticity with the point elasticity in Equation (8.12). Strictly, the demand curve between points 1 and 2 in Fig. 8.26 does not have to be linear for the definition of arc elasticity in Equation (8.25) to hold. The only requirements are that p_m and V_m be defined as average values, that b is defined as the negative slope of the straight line connecting the two points, namely

$$b = -\frac{V_2 - V_1}{p_2 - p_1}$$

and that a is the intercept of that line on the volume axis.

Revenue for the linear demand curve

As for the marginal revenue $\mathrm{d}R/\mathrm{d}V$, if we assume a linear demand curve then the revenue function complies with Equation (8.15) and the marginal revenue (with respect to volume) with Equation (8.16). It follows that maximum revenue is obtained at $p = a/2b$, as given in Equation (8.18), the maximum value being as in Equation (8.19). Also, we note that

when $p = a/2b$ the marginal revenue is zero
$p > a/2b$ the marginal revenue is positive
$p < a/2b$ the marginal revenue is negative

The incremental relative revenue for any demand curve is

$R^* = p^* + V^*(1 + p^*)$

and substituting V^* from (8.24), and simplifying we find that

$$R^* = p^* \frac{(2 + p^*)(1 - \varepsilon_a)}{2 + p^*(1 + \varepsilon_a)} \tag{8.26}$$

A comparison of this result with that of Equation (8.20) for the linear demand curve shows how much simpler the expression in (8.20) is, the difference between the two being caused by the different definitions of elasticity. It is also interesting that while R^* is negative for $\varepsilon_p = 1$ in Equation (8.20), here $R^* = 0$ when $\varepsilon_a = 1$, and the marginal revenue is zero in both cases.

We have already seen from the example in Fig. 8.26 that when many alternative demand curves can be drawn through a given point, it is not possible to determine the value of the point elasticity ε_p at that point and, hence, it is not possible to establish the relationship between the point and arc elasticities. In the case of a specific demand curve, such a relationship can be

Table 8.6
Values of arc elasticity.

p^* (%)	−30	−25	−20	−15	−10	−5	0	5	10	15	20	25	30
$\varepsilon_p = 0.50$	.395	.412	.429	.446	.463	.481	.500	.519	.538	.558	.579	.600	.622
0.75	.573	.600	.628	.657	.687	.718	.750	.783	.818	.854	.892	.931	.972
0.80	.607	.636	.667	.698	.731	.765	.800	.837	.875	.915	.957	1.000	1.045
0.90	.674	.708	.743	.780	.818	.858	.900	.944	.990	1.038	1.088	1.141	1.197
1.00	.739	.778	.818	.860	.905	.951	1.000	1.051	1.105	1.162	1.222	1.286	1.353
1.10	.803	.846	.892	.940	.991	1.044	1.100	1.159	1.222	1.289	1.360	1.435	1.515
1.20	.864	.913	.964	1.018	1.075	1.136	1.200	1.268	1.340	1.418	1.500	1.588	1.683
1.25	.895	.946	1.000	1.057	1.118	1.182	1.250	1.323	1.400	1.483	1.571	1.667	1.769
1.5	1.041	1.105	1.174	1.247	1.326	1.410	1.500	1.597	1.703	1.817	1.941	2.077	2.226
2	1.308	1.400	1.500	1.609	1.727	1.857	2.000	2.158	2.333	2.529	2.750	3.000	3.286

found. When the demand curve is linear, the relative volume increment is given in Equation (8.14) as

$$V^* = -\varepsilon_p p^*$$

while, for the same p^* and V^*, the definition of the arc elasticity is given in Equation (8.23), into which V^* is substituted to obtain

$$\varepsilon_a = \varepsilon_p \frac{2 + p^*}{2 - \varepsilon_p p^*} \tag{8.27}$$

For any given point on the demand curve, the value of ε_p is determined and it is independent of the actual magnitude of p^* (since the definition of ε_p is based on a marginal change in price), whereas ε_a does depend on p^*. The relationship between ε_a and ε_p is tabulated in Table 8.6 for various values of p^*, and some results for the ratio $\varepsilon_a/\varepsilon_p$ are plotted in Fig. 8.28, from which we see that the larger the absolute value of p^* and the larger the value of ε_p, the more ε_a deviates from ε_p.

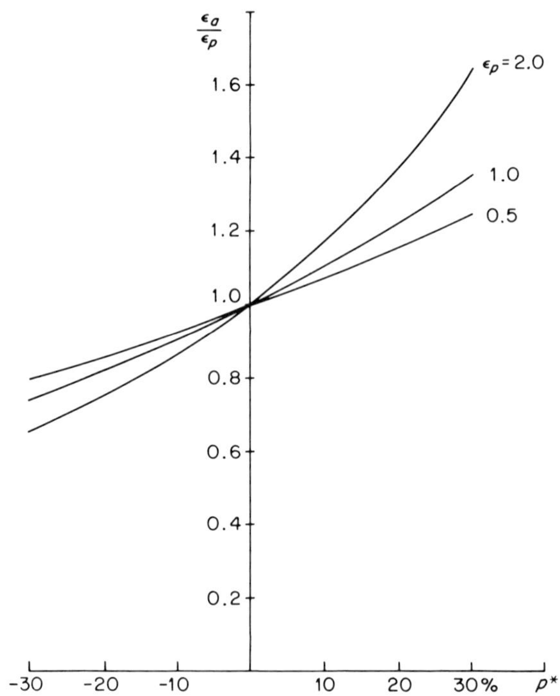

FIG. 8.28. The ratio $\varepsilon_a/\varepsilon_p$.

RELATIVE CHANGE ELASTICITY

Before proceeding with the third definition of elasticity, let us summarize briefly the advantages and disadvantages of the two definitions given above.

Point elasticity

Advantages

Relates to a single point on the curve
It is, therefore, independent of p^*, and also of whether p^* is positive or negative
Easy to fit segments of $\varepsilon_p =$ constant to convex demand functions
$\varepsilon_p = 1$ corresponds to marginal revenue $= 0$
A linear demand function leads to the simple relationship $V^* = -\varepsilon_p p^*$.

Disadvantages

Definition depends on the differential, hence does not apply to functions with discontinuities
Demand function at a given point needs to be known
If used to estimate V^* and R^* when p^* is not very small, then large errors may result
With limited available data (which is usually the case) numerous alternative values of ε_p are equally valid, hence not very useful in most circumstances in practice

Arc elasticity

Advantages

Demand function need not be defined
Function need not be continuous
Elasticity depends on the relative changes of p and V as in Equation (8.23)
The case $\varepsilon_a = 1$ leads to $R^* = 0$

Disadvantages

Relationship between V^* and p^* in Equation (8.24) is cumbersome
Definition relies on the average values p_m and V_m, which are inconvenient to handle.

An alternative definition

As already stated on several occasions, performance in a given period is commonly compared with that in a base period, such as in a historical evaluation of one period with another, or in a postulated performance in a future period compared with the current period. A convenient definition of elasticity is, therefore, based on associating the relative volume increment that results from a relative price increment (as shown in Fig. 8.29) in the expression

$$V^* = -\varepsilon p^* \tag{8.28}$$

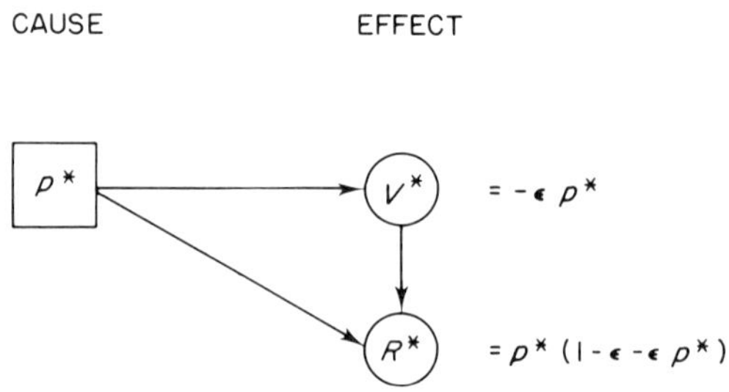

FIG. 8.29. Effect of a change in price.

where ε is the *relative change elasticity*. This definition incorporates many of the advantages while avoiding all the disadvantages of the other two definitions; it does not require the demand curve to be fully defined or to be continuous, it can be applied even when little data are available, and it has the virtue of simplicity which we encountered in Equation (8.14) for the point elasticity in the special case of linear demand. In the main, it accords with the practical intentions behind the definition of the arc elasticity in Equation (8.22) without its cumbersome features. Thus, if a manager is told that

$p^* = 10\%$ is expected to lead to $V^* = -9\%$

while $p^* = -10\%$ is expected to result in $V^* = 12\%$

he immediately concludes that the elasticities are 0.9 and 1.2 respectively, whereas the value of the arc elasticity from (8.22) or (8.23) is less immediately obvious, while the point elasticity from (8.3) cannot be obtained for lack of sufficient data.

Unless stated otherwise, the third definition in Equation (8.28) is adopted throughout this volume, and whenever the term "price elasticity of demand" is used it refers to the relative change elasticity ε. Similar definitions are used in this volume for the effect of advertising

$$V^* = \sigma D^*$$

where σ is the *advertising elasticity of demand*, and for the effect of capital expenditure on fixed and variable costs

$$s^* = -\mu F^*$$

where μ is the *mechanization elasticity*. These expressions are given in Chapter 7.

Relationships between the three elasticities of demand

(1) If the demand curve takes the form

$$V = Kp^{-\varepsilon_p}$$

for which the point elasticity ε_p is constant (as shown earlier), then using the rules of incremental calculus in Chapter 4 we find that

$$V^* = (1+p^*)^{-\varepsilon_p} - 1 \tag{8.29}$$

while we recall that the definition (8.28) for the relative change elasticity is

$$V^* = -\varepsilon p^*$$

hence the relationship between ε and ε_p is

$$\varepsilon = \frac{1}{p^*}[1-(1+p^*)^{-\varepsilon_p}] \tag{8.30}$$

An approximation may be found by expanding the first term in Equation (8.29) by the binomial series to

$$(1+p^*)^{-\varepsilon_p} = 1 - \varepsilon_p p^* + \frac{\varepsilon_p(\varepsilon_p+1)}{2!}p^{*2} - \frac{\varepsilon_p(\varepsilon_p+1)(\varepsilon_p+2)}{3!}p^{*3} + \ldots$$

Equation (8.29) then becomes

$$V^* = -\varepsilon_p p^* + \frac{\varepsilon_p(\varepsilon_p+1)}{2!}p^{*2} - \frac{\varepsilon_p(\varepsilon_p+1)(\varepsilon_p+2)}{3!}p^{*3} + \ldots$$

For the case when $p^* < 1$ and is relatively small, we may use as an approximation only the first two terms, namely

$$V^* \simeq -\varepsilon_p p^*\left(1 - \frac{\varepsilon_p+1}{2}p^*\right)$$

hence, in conjunction with Equation (8.28),

$$\varepsilon \simeq \varepsilon_p \left(1 - \frac{\varepsilon_p + 1}{2} p^*\right) \tag{8.31}$$

so that we get the expected result that the smaller p^* the closer ε gets to ε_p. It is also interesting to note that for this demand curve (and provided $|p^*| \ll 1$)

$$\begin{aligned} \varepsilon &< \varepsilon_p \quad \text{when } p^* > 0 \\ \varepsilon &> \varepsilon_p \quad \text{when } p^* < 0 \end{aligned}$$

The relationship between ε and ε_p for various values of p^* is shown in Fig. 8.30 and a comparison between the result in Equation (8.30) and the approximation in Equation (8.31) is given in Table 8.7

(2) In the case of a linear demand curve, a comparison of Equations (8.14) and (8.28) reveals that

$$\varepsilon = \varepsilon_p$$

so that the elasticity ε can be determined in exactly the same way as suggested in Equation (8.7). As for the arc elasticity, Equations (8.24) and (8.28) suggest

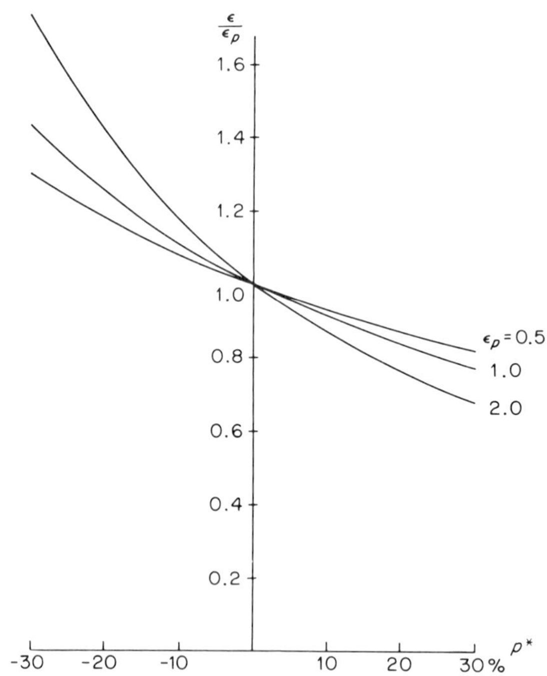

FIG. 8.30. The ratio $\varepsilon/\varepsilon_p$ for the demand function $V = Kp^{-\varepsilon_p}$.

TABLE 8.7

Values of ε for given ε_p for the function $V = K_p^{-\varepsilon_p}$.

p^* (%)	−30	−25	−20	−15	−10	−5	0	5	10	15	20	25	30
$\varepsilon_p = 0.5$													
Eq. (8.30)	.651	.619	.590	.564	.541	.520	.500	.482	.465	.450	.436	.422	.410
App (8.31)	.613	.594	.575	.556	.538	.519	.500	.481	.463	.444	.425	.406	.388
$\varepsilon_p = 1$													
Eq. (8.30)	1.429	1 333	1.250	1.176	1.111	1.053	1.000	.952	.909	.870	.833	.800	.769
App (8.31)	1.300	1.250	1.200	1.150	1.100	1.050	1.000	.950	.900	.850	.800	.750	.700
$\varepsilon_p = 2.0$													
Eq. (8.30)	3.469	3.111	2.813	2.561	2.346	2.161	2.000	1.859	1.736	1.626	1.528	1.440	1.361
App (8.31)	2.900	2.750	2.600	2.450	2.300	2.150	2.000	1.850	1.700	1.550	1.400	1.250	1.100

Eq = Equation; App = Approximation.

that for the same p^* and V^*, and irrespective of the demand curve,

$$\varepsilon_a = \varepsilon \frac{2 + p^*}{2 - \varepsilon p^*} \tag{8.32}$$

as in Equation (8.27). The comparison in Table 8.6 and Fig. 8.28 is applicable here, since $\varepsilon = \varepsilon_p$.

Marginal and relative incremental revenue

The results obtained for the definition of point elasticity for a linear demand function (when Equation (8.14) holds) are relevant here, although as we have seen, there is no need to assume any specific shape for the demand curve in the definition of ε. A straight line connecting two given points has a slope $-b$ and an intercept a on the volume axis, the actual demand function between the two points being quite immaterial. For the marginal revenue, the results in Equations (8.5) and (8.6) apply, except that now $\varepsilon_p = \varepsilon$, i.e.

$$\frac{\mathrm{d}R}{\mathrm{d}V} = p\,\frac{\varepsilon - 1}{\varepsilon}$$

$$\frac{\mathrm{d}R}{\mathrm{d}p} = V(1 - \varepsilon)$$

so that the result obtained earlier, that marginal revenue is zero at $\varepsilon = 1$ also applies; in general,

$$\frac{\mathrm{d}R}{\mathrm{d}p} \gtrless 0 \quad \text{or} \quad \frac{\mathrm{d}R}{\mathrm{d}V} \lessgtr 0 \text{ when } \varepsilon \lessgtr 1 \tag{8.33}$$

The relative incremental revenue is shown in Fig. 8.29 and is as in (8.20), namely

$$R^* = p^*(1 - \varepsilon - \varepsilon p^*) \tag{8.34}$$

At $\varepsilon = 1$ the relative incremental revenue is negative, i.e. revenue declines. Thus, if we are told that an $x\%$ price change causes a change of $y\%$ in volume in the opposite direction, then such a change must result in a fall in revenue for $\varepsilon \geqslant 1$. From the last equation it is clear that if the price increases ($p^* > 0$), the revenue will rise provided

$$1 - \varepsilon - \varepsilon p^* > 0$$

or

$$p^* < \frac{1 - \varepsilon}{\varepsilon} \tag{8.35}$$

and, similarly, when $p^* < 0$ then $R^* > 0$ if the inequality in (8.35) reads the other way.

CONCLUSION

Price is the amount of money that a buyer is prepared to pay and that a seller is prepared to accept in exchange for goods or services. Price is affected by the value systems of buyers and sellers, by competition, by fashion and customer preferences, by the amount of discretionary expenditure at his disposal and by a balance between supply and demand. When supply exceeds demand in a free market, prices tend to fall, and when goods are scarce, prices tend to rise. An equilibrium is obtained at the intersection of the demand and supply curves, as in Fig. 8.1.

Elasticity of demand is a measure of market response to a change in price. Three definitions of elasticity of demand are given in this chapter:

ε_p = *point elasticity*, defined for *a given point* on the demand function and relies on the derivative of the function at that point

ε_a = *arc elasticity*, defined for *the mid-point* of an arc connecting two points, irrespective of the shape of the demand function

ε = *relative change elasticity*, defined *for two given points* as minus the ratio of the relative volume increment to the relative price increment, and this is the definition adopted in this volume.

Of the three, ε_p has the merit that marginal revenue is zero at $\varepsilon_p = 1$. It is also convenient when curves with ε_p = constant can be fitted to price-demand data. In practice, apart from the fact that ε_p is often difficult to determine, management is mainly concerned with discrete increments and not marginal changes. In this respect ε_a is more useful, but ε is preferable to both because relative increments refer to a given base period (instead of a midpoint, as in the case of ε_a), and because of the simplicity of its definition and application. The main features of the three elasticities are summarized in Table 8.8.

Other noteworthy results are:

(1) For any demand curve,

$$\varepsilon_a = \varepsilon \frac{2+p^*}{2-\varepsilon p^*}$$

A general relationship between ε_p and the other two elasticities is not possible to establish.

(2) For a demand curve of the type

$$V = Kp^{-\varepsilon_p}$$

and when p^* is relatively small, then

$$\varepsilon \simeq \varepsilon_p[1 - \tfrac{1}{2}p^*(\varepsilon_p + 1)]$$

and the relationship between ε_a and ε_p is obtained by substituting into the expression of ε_a above.

TABLE 8.8
Summary of definitions and results.

	Point elasticity ε_p	Arc elasticity ε_a	Relative change elasticity ε
Definitions	$-\frac{dV}{dp}\frac{p}{V}$	$-\frac{V_2-V_1}{V_2+V_1}\Big/\frac{p_2-p_1}{p_2+p_1}$	$-\frac{V^*}{p^*}$
for (ii) (see footnote)	$\frac{a}{V}-1$	$b\frac{p_m}{V_m}=\frac{a}{V_m}-1$	$\frac{a}{V}-1$
Revenue	pV (i) $Kp^{1-\varepsilon_p}$ (ii) $ap-bp^2$	pV $ap-bp^2$	pV $ap-bp^2$
$\frac{dR}{dV}=p+V\frac{dp}{dV}$	$p\frac{\varepsilon_p-1}{\varepsilon_p}$		
(ii) $\frac{dR}{dV}=2p-\frac{a}{b}$	$\frac{a}{b}\frac{\varepsilon_p-1}{\varepsilon_p+1}$		$\frac{a}{b}\frac{\varepsilon-1}{\varepsilon+1}$
$\frac{dR}{dp}=V+p\frac{dV}{dp}$	$V(1-\varepsilon_p)$		
(ii) $\frac{dR}{dp}=a-2bp$	$a\frac{1-\varepsilon_p}{1+\varepsilon_p}$		$a\frac{1-\varepsilon}{1+\varepsilon}$
V^*	(i) $(1+p^*)^{-\varepsilon_p}-1$ (ii) $-\varepsilon_p p^*$	$-\frac{2\varepsilon_a p^*}{2+p^*+\varepsilon_a p^*}$	$-\varepsilon p^*$

R^*	(i) $(1+p^*)^{1-\varepsilon_p}-1$ (ii) $p^*(1-\varepsilon_p-\varepsilon_p p^*)$	$p^*\frac{(2+p^*)(1-\varepsilon_a)}{2+p^*(1+\varepsilon_a)}$	$p^*(1-\varepsilon-\varepsilon p^*)$
$R^*=0$		if $\varepsilon_a=1$	
Elasticity conversion		$\varepsilon_a=\varepsilon\frac{2+p^*}{2-\varepsilon p^*}$	$\varepsilon=\frac{2\varepsilon_a}{2+p^*(\varepsilon_a+1)}$ (i) $\varepsilon\simeq\varepsilon_p\left(1-\frac{\varepsilon_p+1}{2}p^*\right)$ (ii) $\varepsilon=\varepsilon_p$

Case (i): $\varepsilon_p=$ const, hence $V=Kp^{-\varepsilon_p}$
Case (ii): linear demand function, where $V=a-bp$.

(3) For the linear demand curve, from Equation (8.32),

$$\varepsilon = \varepsilon_p = \frac{2-\varepsilon p^*}{2+p^*}\,\varepsilon_a$$

$$R_{\max} = \frac{a^2}{4b} \qquad \text{obtained at } p_{\text{opt}} = \frac{a}{2b}$$

and

$$R^* = p^*(1-\varepsilon-\varepsilon p^*)$$

the maximum value of R^* being at

$$p^* = \frac{1-\varepsilon}{2\varepsilon}$$

In the special case $\varepsilon = 1$ we have

$$\frac{\mathrm{d}R}{\mathrm{d}V} = \frac{\mathrm{d}R}{\mathrm{d}p} = 0$$

but

$$R^* = -p^{*2}$$

so that *an increment change in price (positive or negative) has the effect of reducing revenue.*

REFERENCES

[1] Bach, G. L. (1968). "Economics". Prentice Hall.
[2] Brigham, E. F. and Pappas, J. L. (1972). "Managerial Economics". The Dryden Press.
[3] Dorfman, R. (1965). "The Price System". Prentice Hall.
[4] Stigler, G. J. (1952). "The Theory of Price". Macmillan, New York.

EXERCISES

1. If the unit cost c is a linear function of the volume V, prove that as V increases it is not possible to maintain both $z = \text{constant}$ and $\alpha = \text{constant}$.
2. Suppose that

$$c = 10 - 0.01\,V \qquad \text{for } V \leqslant 500$$
$$c = 5 \qquad \text{for } V \geqslant 500$$

Calculate the unit price p and the unit profit z as a function of V for the following alternative policies:

(a) $z = 0.6$ irrespective of volume
(b) $\alpha = 10\%$ irrespective of volume.

3. In Table 8.1 it is suggested that when ε_p = constant then a price increase by $x\%$ has a lesser absolute effect on volume (in percentage terms) than a price reduction of $x\%$. Prove that this result generally holds for ε_p = constant.
4. For the demand function

$$V = 100p^{-\varepsilon_p}$$

The revenue is shown in Fig. 8.16 for three values of ε_p and the marginal revenue dR/dV in Fig. 8.17. Calculate and plot the marginal revenue with respect to price dR/dp.
5. If the current price is 13 and the demand curve is given by

$$V = 100 - 4p$$

Compute and plot:

(a) the point elasticity for all values of V
(b) the expected relative change in volume
(c) the volume and revenue for different values of p
(d) the marginal revenue for different values of p.

6. For any demand curve, prove the following:

(a) $R = \left(\frac{dR}{dV} - p\right)\left(\frac{dR}{dp} - V\right)$

(b) $R = -\frac{dR}{dV}\frac{dR}{dp}\frac{\varepsilon_p}{(1-\varepsilon_p)^2} \qquad (\varepsilon_p \neq 1)$

7. The demand function is

$$V = 10 + \frac{200}{p}$$

for the range $10 \leqslant p \leqslant 20$. Compute:

(a) the revenue function
(b) the marginal revenue in relation to V and p respectively
(c) the three elasticities defined in this chapter, when for any given p an increase of 10% in price is considered.

8. Prove that if

$$V = K_0 + Kp^{-1}$$

for

$$K_0, K > 0$$

then

$$\varepsilon_p < 1$$

9. Maximum revenue for the linear demand curve is

$$R_{\max} = \frac{a^2}{4b}$$

Prove that at this point $\varepsilon_p = 1$.

10. A linear demand function

$$V = 1000 - 0.8p$$

with a current volume of 600 at a unit price of 500 is expected to have a positive shift to a new line

$$V = 1100 - 0.8p$$

A change in price is contemplated (in addition to the shift of the curve) for two cases

(a) $p^* = 10\%$
(b) $p^* = -10\%$

Determine how each of the three price elasticities of demand will change from the old demand line to the new line at $p = 500$ and $V = 600$ respectively.

11. The demand function may be represented by two linear segments as follows:

	p	V
1st segment	45	100
	40	300
2nd segment	20	500

(The 1st segment is bracketed over the rows p = 45 and 40; the 2nd segment over p = 40 and 20.)

Current operations are at $V = 400$. If price is alternatively increased by 10%, 20%, . . . , 50%, calculate the three elasticities in each case.

12. The following limited data are available for the relationship between price and volume:

Price	Volume
40	1000
32	1500
20	3000
10	5000

Present operations are at $V = 3000$ and management would like to know what effect a change in price of $+20\%$ and -20% respectively will have on elasticity and on revenue. There are several alternatives for fitting a demand curve to the data and it is suggested that you try the following:

(a) An exponential curve of the type $Kp^{-\varepsilon_p}$
(b) A linear function
(c) Two linear segments for $1000 \leqslant V \leqslant 3000$ and $3000 \leqslant V \leqslant 5000$
(d) Linear segments between the given points

13. (a) It is stated in this chapter that when $\varepsilon_a = 1$ then

$$V^* = -\frac{p^*}{1+p^*}$$

and

$$R^* = 0$$

Prove these results

(b) Show that when $p^* \ll 1$ then

$$R^* \simeq \frac{2p^*(1-\varepsilon_a)}{2+\varepsilon_a p^*}$$

14. Prove that when a price increase takes place, an upper bound for R^* is given by $p^*(1-\varepsilon_a)$.

9
Revenue and Profit Models

Themes in this chapter

The basic functions of incremental changes in costs, unit cost, revenue and profit
Analysis of circumstances leading to a profit increase
Iso-Z^* charts
Comparison of the relative effects of
- **increasing unit price**
- **decreasing unit cost**
- **increasing volume**

Effect of price elasticity on profit and revenue
Price changes that maximize profit and revenue
The concept of dominance for multi-objectives
The effect of increased advertising expenditure

THE EFFECT OF THREE BASIC VARIABLES

Apart from the analysis of past performance, management is anxious to look at the future and to anticipate the possible effects of changes in the values of key variables in the system on revenue and profit. If the major variables are identified as unit price p, unit cost c and volume V, then a summary of the possible consequences of changes is given in Tables 9.1 and 9.2.

Table 9.1 shows the effect on revenue: if both unit price and volume increase,

TABLE 9.1
Effect on revenue.

	V ↑	V ↓
p ↑	↑	?
p ↓	?	↓

TABLE 9.2
Effect on profit.

p	c	$z=p-c$	V ↑	V ↓
↑	↑	?	?	?
↑	↓	↑	↑	?
↓	↑	↓	?	↓
↓	↓	?	?	?

then revenue increases, and if both decline then revenue declines. However, if one increases and the other decreases, the outcome is uncertain. Table 9.2 consists of two parts: The left-hand side shows the effect on the unit profit z for four combinations of changes in p and c. If both increase or decrease, the outcome for z is uncertain; if p increases and c decreases then z increases; and if p decreases and c increases then z decreases. The last two columns of Table 9.2 show the outcome for the profit Z, which is given by

$$Z = (p-c)V = zV$$

As we can see, of the eight combinations shown on the right-hand side, the results of only two are clear cut (one case is when p and V increase coupled with a decline in c, then Z increases; the other is when all the changes are in the opposite direction), while in six cases the outcome is uncertain, and a general qualitative answer cannot be given. The result then depends on the magnitude of the changes that the variables are subjected to. Even in the two clear cut cases, a qualitative answer may not be adequate, since management often wishes to know by *how much* the profit is likely to be affected for a given combination of changes. In all these cases, therefore, a general quantitative model needs to be developed.

THE BASIC FUNCTIONS

As shown in earlier chapters, the basic relationships for total cost, revenue and profit may be summarized as follows (see also Fig. 5.9):

COST $\quad C = cV \qquad (9.1)$

$C^* = c^* + V^* + c^* V^* \qquad (9.2)$

REVENUE $\quad R = pV \qquad (9.3)$

$R^* = p^* + V^* + p^* V^* \qquad (9.4)$

PROFIT
$$Z = (p-c)V = zV \tag{9.5}$$
$$Z^* = z^* + V^* + z^*V^* \tag{9.6}$$

These functions have a common format, in that they refer to the unit cost c, the unit price p and the unit profit z respectively, and in all cases to the volume V. In order to relate profit to cost and revenue, we turn to the simple relationship

$$Z = R - C \tag{9.7}$$

so that the relative incremental profit (as depicted in the simple network in Fig. 9.1) is

$$Z^* = \frac{R}{Z}R^* - \frac{C}{Z}C^*$$

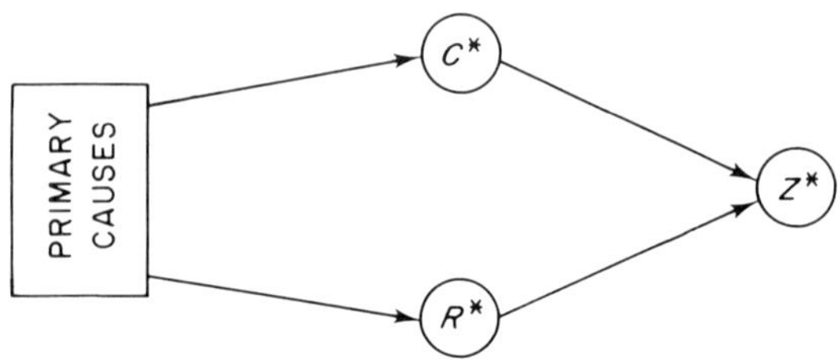

FIG. 9.1. The basic effects on changes in profit.

Using the definition for the profit margin in the base period

$$\alpha = \frac{p-c}{p} = \frac{R-C}{R} = \frac{Z}{R} \tag{9.8}$$

we can substitute

$$\frac{R}{Z} = \frac{1}{\alpha}$$

$$\frac{C}{Z} = \frac{1-\alpha}{\alpha}$$

to obtain (derived in Chapter 4, Equation (4.18))

$$Z^* = \frac{1}{\alpha}[R^* - (1-\alpha)C^*] \tag{9.9}$$

As suggested in Fig. 1.3, an array of variables in the enterprise is subject to various changes; some relating to external factors, such as changes in market conditions or factor prices, others may be the outcome of internal changes, including working practices and the use of resources. The effect of these changes is reflected either in the total cost or in the revenue, or both, and in turn they contribute to a change in the level of profit. But as Equation (9.9) clearly states, in addition to R^* and C^* the relative profit increment is affected by the profit margin α, to which it becomes very sensitive when the profit margin is close to zero. This is illustrated by the following example: Suppose that both revenue and cost increase by 15% and 12% respectively, i.e.

$$R^* = 0.15 \qquad C^* = 0.12$$

If $\alpha = 10\%$ then $Z^* = 42\%$, but if $\alpha = 5\%$ then $Z^* = 72\%$ and if $\alpha = 1\%$ then $Z^* = 312\%$. Similarly, the change in profit is sensitive to changes in revenue and cost, and in particular to whether these changes are in the same direction (in which case they cancel each other out to some extent) or in the opposite direction (thereby amplifying the effect on profit).

This is demonstrated in Table 9.3 for the case where the profit margin is $\alpha = 10\%$ and each of the revenue and cost elements rises or declines by 10%. The results show that if both revenue and cost increase (decrease) by 10%, profit increases (decreases) by 10%, but if a 10% increase in revenue is coupled with a 10% decrease in cost, the result is 190% rise in profit, while a decline in revenue coupled with a cost increase has disastrous consequences for the level of profit. Further examples to illustrate this point are given in Table 9.4 for various values of C^* and R^* and for three levels of profit margin. Such examples explain why company profits can fluctuate so wildly from year to year.

TABLE 9.3
Effect on profit (in %) of 10% increase/decrease in revenue and cost.

		Cost C	
		+10	−10
Revenue R	+10	+10	+190
	−10	−190	−10

RISING AND DECLINING PROFIT

Equation (9.9) reveals the following obvious result: If revenue and cost change by the same percentage, then

(1) profit also changes by that percentage if $\alpha > 0$ (see examples in Table 9.4)
(2) profit changes by that percentage in the opposite direction if $\alpha < 0$

To prove this result, substitute $R^* = C^*$ into Equation (9.9), to get

$$Z^* = R^* = C^*$$

TABLE 9.4
The value of Z^* for given C^*, R^* and α (all values in %).

	R^*	−20	−10	0	10	20
$\alpha = 5\%$;	$C^* = -20$	−20	180	380	580	780
	−10	−210	−10	190	390	590
	0	−400	−200	0	200	400
	10	−590	−390	−190	10	320
	20	−780	−580	−380	−180	20
$\alpha = 10\%$;	$C^* = -20$	−20	80	180	280	380
	−10	−110	−10	90	190	290
	0	−200	−100	0	100	200
	10	−290	−190	−90	10	110
	20	−380	−280	−180	−80	20
$\alpha = 20\%$;	$C^* = -20$	−20	30	80	130	180
	−10	−60	−10	40	90	140
	0	−100	−50	0	50	100
	10	−140	−90	−40	10	60
	20	−180	−130	−80	−30	20

which holds irrespective of the value of α (provided $\alpha \neq 0$). Now, if $\alpha > 0$, which means (from the definition in (9.8)) that the profit in the base period is positive (since R is always assumed to be positive), and as $Z^* = \delta Z/Z$ it follows that the profit increment is

$$\delta Z = ZZ^*$$

so that if Z^* is positive then δZ is positive, and if Z^* is negative the increment δZ is also negative. If, however, $\alpha < 0$ then $Z < 0$, and in this case the profit increment is negative if $Z^* > 0$ and positive if $Z^* < 0$. These relationships are summarized in Table 9.5.

This result is of particular interest when unit cost and unit price can be kept constant and unaffected by a change in volume. In this case, $c^* = p^* = 0$ and

TABLE 9.5
The increment profit δZ for positive and negative values of Z and Z^*.

		Z (or α) >0	Z (or α) <0
Z^*	>0	>0	<0
	<0	<0	>0

Equations (9.2) and (9.4) yield

$$C^* = R^* = V^*$$

Thus, the only way to increase revenue is by increasing volume, but consequently the total cost also increases. What, then, is the outcome with respect to profit?

The obvious answer is that, provided the operation is profitable to start with (i.e. $\alpha > 0$), then increasing volume ($V^* > 0$) is beneficial, but if the operation is run at a loss ($\alpha < 0$), then an increase in volume will make matters worse. Two simple numerical examples are given in Table 9.6. In both cases revenue and cost increase by 10% so that Z^* is also 10%. But in example (a) the profit margin is positive, so that *the profit increases by* 10%, whereas in (b) the margin is negative and *the loss increases by* 10%.

TABLE 9.6
Two examples to illustrate effect on profit.

	Example (a): $\alpha > 0$			Example (b): $\alpha < 0$		
	R	C	Z	R	C	Z
Period 1	100	80	20	100	110	−10
Period 2	110	88	22	110	121	−11
Change	10%	10%	$Z^* = 10\%$ $\delta Z = 2$	10%	10%	$Z^* = 10\%$ $\delta Z = -1$

Condition for $Z^* \geqslant 0$

In normal circumstances, management is anxious to ensure that profit does not decline. Unless stated otherwise, we assume that $\alpha > 0$. From Equation (9.9)

$$Z^* \geqslant 0 \quad \text{if } R^* \geqslant (1-\alpha)C^*$$

and if $C^* > 0$ this condition becomes

$$\frac{R^*}{C^*} \geqslant 1-\alpha \tag{9.10}$$

while if $C^* < 0$ the inequality sign should read the other way. This condition is shown in Fig. 9.2, where the straight line represents the function $1-\alpha$ and stands for $Z^* = 0$; the domain above it corresponds to $Z^* > 0$ and below it to $Z^* < 0$. Values for $R^*/C^* = 1$ fall above the line when α is positive and ensure an increase in profit, but even when the increase in cost is greater in percentage terms than the increase in revenue, it is possible for profit to increase, particularly when α is relatively high. For example, if $\alpha = 0.3$ and if a revenue

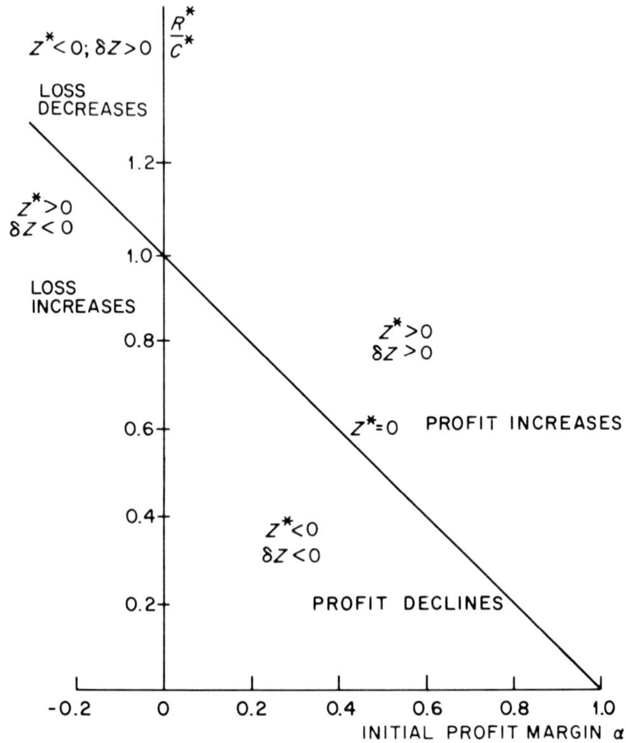

FIG. 9.2. Effect on Z^* and on profit for $C^* > 0$.

increase of 10% is coupled with a cost increase of 14%, the condition in (9.10) is met:

$$\frac{0.10}{0.14} = 0.71 > 0.7 = 1 - \alpha$$

and by Equation (9.9) the change in profit is

$$Z^* = \frac{1}{0.3}(0.10 - 0.7 \times 0.14) = 0.007 \ (0.7\%)$$

Notice that although the profit increases by 0.7%, the profit margin declines, since the new margin is

$$\alpha_{\text{new}} = \frac{Z + \delta Z}{R + \delta R} = \frac{Z(1 + Z^*)}{R(1 + R^*)}$$

$$= \alpha_{\text{old}} \frac{1 + Z^*}{1 + R^*} \tag{9.11}$$

hence

$$\alpha_{\text{new}} = 0.30\,\frac{1.007}{1.10} = 0.275$$

i.e. it falls from 30% to 27.5%. A further increase in revenue by 10%, coupled with a cost increase of 14%, will not improve the profit any further, since now $1-\alpha = 0.725$ and the condition in (9.10) is no longer satisfied.

If the profit margin in the base year is negative ($\alpha < 0$), the line $1-\alpha$ in Fig. 9.2 still denotes $Z^* = 0$ but now $Z^* > 0$ corresponds to

$$R^* < (1-\alpha)C^*$$

and if $C^* > 0$ the domain $Z^* > 0$ is below the line, resulting in a negative value for δZ (from Table 9.5), so that the loss increases. Above the line, however, $Z^* < 0$, so that δZ is positive and the loss declines. If performance can be attained at a point well above the line, where the increase in revenue is sufficiently high, then a loss situation may be transformed into one of profit. Thus, whether the initial profit margin α is positive or negative, the domain above the line implies $\delta Z > 0$, whereas below the line we have $\delta Z < 0$.

Incremental revenue involving incremental cost

Management often has to face the following question: is it worthwhile striving for an additional revenue δR if it involves an incremental cost δC? Similarly, should a reduction in revenue, coupled with a reduction in cost, be considered?

The answer is in the affirmative if

$$\delta R - \delta C > 0$$

or (since $\delta R = R^*R$ and $\delta C = C^*C$)

$$R^*R - C^*C > 0$$

and as $C/R = 1-\alpha$ it follows that we require

$$R^* > (1-\alpha)C^*$$

which for $C^* > 0$ is the condition in (9.10) for profit to increase.

The various combinations of values of δR and δC are depicted in Fig. 9.3. The line at 45° denotes no change in profit ($Z^* = 0$), i.e. any positive increment in revenue is totally expended, and any cost saving is matched by an equal fall in revenue. The domain above the line is "good", in the sense that $\delta R > \delta C$, and below the line is "bad" because $\delta R < \delta C$. The six regions in Fig. 9.3 may be characterized as:

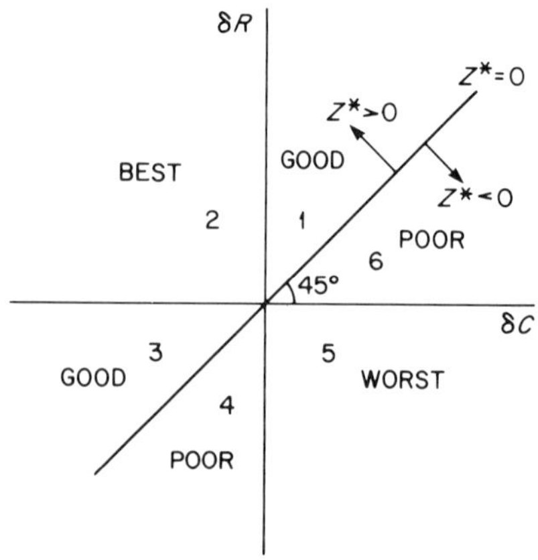

FIG. 9.3. Effect of δR and δC on profit.

Region 1—good, profit increases
Region 2—excellent, revenue rises and cost falls, the ideal combination for a dramatic increase in profit
Region 3—as region 1
Region 4—poor, fall in revenue exceeds cost saving, and profit declines
Region 5—revenue falls and cost increases, the worst possible combination
Region 6—poor

Iso-Z^* chart

From Equation (9.9), or from the sum $R = Z + C$, we derive the function for R^*

$$R^* = \alpha Z^* + (1-\alpha)C^* \tag{9.12}$$

which can be used as a basis for drawing a chart of iso-Z^* lines, i.e. lines for which Z^* is constant. Note the following special cases:

(1) *The break-even point is arrived at* $Z^* = -1$, since by definition, the new profit value is zero, which means that if a plant moves from a position of profit ($Z > 0$) to breakeven, then the profit increment is $\delta Z = -Z$, leading to $Z^* = -1$. Substituting into Equation (9.12), this point occurs at

$$R^* = (1-\alpha)C^* - \alpha$$

This linear relationship between R^* and C^* is shown for various values of α in Fig. 9.4.

(2) *Profit is maintained when* $Z^* = 0$, and that occurs when

$$R^* = (1-\alpha)C^*$$

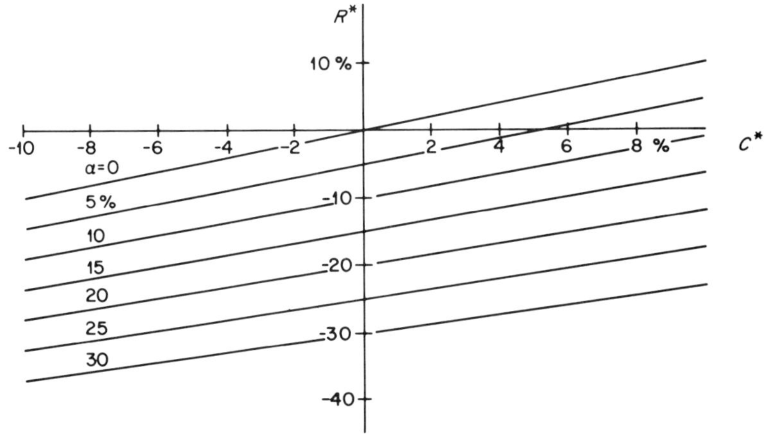

FIG. 9.4. The condition for breakeven ($Z^* = -1$).

Example

Consider the case where $\alpha = 12.5\%$. Hence from Equation (9.12)

$$R^* = 0.125Z^* + 0.875C^*$$

An iso-Z^* chart for this case is shown in Fig. 9.5, so that for any combination of two of the three relative increments R^*, C^* and Z^* the consequences for the third can be determined graphically, and this simple presentation makes the chart a convenient tool for managerial purposes. An alternative presentation is given in Table 9.7, where the outcome for Z^* is tabulated for various values of R^* and C^*. Table 9.7 is divided into three parts (assuming that initially $Z > 0$): the first is where profit is expected to rise; the second corresponds to a decline in profit, but not to the extent that a loss is incurred (i.e. $-1 < Z^* < 0$); the third involves a loss ($Z^* < -1$). Again, the wide range of values for the outcome Z^*, even for relatively moderate changes in revenue and cost, is evident both from Fig. 9.5 and from Table 9.7.

The lines in the iso-Z^* chart in Fig. 9.5 are all parallel and their slope is $1-\alpha$, as indicated by Equation (9.12). The outcome $Z^* = 0$, signifying no change in profit, is shown in Fig. 9.6 for various values of α. In each case the domain above the line corresponds to a profit increase, and it is obvious that when the percentage incremental revenue matches the percentage incremental cost (i.e. $R^* = C^*$, given by the 45° line), profit increases provided $\alpha > 0$. In the particular case where $\alpha = 0$ (i.e. the profit is initially zero), the line for $Z^* = 0$ coincides with that of $R^* = C^*$, and any point on this line means that profit will remain at zero. Lines representing negative values of α are not shown in Fig. 9.6, and if drawn they would lie above the line for $\alpha = 0$.

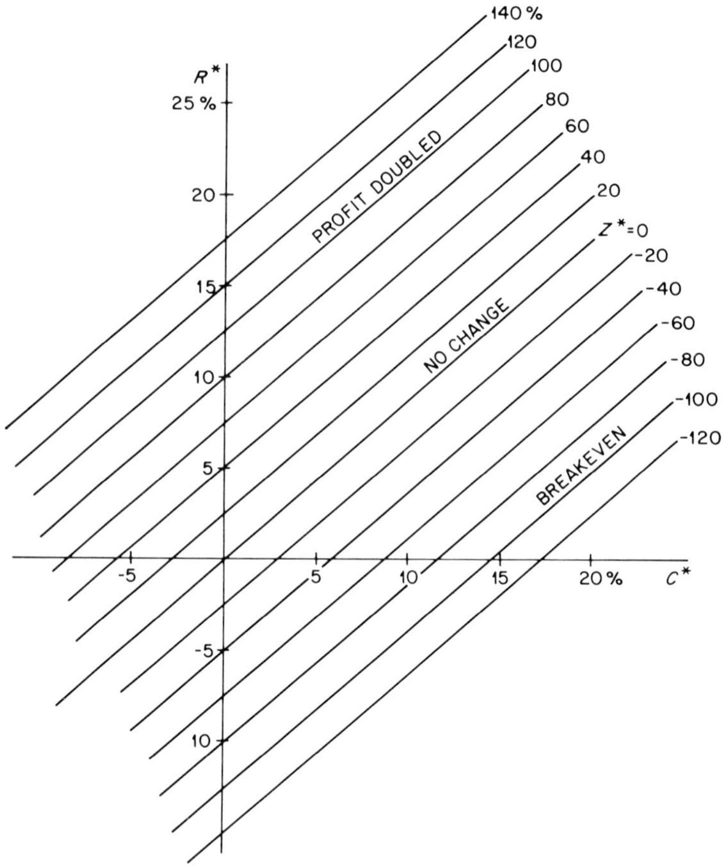

FIG. 9.5. An iso-Z^* chart (for $\alpha = 12.5\%$).

EFFECT OF UNIT COST, UNIT PRICE AND VOLUME

Equation (9.9) expresses the relative incremental profit as a function of the relative increments in revenue and total cost. If we now substitute R^* and C^* from (9.4) and (9.2), then the effect of the incremental unit cost, unit price and volume is obtained. Alternatively, the incremental unit profit is found from

$$z^* = (p - c)^*$$

$$= \frac{p}{z} p^* - \frac{c}{z} c^*$$

TABLE 9.7

Expected value of Z^* (in %) for $\alpha = 12.5\%$.

C^* (%) =	0	2	4	6	8	10	12	14	16	18	20
R^* (%) = 0	0	−14	−28	−42	−56	−70	−84	−98	−112	−126	−140
2	16	2	−12	−26	−40	−54	−68	−82	−96	−110	−124 ←Loss
4	32	18	4	−10	−24	−38	−52	−66	−80	−94	−108
6	48	34	20	6	−8	−22	−36	−50	−64	−78	−92
8	64	50	36	22	8	−6	−20	−34	−48	−62	−76
10	80	66	52	38	24	10	−4	−18	−32	−46	−60

↑ Profit will increase

↑ Profit will decline but operation still profitable

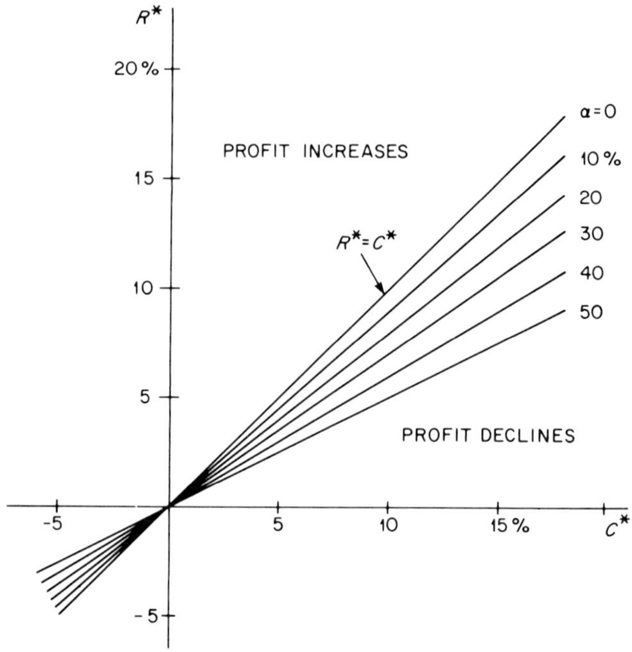

FIG. 9.6. The condition $Z^* = 0$ for various values of α.

Substituting

$$\frac{p}{z} = \frac{1}{\alpha}$$

$$\frac{c}{z} = \frac{1-\alpha}{\alpha}$$

the result is

$$z^* = \frac{1}{\alpha}\left[p^* - (1-\alpha)c^*\right] \tag{9.13}$$

Notice the similarity between this expression and that in Equation (9.9). When the value of z^* in (9.13) is substituted into (9.6), then

$$Z^* = \frac{1+V^*}{\alpha}\left[p^* - (1-\alpha)c^*\right] + V^* \tag{9.14}$$

Here Z^* is a function of the three variables p^*, c^* and V^*, in addition to the initial profit margin α. The following conclusions may be drawn from this equation (assuming the initial profit Z is positive):

(1) An increase in unit price on its own ($p^* > 0$) has a beneficial effect on profit. If $c^* = V^* = 0$ then

$$Z^* = \frac{p^*}{\alpha}$$

Since $0 < \alpha < 1$, the percentage increase in profit exceeds that of the unit price. Clearly, the larger the price increase and the smaller the original profit margin, the greater is the impact on Z^*.

(2) A reduction in unit cost on its own ($c^* < 0$) is also beneficial, but to a slightly lesser extent. If $p^* = V^* = 0$ then

$$Z^* = -\frac{1-\alpha}{\alpha}c^*$$

and if c^* is negative then Z^* is positive. Notice that when the initial profit margin α is positive then

$$\frac{1}{\alpha} > \frac{1-\alpha}{\alpha}$$

so that a rise of $x\%$ in the unit price on its own is bound to improve profit more than a fall of $x\%$ in the unit cost, and this result is further demonstrated by the functions $1/\alpha$ and $(1-\alpha)/\alpha$ in Fig. 9.7, where the

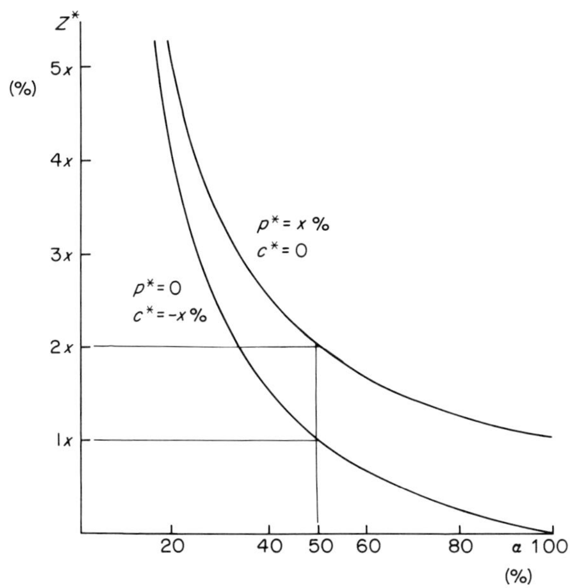

FIG. 9.7. Effect of changing unit price and unit cost by $x\%$ (when $V^* = 0$).

effects of changes in unit price and unit cost are explored. The upper curve corresponds to an increase of $x\%$ in the unit price for $c^* = V^* = 0$, whereas the lower curve corresponds to a reduction of $x\%$ in the unit cost while $p^* = V^* = 0$. The two curves are derived by the functions $1/\alpha$ and $(1-\alpha)/\alpha$ respectively. Notice further that

$$\frac{1-\alpha}{\alpha} > 1$$

when $\alpha < 0.5$, which means that a reduction in unit cost by $x\%$ causes profit to rise by more than $x\%$. However, the relative effect declines as α increases; when $\alpha = 0.5$ a reduction in unit cost by $x\%$ leads to a profit increase by $x\%$, but when $\alpha > 0.5$ the profit increase is less than $x\%$. On the other hand, at $\alpha = 0.5$ an increase in the unit price of $x\%$ results in twice that increase in profit, and throughout the range $0 < \alpha < 1$ profit improves by more than $x\%$.

(3) An increase in volume on its own ($V^* > 0$) is less effective than a price increase and usually less effective than a unit cost decrease (unless α is relatively high, i.e. $\alpha > 0.5$, when the effect of a reduction in unit cost is muted, as shown in Fig. 9.7). If $p^* = c^* = 0$ then

$$Z^* = V^*$$

i.e. the percentage increase in profit matches the percentage increase in volume.

THE EFFECTS OF p^*, c^* AND V^*

These results and others are demonstrated [1] in Table 9.8, where 12 cases are listed for $\alpha = 10\%$, with various combinations of changes in p, c and V, for which the resultant Z^* is noted. Some comments on these results are given below.

1. The relative importance of p^*, c^* and V^*

In *cases 1, 2 and 3* a 1% change in the "right" direction (i.e. to benefit profit) is stipulated for p, c and V respectively. The greatest effect on Z^* is recorded for an increase in the unit price, a slightly smaller effect is caused by a fall in unit cost, and the effect of an increase in volume is relatively muted. These results explain why so much effort is devoted in industry to improving, or at least maintaining, the unit profit, either by increasing the unit price or reducing the unit cost or both. As seen from *case 4* in Table 9.8, the combined effect of $p^* = 1\%$ and $c^* = -1\%$ is equal to the sum of the results in cases 1 and 2.

TABLE 9.8
Effect on Z^* for given values of p^*, c^* and V^* (for $\alpha = 10\%$) (all figures in %).

Case	p^*	c^*	V^*	Z^*
1	1	0	0	10.0
2	0	−1	0	9.0
3	0	0	1	1.0
4	1	−1	0	19.0
5	1	0	1	11.1
6	0	−1	1	10.1
7	1	−1	1	20.2
8	1	0	−1	8.9
9	1	1	−1	−0.01
10	0	1	1	−8.1
11	5	−5	5	104.8
12	−5	5	−5	−95.2

2. The unit profit

From Equation (9.13) it is clear that unit profit will not decline provided

$$p^* \geqslant (1-\alpha)c^* \qquad \text{for } \alpha > 0$$

This result also applies for $Z^* \geqslant 0$ when the level of output does not change ($V^* = 0$), as is clear from Equation (9.14).

3. Interaction of V^* with p^* and c^*

In *case 5* in Table 9.8, p and V rise by 1% each, and in *case 6* c declines by 1% and V rises by 1%. Notice that the result for case 5 is greater than the sum of cases 1 and 3, and similarly case 6 is better than the sum of the results for cases 2 and 3. The reason for cases 5 and 6 behaving differently in this respect from case 4 is that V^* interacts with p^* and c^* in Equation (9.14), where p^* and c^* appear in separate terms in that equation.

4. Combined favorable changes

In *case 7* all three variables change simultaneously in the "right" direction, and the result is better than the sum of the individual effects in cases 1–3. *Case 11* is similar to case 7, except that the relative changes are 5% instead of 1% in magnitude, and it is interesting to note that the effect is greater than fivefold.

5. Effect of a "mixed bag" of changes

When changes that take place are not all in the "right" direction, the combined effect obviously suffers as a result. In *case 8* a 1 % price increase is accompanied by a 1 % decline in volume. The overall outcome is still beneficial to profit. In *case 9* there is, in addition to case 8, a 1 % increase in the unit cost, with the result that the improvement in profit is completely wiped out. In *case 10* the benefit of increased volume is not sufficient to counterbalance the adverse effect of increased unit cost, and the profit therefore declines.

6. Profit can be sensitive even to relatively small changes in its determinant variables

Case 11 comprises changes, which are all in the "right" direction, and it is interesting that although each of the changes is relatively modest, the combined effect is a doubling of the profit level. In *case 12* all the changes are in the "wrong" direction, and again they are relatively small, but because they are all detrimental to performance, their combined effect is almost to wipe out the entire profit. Notice also the asymmetry of the results in cases 11 and 12.

7. Condition for $Z^* > 0$ when $p^* = 0$

In the absence of a price change ($p^* = 0$), Equation (9.14) is reduced to

$$Z^* = -(1+V^*)\frac{1-\alpha}{\alpha}c^* + V^* \qquad \text{(for } p^* = 0\text{)} \tag{9.15}$$

The first term shows the detrimental effect of an increase in unit cost on profit (for $\alpha > 0$). An increase in volume has a beneficial effect, as indicated by the second term, but an adverse effect (if c^* and α are positive) shown in the first term. Notice that in the special case where

$$c^* = \frac{\alpha}{1-\alpha}$$

the expression for Z^* is reduced to $Z^* = -1$, i.e. the profit vanishes altogether and performance is reduced to a break-even point operation. To ensure that the operation remains in profit (i.e. $Z^* > -1$) Equation (9.15) yields

$$\left(1 - c^*\frac{1-\alpha}{\alpha}\right)(1+V^*) > 0$$

i.e.

$$c^* < \frac{\alpha}{1-\alpha} \tag{9.16}$$

for which a sufficient condition is given by

$$c^* < \alpha$$

Equation (9.15) also provides a relationship between c^* and V^* to ensure that profit does not fall. In that case $Z^* \geqslant 0$, provided

$$c^* \leqslant \frac{\alpha}{1-\alpha} \frac{V^*}{1+V^*} \tag{9.17}$$

or, for any given value of c^*, the minimum required increase in volume is given by (assuming $\alpha > 0$)

$$V^* \geqslant \frac{(1-\alpha)c^*}{\alpha - (1-\alpha)c^*}$$

Clearly, the condition (9.16) is less stringent than (9.17) for $V^* > 0$, since the former ensures that profit does not evaporate, whereas the latter sets the scene for profit not to fall.

Consider the following example: $\alpha = 0.20$ and $c^* = 0.12$. Here $c^* < \alpha$, hence (9.16) is met and the operation will remain in profit irrespective of V^*. The minimum increase in volume needed to guard against a fall in profit is obtained from the last expression:

$$V^* \geqslant \frac{0.8 \times 0.12}{0.2 - 0.8 \times 0.12} = 0.923$$

i.e. volume will have to increase by 92.3% to offset the effect of 12% unit cost increase.

Figure 9.8 shows the minimum required level of V^* plotted against c^* for three cases of α. Each curve stands for $Z^* = 0$, the domain above it corresponds to $Z^* > 0$ and the domain below to $Z^* < 0$. The slope of the curve indicates how sensitive the need for a volume increase is when the unit cost rises, and this slope would have been more prominent had the other two axes been drawn to the same scale. Results for several other values of α are given in Table 9.9, from which it is clear that the relationship between c^* and V^* is very sensitive when the profit margin is low. For example, when $\alpha = 5\%$, an increase of 5% in the unit cost requires a 19-fold increase in volume for the profit level to be maintained! When $c^* \geqslant 5.3\%$ the condition in (9.16) is no longer satisfied for $\alpha = 5\%$, which means that no amount of volume increase can prevent the occurrence of a loss. The blanks in Table 9.9 signify that profit cannot be maintained. When a reduction in unit cost takes place, volume may decline by up to the amounts shown in Table 9.9, if profit is not to fall, and again the sensitivity to a change in unit cost is evident when the profit margin is low.

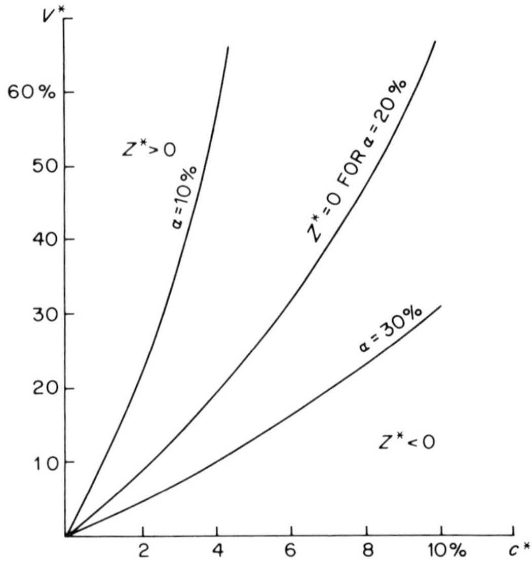

FIG. 9.8. Condition for $Z^* \gtrless 0$ when $p^* = 0$.

8. The case $p^* \neq 0$

In the more general case where $p^* \neq 0$ the condition to ensure that the profit level does not fall is derived from Equation (9.14), assuming $\alpha > 0$, namely

$$(1+V^*)[p^* - (1-\alpha)c^*] + \alpha V^* \geqslant 0$$

leading to the general condition

$$p^* + \frac{\alpha V^*}{1+V^*} \geqslant (1-\alpha)c^* \tag{9.18}$$

from which (9.17) is derived as a special case when $p^* = 0$, so that for any given values of p^* and c^* the lower bound for V^* now becomes

$$V^* \geqslant \frac{(1-\alpha)c^* - p^*}{p^* + \alpha - (1-\alpha)c^*} \tag{9.19}$$

from which the special case for $p^* = 0$ is easily derived. Again, to ensure that the operation remains in profit ($Z^* > -1$), there is an upper bound that c^* must not exceed, derived from Equation (9.14) and given by

$$c^* < \frac{p^* + \alpha}{1-\alpha} \tag{9.20}$$

TABLE 9.9
Change in volume needed to compensate for a change in unit cost to maintain profit when $p^* = 0$ (figures in %).

$\alpha =$	5	10	15	20	25	30
$c^* = 0$	0	0	0	0	0	0
1	23.5	9.9	6.0	4.2	3.1	2.4
2	61.3	22.0	12.8	8.7	6.4	4.9
3	132.6	37.0	20.5	13.6	9.9	7.5
4	317	56.3	29.3	19.0	13.6	10.3
5	1900	81.8	39.5	25.0	17.6	13.2
6		117.4	51.5	31.6	22.0	16.3
7		170.3	65.7	38.9	26.6	19.5
8		257	82.9	47.1	31.6	23.0
9		426	104.1	56.3	37.0	26.6
10		900	130.8	66.7	42.9	30.4
−1	−16.0	−8.3	−5.4	−3.8	−2.9	−2.3
−2	−27.5	−15.3	−10.2	−7.4	−5.7	−4.5
−3	−36.3	−21.3	−14.5	−10.7	−8.3	−6.5
−4	−43.2	−26.5	−18.5	−13.8	−10.7	−8.5
−5	−48.7	−31.0	−22.1	−16.7	−13.0	−10.4
−6	−53.3	−35.1	−25.4	−19.4	−15.3	−12.3
−7	−57.1	−38.7	−28.4	−21.9	−17.4	−14.0
−8	−60.3	−41.9	−31.2	−24.2	−19.4	−15.7
−9	−63.1	−44.8	−33.8	−26.5	−21.3	−17.4
−10	−65.5	−47.4	−36.2	−28.6	−23.1	−18.9

from which (9.16) is obtained as a special case when $p^* = 0$. If in (9.19) the inequality sign is replaced by an equality, then the result is no change in profit ($Z^* = 0$). The expression (9.20) with an equality sign is shown graphically in Fig. 9.9 for several values of α. For a given profit margin, the condition (9.20) is a linear relationship between p^* and c^*, but it should be noted that the lines in Fig. 9.9 are not parallel. We conclude that as condition (9.18) ensures $Z^* \geqslant 0$ while (9.20) is designed for $Z^* > -1$, the former must be more demanding on the upper bound allowable for an increase in the unit cost, so that satisfying the condition in (9.20) is *necessary but not sufficient to guard against a fall in profit.*

In addition to meeting (9.20), the rise in volume must be adequate to lead to a rise in profit, and the minimum level required for a volume increase is given by (9.19). It is clear from expression (9.19) that if the unit price is allowed to increase, the lower bound for V^* shown in Fig. 9.8 is ameliorated. This is demonstrated in Fig. 9.10, where the top curve (corresponding to $p^* = 0$) gives the minimum increase in volume required, for any given increase in the unit cost, to ensure that profit does not decline. This curve is the same as the one shown in Fig. 9.8 for $\alpha = 20\%$. Two other curves in Fig. 9.10 reveal the effect of

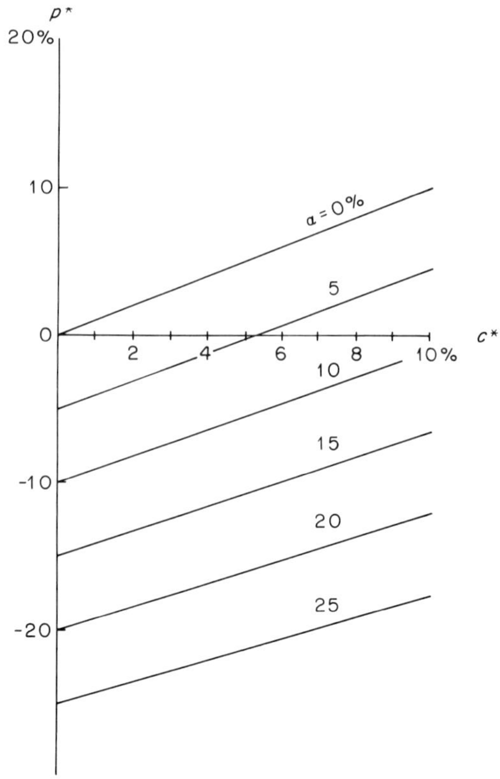

FIG. 9.9. The condition for breakeven ($Z^* = -1$). The operation remains profitable ($Z^* > -1$) above the corresponding line.

increasing the unit price and, evidently, even a relatively small price increase helps to relax significantly the requirement for V^*. In each case the curve signifies no change in profit, while profit increases above the curve. Note that provided c^* is not excessive, it is even possible for volume to decline (as shown by the shaded area in Fig. 9.10) and yet for profit to rise, since the price increase is sufficient here to compensate for a rise in the unit cost.

9. Decomposition of the unit cost

If we assume that the unit cost function consists of fixed and variable cost elements, as discussed in Chapter 7, then

$$c = s + \frac{F}{V}$$

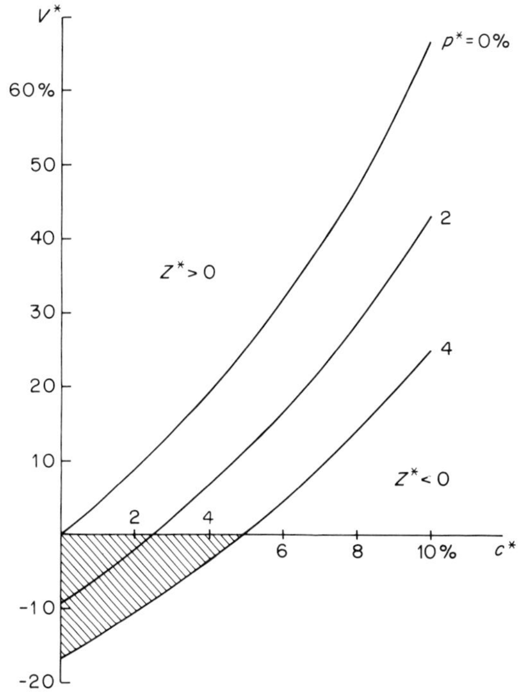

FIG. 9.10. Effect of p^* on the lower bound of V^* when $\alpha = 20\%$.

No separate account is taken here of the interest charges J and the effect of working capital, which is considered further in Chapter 11. If we also assume that $s^* = F^* = 0$, then

$$c^* = -\frac{fV^*}{1+V^*} \tag{9.21}$$

and we recall that here $f = f_2 = F/C$. Substituting this value of c^* into Equation (9.14) we get

$$Z^* = \frac{1}{\alpha}[p^*(1+V^*) + (\alpha + f - f\alpha)V^*]$$

Denote

$$\beta = \alpha + f - f\alpha \tag{9.22}$$

It will be shown in the next chapter that β is the gross profit margin, defined by

$$\beta = \frac{p-s}{p}$$

If the definition in (9.22) is substituted into the last equation for Z^* we obtain

$$Z^* = \frac{1}{\alpha}\left[p^*(1+V^*)+\beta V^*\right] \tag{9.23}$$

and for the case where V^* is relatively small, i.e. $V^* \ll 1$, the following approximation holds

$$Z^* \simeq \frac{1}{\alpha}(p^* + \beta V^*) \tag{9.24}$$

If the original profit margin α is positive, the condition for profit not to decline is derived from Equation (9.23) as

$$Z^* \geqslant 0 \qquad \text{if } p^* \geqslant \frac{-\beta V^*}{1+V^*} \tag{9.25}$$

This result is depicted in Fig. 9.11 for various values of β, where each curve represents $Z^* = 0$ and the domain above it corresponds to an increase in profit.

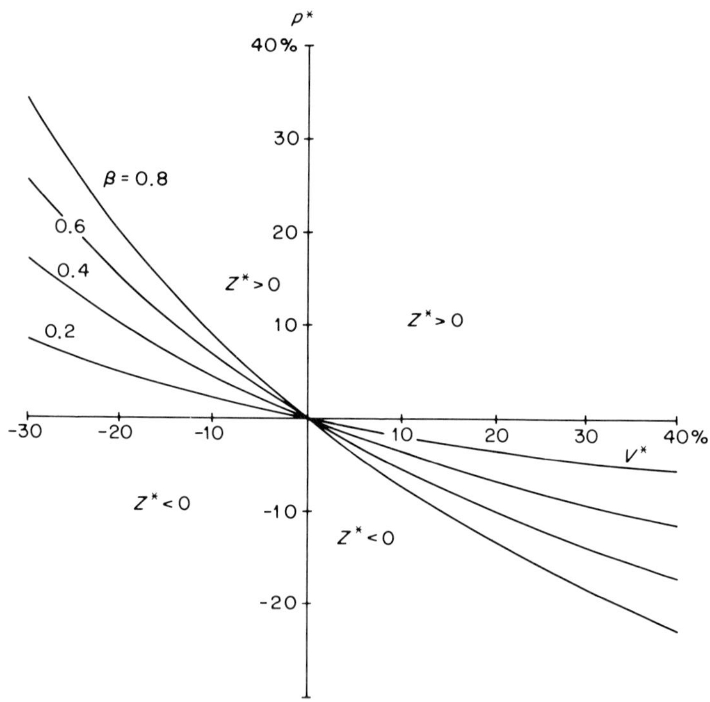

Fig. 9.11. Conditions for no change in profit ($Z^* = 0$).

10. The effect of p^* on V^*

The results in Table 9.8, and others discussed hitherto, are cited for various combinations of data for p^*, c^* and V^*, without stipulating what connection there might be between them. For example, case 1 involves a price increase on its own, and yet, as discussions in the previous chapter suggest, such an increase is expected to lead to a decline in volume, which in turn should affect the unit cost. The network of relationships that would then apply is shown in Fig. 9.12, and all the secondary effects need to be accounted for in determining the overall outcome for Z^*. In this respect, therefore, many of the cases in Table 9.8 may not be realistic, but they do serve to indicate the relative importance of the variables in question, and the consequences of primary causes of change when the secondary effects are ignored in the first instance. A more detailed account of these effects will now follow.

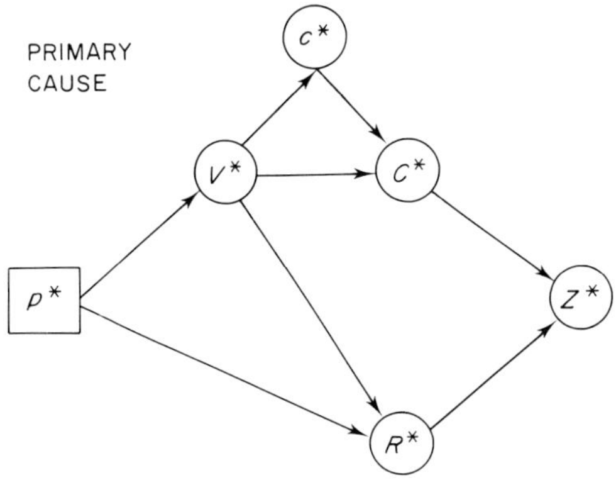

FIG. 9.12. Effect of changing the unit price.

EFFECT OF A PRICE CHANGE

If we assume a change in price as the primary cause, then the sequence of events is shown in Fig. 9.12:

- p^* causes a change in volume (we assume here that this is the only reason for V^*)
- p^* and V^* affect the revenue
- V^* affects the unit cost

- V^* and c^* affect the total cost
- R^* and C^* affect the profit level

The relationship between p^* and V^* is described by the use of the relative change elasticity of demand ε, as discussed in the previous chapter, namely (assuming V^* is solely due to p^*)

$$V^* = -\varepsilon p^* \tag{9.26}$$

Effect on revenue

When this value of V^* is substituted into Equation (9.4) for the relative incremental revenue we get

$$\begin{aligned} R^* &= p^* - \varepsilon p^* - \varepsilon p^{*2} \\ &= p^*(1 - \varepsilon - \varepsilon p^*) \end{aligned} \tag{9.27}$$

for which the following observations may be made:

(1) Since R^* is a quadratic function of p^* we can find, for any given value of ε (i.e. assuming ε remains constant), the value of p^* that maximizes the increase in revenue, namely at

$$\frac{\mathrm{d}R^*}{\mathrm{d}p^*} = 1 - \varepsilon - 2\varepsilon p^* = 0$$

or

$$p^*_{\mathrm{opt}}(R^*) = \frac{1-\varepsilon}{2\varepsilon} \tag{9.28}$$

The notation $p^*_{\mathrm{opt}}(R^*)$ denotes the optimum value of p^* to maximize R^*. That this is a maximum point is obvious from the second derivative being negative:

$$\frac{\mathrm{d}^2 R^*}{\mathrm{d}p^{*2}} = -2\varepsilon < 0 \qquad \text{for } \varepsilon > 0$$

(2) The result in Equation (9.28) is obtained on the assumption that ε is constant, though in reality (as we saw in the last chapter) ε may well depend on the magnitude of the price change. If the functional relationship between ε and p^* is known, it can be substituted into Equation (9.27) for the maximum value of R^* to be ascertained.

(3) For any given value of $p^* > 0$ it follows from (9.27) that R^* declines as ε increases, since the derivative of R^* with respect to ε is

$$\frac{\mathrm{d}R^*}{\mathrm{d}\varepsilon} = -p^*(1 + p^*) < 0 \qquad \text{for } p^* > 0$$

However, when the price is reduced, i.e. if $p^* < 0$, the derivative becomes positive and R^* then increases with increasing ε.

(4) From equations (9.27) and (9.28) it is further clear that if $p^* > 0$, there is a limit to the level to which p^* may be allowed to rise in order to achieve an increase in revenue, i.e.

$$0 < p^* < \frac{1-\varepsilon}{\varepsilon}$$

But since we only consider positive values of ε, then for $p^* > 0$ it follows that $\varepsilon < 1$. Similarly, if the price is reduced, i.e. $p^* < 0$, then the condition for $R^* > 0$ from Equation (9.27) is

$$0 > p^* > \frac{1-\varepsilon}{\varepsilon}$$

where now ε must exceed 1. These results may be summarized as follows:

$$R^* > 0 \begin{cases} \text{if } 0 < p^* < \dfrac{1-\varepsilon}{\varepsilon} & \varepsilon < 1 \\ \text{or } \dfrac{1-\varepsilon}{\varepsilon} < p^* < 0 & \varepsilon > 1 \end{cases} \tag{9.29}$$

i.e. *if* $0 < \varepsilon < 1$ *a price increase up to a point* (*as given in* (9.29)) *causes revenue to rise, but if* $\varepsilon > 1$ *then a price increase causes it to decline, while at* $\varepsilon = 1$ *a price change* (*increase or decrease*) *results in a fall in revenue.*

(5) The conclusions drawn from the conditions (9.29) are shown in Fig. 9.13, where the solid line (given by the function $(1-\varepsilon)/\varepsilon$) represents $R^* = 0$. For the case of a price increase, the domain below the curve corresponds to an increase in revenue and above the curve to a decline in revenue. Notice how steep the curve is, particularly for low values of ε, implying that substantial price increases may be considered without detrimental effects on revenue. For example, if $\varepsilon = 0.6$ the price can be increased by up to 66.7 % before the revenue level falls below that prevailing in the base period, and if $\varepsilon = 0.5$ this bound for a price increase rises to 100 %. The path which corresponds to the maximum value of R^* is shown by the broken line in Fig. 9.13. Further results for the bound for p^* to ensure that revenue does not decline are given in Table 9.10. For example, if $\varepsilon = 0.92$ the upper bound for p^* is 8.7 %.

(6) The line for $R^* = 0$ is also shown in Fig. 9.14, where Equation (9.27) is used to draw a contour map of iso-R^* lines. Such a map helps to assess the degree of sensitivity of a change in revenue to a change in price, for any given price elasticity. Notice that the terrain in the map has the shape of a saddle: starting at the point $p^* = 0$ and $\varepsilon = 1$, the value of R^* declines

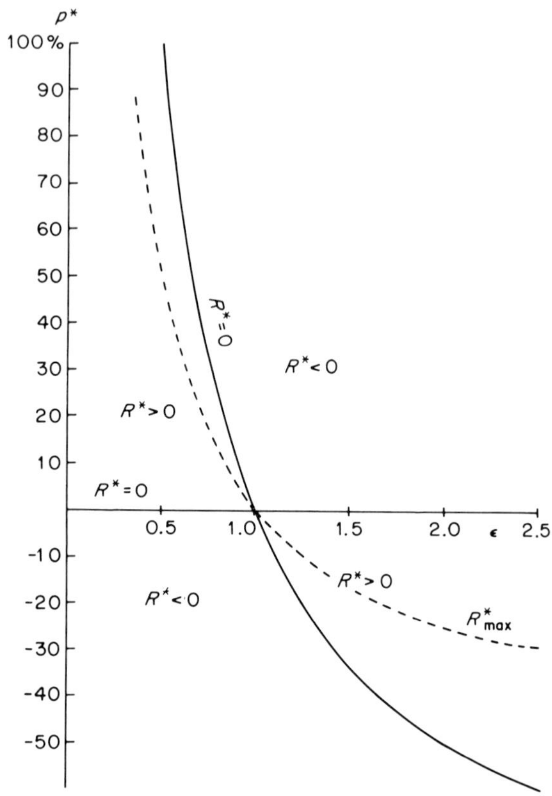

FIG. 9.13. Condition for revenue to increase or decrease.

when we proceed either in a north-easterly or south-westerly direction, but R^* increases when proceeding along the north-west or south-east axis.

(7) Thus, when $\varepsilon < 1$ and price is increased, revenue increases as well, until the optimum point given by Equation (9.28) is reached, beyond which the revenue begins to decline. The maximum level of revenue is obtained when p^*_{opt} from Equation (9.28) is substituted into (9.27) to yield

$$R^*_{max} = \frac{(1-\varepsilon)^2}{4\varepsilon} = \varepsilon[p^*_{opt}(R^*)]^2 \tag{9.30}$$

The magnitude of R^*_{max} is tabulated in Table 9.11 for various values of ε and it is evident that the scope for increasing revenue by manipulation of the unit price is somewhat limited, when the elasticity ε is the range of, say, 0.6 to 1.6, and particularly in the vicinity of $\varepsilon = 1$.

TABLE 9.10

Bound for p^* (in %) to avoid a decline in revenue.

$\varepsilon =$	0.00	0.01	0.02	0.03	0.04	0.05	0.06	0.07	0.08	0.09
0.5	$p^* =$ 100.0	96.1	92.3	88.7	85.2	81.8	78.6	75.4	72.4	69.5
0.6	66.7	63.9	61.3	58.7	56.3	53.8	51.5	49.3	47.1	44.9
0.7	42.9	40.8	38.9	37.0	35.1	33.3	31.6	29.9	28.2	26.6
0.8	25.0	23.5	22.0	20.5	19.0	17.6	16.3	14.9	13.6	12.4
0.9	11.1	9.9	8.7	7.5	6.4	5.3	4.2	3.1	2.0	1.0
1.0	0	−1.0	−2.0	−2.9	−3.8	−4.8	−5.7	−6.5	−7.4	−8.3
1.1	−9.1	−9.9	−10.7	−11.5	−12.3	−13.0	−13.8	−14.5	−15.3	−16.0
1.2	−16.7	−17.4	−18.0	−18.7	−19.4	−20.0	−20.6	−21.3	−21.9	−22.5
1.3	−23.1	−23.7	−24.4	−24.8	−25.4	−25.9	−26.5	−27.0	−27.5	−28.1
1.4	−28.6	−29.1	−29.6	−30.1	−30.6	−31.0	−31.5	−32.0	−32.4	−32.9
1.5	−33.3	−33.8	−34.2	−34.6	−35.1	−35.5	−35.9	−36.3	−36.7	−37.1
1.6	−37.5	−37.9	−38.3	−38.7	−39.0	−39.4	−39.8	−40.1	−40.5	−40.8
1.7	−41.2	−41.5	−41.9	−42.2	−42.5	−42.9	−43.2	−43.5	−43.8	−44.1
1.8	−44.4	−44.8	−45.1	−45.4	−45.7	−45.9	−46.2	−46.5	−46.8	−47.1
1.9	−47.4	−47.6	−47.9	−48.2	−48.5	−48.7	−49.0	−49.2	−49.5	−49.7
2.0	−50.0									

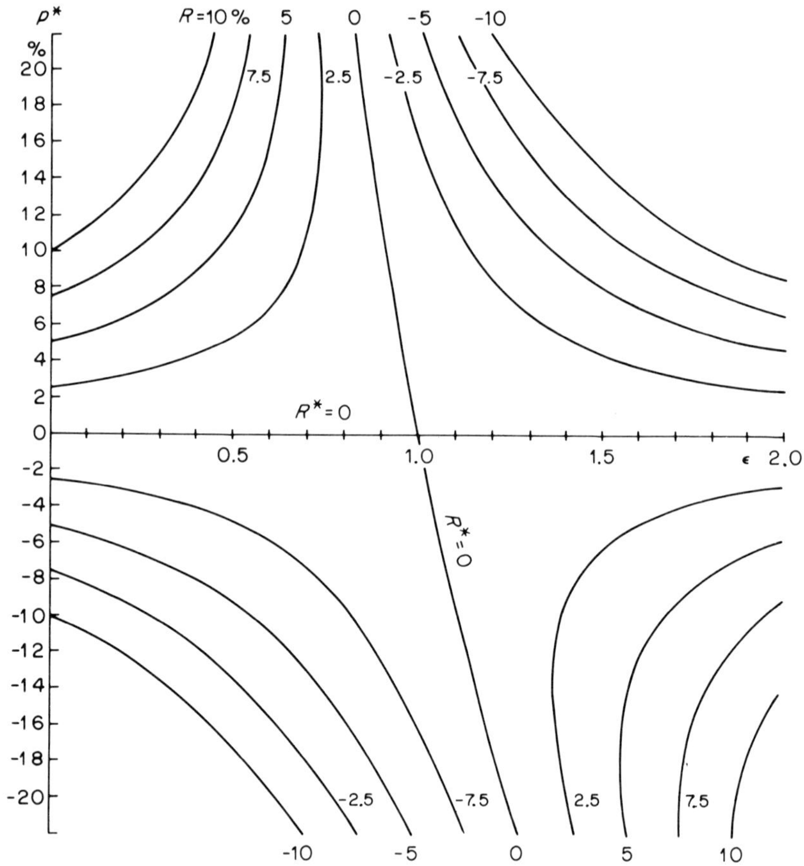

FIG. 9.14. Iso-R^* map.

(8) Notice the fact that p^*_{opt} in Equation (9.28) is exactly half the bound in (9.29). To illustrate how the relative incremental revenue changes with price compare the two examples $\varepsilon = 0.5$ and $\varepsilon = 0.75$, for which the results for R^* are given in Fig. 9.15. The effect of ε is clearly demonstrated by the relative magnitude of the two R^* functions.

Effect on profit

If we substitute Equation (9.26) into expression (9.18), the condition for $Z^* \geqslant 0$ (when $\alpha > 0$) becomes

$$(1 - \varepsilon p^* - \alpha\varepsilon)p^* \geqslant (1 - \alpha)(1 - \varepsilon p^*)c^* \tag{9.31}$$

TABLE 9.11
Maximum R^* for given elasticity ε.

ε	0.1	0.2	0.3	0.4	0.5	0.6	0.7	0.8	0.9	1.0
p^*_{opt} (%)	450	200	116.7	75	50	33.3	21.4	12.5	5.6	0
R^* (%)	202.5	80.0	40.8	22.5	12.5	6.7	3.2	1.3	0.3	0
ε	1.1	1.2	1.3	1.4	1.5	1.6	1.7	1.8	1.9	2.0
p^*_{opt} (%)	−4.6	−8.3	−11.5	−14.3	−16.7	−18.8	−20.6	−22.2	−23.7	−25.0
R^* (%)	0.2	0.8	1.7	2.9	4.2	5.6	7.2	8.9	10.7	12.5

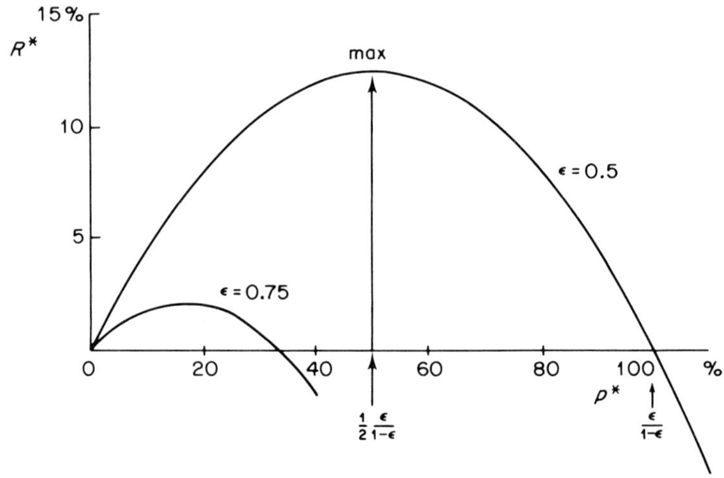

FIG. 9.15. Effect of a price increase on revenue for two values of ε.

For a relatively small value of p^* it may be assumed that

$$1 - \varepsilon p^* \simeq 1$$

and (9.31) may then be replaced by the approximation

$$(1 - \alpha\varepsilon)p^* \geqslant (1 - \alpha)c^* \tag{9.32}$$

If the original profit is rather small ($\alpha \ll 1$), or if $\varepsilon \simeq 1$, then this condition is reduced to

$$p^* \geqslant c^*$$

i.e. the relative increment of the unit price must exceed that of the unit cost for profit to increase.

How good the approximation (9.32) is can be judged from Table 9.12, which shows some results for selected values of α and ε. The approximation is clearly quite reasonable, except in circumstances which combine relatively high values of p^*, ε and α.

Substituting $V^* = -\varepsilon p^*$ into Equation (9.23), the following simple expression for Z^* is derived:

$$Z^* = \frac{p^*}{\alpha}(1 - \varepsilon p^* - \varepsilon\beta) \tag{9.33}$$

This is a parabolic function of p^*. Thus, if there is a price increase ($p^* > 0$) and if $\alpha > 0$ then profit will rise ($Z^* > 0$) provided

$$1 - \varepsilon p^* - \varepsilon\beta > 0$$

TABLE 9.12
The bound for c^* (in %) by approximation (9.32) compared with (9.31).

p^* (%)	−20	−15	−10	−5	0	5	10	15	20
$\varepsilon = 0.5$									
$\alpha = 0.1$ appr (9.32)	−21.1	−15.8	−10.6	−5.3	0	5.3	10.6	15.8	21.1
(9.31)	−21.1	−15.9	−10.6	−5.3	0	5.3	10.5	15.8	21.0
$\alpha = 0.2$ appr (9.32)	−22.5	−16.9	−11.3	−5.6	0	5.6	11.3	16.9	22.5
(9.31)	−22.7	−17.0	−11.3	−5.6	0	5.6	11.2	16.7	22.2
$\varepsilon = 1.0$									
$\alpha = 0.1$ appr (9.32)	−20.0	−15.0	−10.0	−5.0	0	5.0	10.0	15.0	20.0
(9.31)	−20.4	−15.2	−10.1	−5.0	0	5.0	9.9	14.7	19.4
$\alpha = 0.2$ appr (9.32)	−20.0	−15.0	−10.0	−5.0	0	5.0	10.0	15.0	20.0
(9.31)	−20.8	−15.5	−10.2	−5.1	0	4.9	9.7	14.3	18.8
$\varepsilon = 2.0$									
$\alpha = 0.1$ appr (9.32)	−17.4	−13.3	−8.9	−4.4	0	4.4	8.9	13.3	17.8
(9.31)	−19.0	−14.1	−9.3	−4.5	0	4.3	8.3	11.9	14.8
$\alpha = 0.2$ appr (9.32)	−15.0	−11.3	−7.5	−3.8	0	3.8	7.5	11.3	15.0
(9.31)	−17.9	−13.0	−8.3	−4.0	0	3.5	6.3	8.0	8.3

appr = approximation.

or when

$$p^* < \frac{1}{\varepsilon} - \beta$$

where β must be smaller than $1/\varepsilon$ to allow p^* to become positive, otherwise no price increase with the intention of improving profit is possible. If a price decrease is considered then $Z^* > 0$ when

$$p^* > \frac{1}{\varepsilon} - \beta$$

and here $\beta > 1/\varepsilon$ to allow a price decrease. These conclusions are summarized as follows:

$$Z^* > 0 \qquad \text{for } \alpha > 0$$

if

$$\left.\begin{array}{ll} 0 < p^* < \dfrac{1}{\varepsilon} - \beta & \text{when } \beta < \dfrac{1}{\varepsilon} \\ \text{or} & \\ \dfrac{1}{\varepsilon} - \beta < p^* < 0 & \text{when } \beta > \dfrac{1}{\varepsilon} \end{array}\right\} \tag{9.34}$$

These results are presented graphically in the $\beta - p^*$ chart in Fig. 9.16, where it is assumed that $\alpha > 0$. The horizontal axis denotes β which has a value in the range 0–1. The price change p^* is given by the vertical axis, so that any point in this chart is defined by the coordinates β, and p^*. Now, the value $1/\varepsilon$ is marked off on a vertical axis and a line declining at a slope of 45° is drawn, so that the

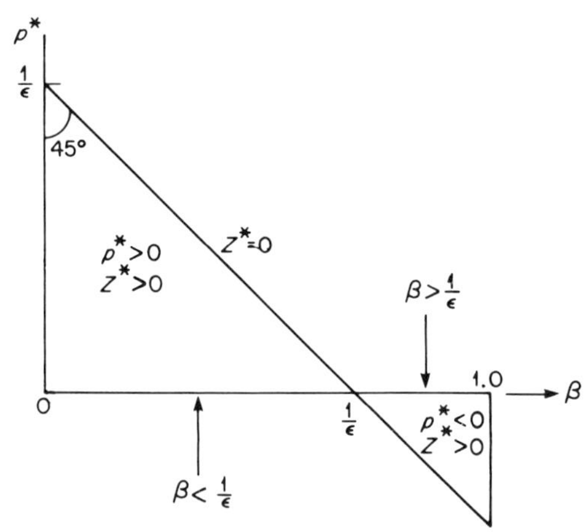

FIG. 9.16. Condition for a price change to increase profit (where $\alpha > 0$).

ordinate of this line for any given value of β is $(1/\varepsilon)-\beta$. The line stands for $Z^* = 0$ and it intersects the horizontal axis at $\beta = 1/\varepsilon$, so that the horizontal axis is cut into two parts, one where $\beta < 1/\varepsilon$ and the other where $\beta > 1/\varepsilon$. A profit increase occurs within the two triangles: in the first p^* is positive and the domain $Z^* > 0$ lies below the $Z^* = 0$ line, and in the second p^* is negative and $Z^* > 0$ lies above the line. Outside the two triangles, profit must decline. The $Z^* = 0$ line in the chart is based on the assumption that the elasticity ε is independent of the magnitude of the contemplated change p^*. Needless to say, when a relationship between ε and p^* is known, or can be postulated, it can be incorporated into Equation (9.33) and in the conditions in (9.34), and the line in Fig. 9.16 would have to be adjusted accordingly.

Following the same assumption, the general $\beta - p^*$ chart for various values of elasticity is shown in Fig. 9.17, indicating how the first triangle expands as ε

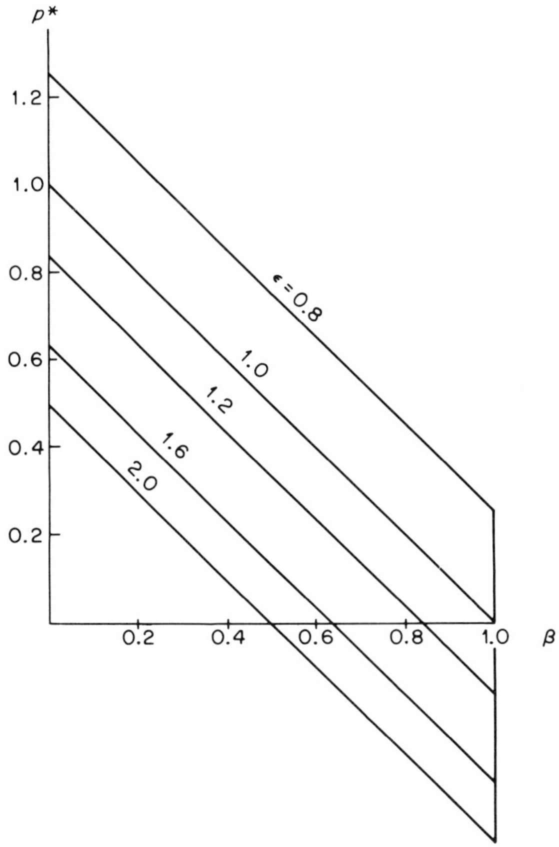

FIG. 9.17. The β-p^* chart.

decreases, thereby covering a wide range of circumstances where increasing profit by a price rise is feasible. The second triangle contracts simultaneously, and at $\varepsilon = 1$ it vanishes altogether, so that when $\varepsilon \leqslant 1$ there is no scope for increasing profit as a consequence of reduction in unit price, a conclusion arrived at earlier in our discussion.

The optimum price increase

From Equation (9.33) the optimum price increase to maximize the relative profit increment Z^* can be found from

$$\frac{\mathrm{d}Z^*}{\mathrm{d}p^*} = \frac{1}{\alpha}(1 - 2\varepsilon p^* - \varepsilon\beta) = 0$$

where we assumed that ε is independent of p^*. Hence the optimum for Z^* is obtained at

$$p^*_{\text{opt}}(Z^*) = \tfrac{1}{2}\left(\frac{1}{\varepsilon} - \beta\right) \tag{9.35}$$

and the maximum profit increase (assuming $\alpha > 0$) can be shown as

$$Z^*_{\max} = \frac{\varepsilon}{4\alpha}\left(\frac{1}{\varepsilon} - \beta\right)^2 = \frac{\varepsilon}{\alpha}[p^*_{\text{opt}}(Z^*)]^2 \tag{9.36}$$

It is interesting to note that p^*_{opt} lies exactly half way along the range given by the bound in the condition (9.34), similar to the result for p^*_{opt} in Equation (9.28) which maximizes revenue, when that p^*_{opt} is compared with the condition for p^* in expression (9.29). We find, however, that the value of p^*_{opt} is different for the two cases. *Hence, a price increase that maximizes an increase in revenue generally does not maximize an increase in profit, and vice versa*, except when

$$p^*_{\text{opt}}(Z^*) = p^*_{\text{opt}}(R^*)$$

or

$$\tfrac{1}{2}\left(\frac{1}{\varepsilon} - \beta\right) = \frac{1 - \varepsilon}{2\varepsilon}$$

namely in the extreme case when $\beta = 1$ (i.e. when the marginal cost $s = 0$). Normally, however, $\beta < 1$ from which it follows that

$$p^*_{\text{opt}}(Z^*) > p^*_{\text{opt}}(R^*) \tag{9.37}$$

Furthermore, note that if $\varepsilon > 1$, a price increase cannot lead to an increase in revenue, since then $R^* < 0$ by Equation (9.27), whereas an increase in profit is possible, provided $\alpha > 0$ and $\beta < 1/\varepsilon$, as can be seen from Equations (9.33) and (9.35).

As the point at which the maximum value of Z^* is attained is half the bound

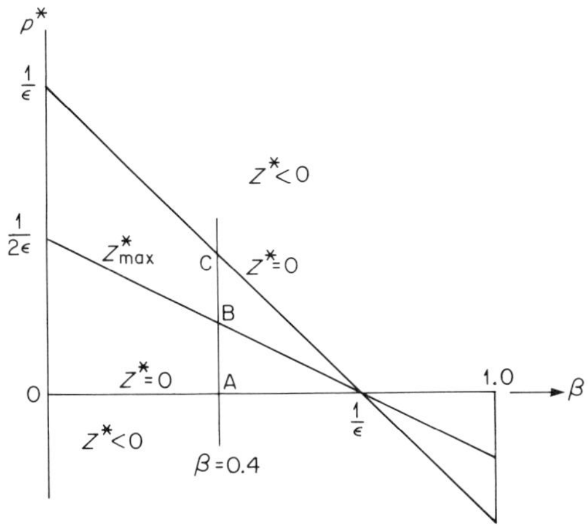

FIG. 9.18. The line for Z^*_{max}.

of p^* given in (9.34), the locus of all the points of Z^*_{max} on the $\beta - p^*$ chart is shown in Fig. 9.18, so that for any given value of β (say, $\beta = 0.4$), the maximum Z^* is obtained at B where AB = BC, and the line for Z^*_{max} cuts the vertical axis at $1/(2\varepsilon)$.

If we visualize Z^* on a third coordinate perpendicular to the $\beta - p^*$ plane, the three-dimensional presentation of β, p^* and Z^* is given in Fig. 9.19. The Z^* envelope can be seen as the ceiling of two sections of a kind of tunnel with a common vertex at $\beta = 1/\varepsilon$. The projection of the tunnel on the $\beta - p^*$ plane yields the two triangles shown in Fig. 9.16. For any given value of β the Z^* function is derived as the parabolic ceiling of the tunnel, obtained as a cross-section perpendicular to the $\beta - p^*$ plane, a large opening of one section being at $\beta = 0$ while the cross-section at the other end is at $\beta = 1$. The horizontal projection of the tunnel on the $\beta - p^*$ plane consists of two right angled isoceles triangles (see Fig. 9.16) meeting at the point $\beta = 1/\varepsilon$, where the height of the ceiling is $Z^* = 0$. For the example $\beta = 0.4$, marked on Fig. 9.18, the corresponding Z^* function is shown in Fig. 9.19, with Z^*_{max} occurring at point B, which is halfway between A and C.

The effect of ε on Z^* is evident from Equation (9.33). The larger the value of ε, the smaller is the impact of a price increase on profit. Figure 9.20 shows a contour map of iso-Z^* lines for the case where $\alpha = 0.2$ and $\beta = 0.5$. Here no change in profit (i.e. $Z^* = 0$) occurs when, from Equation (9.33), $p^* = 0$ and

$$1 - \varepsilon(p^* + \beta) = 0$$

or, after substituting $\beta = 0.5$

$$p^* = \frac{1}{\varepsilon} - 0.5$$

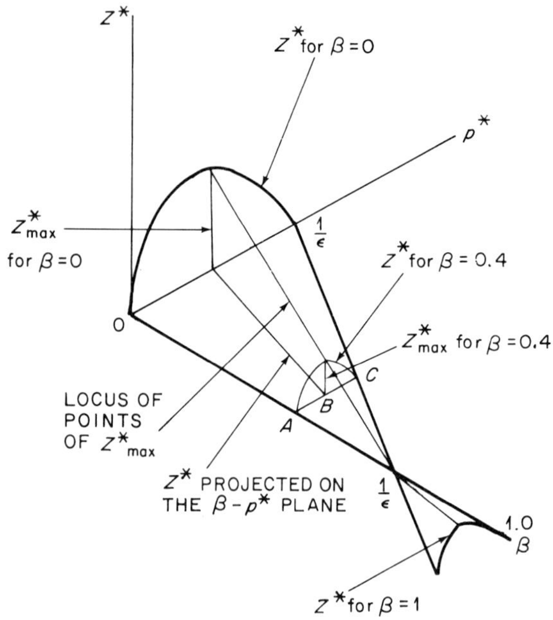

FIG. 9.19. The Z^* envelope.

Similar expressions for other values of Z^* can be easily derived and the map constructed accordingly. Like the R^* map in Fig. 9.14, we have here a saddle-shaped terrain, except that part of it in Fig. 9.20 (in the south-easterly direction) can be revealed only when the map is extended to cover larger values of ε.

Example

To illustrate the functions R^* and Z^*, consider the following example:

$$\varepsilon = 0.75$$
$$\alpha = 0.2$$
$$\beta = 0.4$$

Hence, the relative incremental revenue, by Equation (9.27), is

$$R^* = p^*(0.25 - 0.75\,p^*)$$

The maximum incremental revenue is obtained at

$$p^*_{\text{opt}}(R^*) = \frac{1-\varepsilon}{2\varepsilon} = \frac{0.25}{1.50} = 0.167 \quad (16.7\,\%)$$

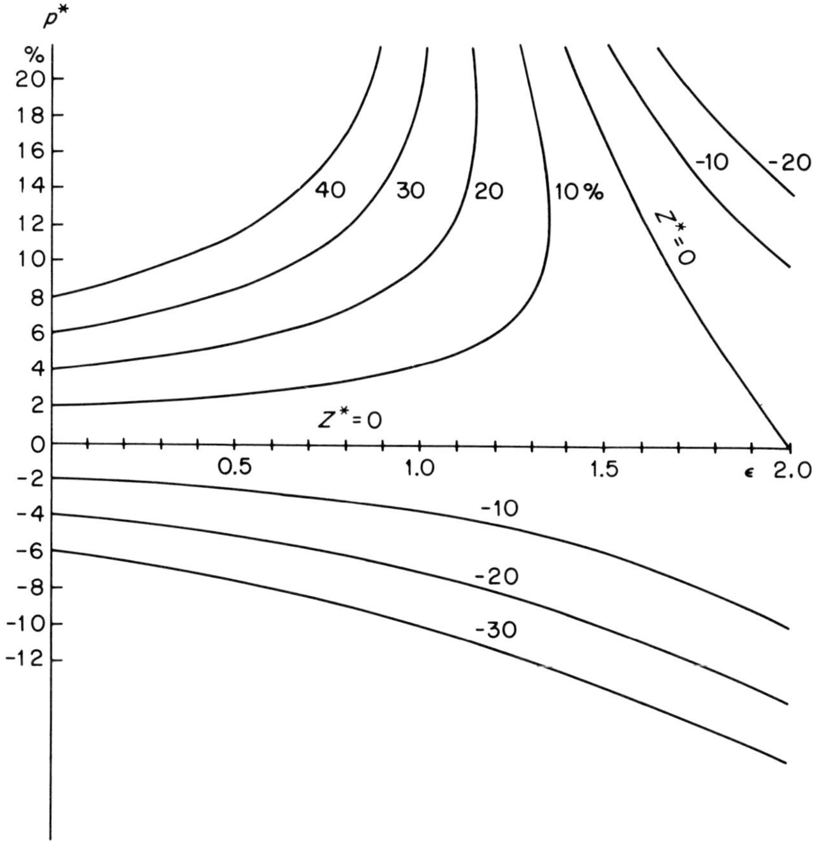

FIG. 9.20. Iso-Z^* map for the case $\alpha = 0.2$, $\beta = 0.5$.

where

$$R^*_{\max} = \varepsilon(p^*_{\text{opt}})^2 = \frac{0.75}{36} = 0.021 \ \ (2.1\,\%)$$

The relative incremental profit, by Equation (9.33), is

$$Z^* = 5p^*(0.7 - 0.75p^*)$$

having a maximum at

$$p^*_{\text{opt}}(Z^*) = \tfrac{1}{2}\left(\frac{1}{\varepsilon} - \beta\right) = \tfrac{1}{2}(1.333 - 0.4) = 0.467 \ \ (46.7\,\%)$$

where

$$Z^*_{\max} = \frac{\varepsilon}{\alpha}(p^*_{\text{opt}})^2 = \frac{0.75}{0.2}0.218 = 0.817 \ \ (81.7\,\%)$$

The Z^* function is the cross-section of Z^* envelope in Fig. 9.19 at $\beta = 0.4$. The results are shown in Fig. 9.21 and some of the salient points are summarized in Table 9.13. It is clear from these results that while the R^* function is relatively flat when the price changes from -10% to $+50\%$, the Z^* function is very sensitive to a price change. As the price increases, revenue rises slowly, attaining a maximum increase of 2.1 % while profit shoots up 47.9 % at that point. Then the increase in revenue begins to fall and at $p^* = 33.3$ it is back to its initial level, while profit continues to rise. At $p^* = 46.7\%$ a profit increase of 81.7 % is obtained, even though at that point volume is down 35 % and revenue

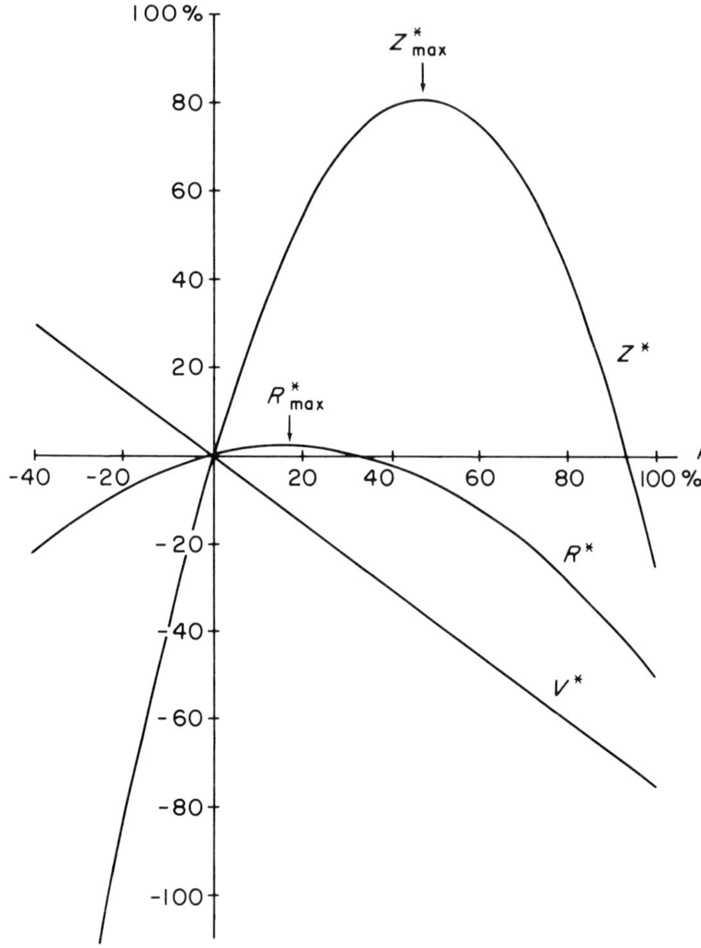

FIG. 9.21. R^* and Z^* for the example $\varepsilon = 0.75$, $\alpha = 0.2$, $\beta = 0.4$.

TABLE 9.13
Selected results for the example $\varepsilon = 0.75$, $\alpha = 0.2$, $\beta = 0.4$ (all figures in %).

p^*	V^*	R^*	Z^*	Note
0	0	0	0	Initial point
16.7	−12.5	2.1	47.9	Maximum R^*
33.3	−25.0	0	75.0	Revenue unchanged, profit increased
46.7	−35.0	−4.7	81.7	Maximum Z^*, revenue reduced
93.3	−70.0	−42.0	0	Profit unchanged
−22.9	17.2	−9.7	−99.8	Breakeven point virtually obtained
−23.0	17.3	−9.7	−100.3	Operating at a loss

is lower than its value at the baseline. At $p^* = 93.3\%$ profit is back to its starting point and revenue is substantially reduced. A change in price in the opposite direction ($p^* < 0$) causes both revenue and profit to fall (the latter falling quite sharply) even though volume increases.

THE DUAL OBJECTIVE OF INCREASING REVENUE AND PROFIT

We see, therefore, that *increasing revenue and maximizing profit may not be compatible objectives*. Another commonly proclaimed objective, that of increasing market share, may not be compatible with increasing profit either. There are two ways in which market share may be measured:

(1) *as a proportion of the total revenue*, in which case an increase in market share demands maximizing the revenue of the firm, but (as we have just noted) revenue may well have to decline for a higher level of profit to be attained;
(2) *as a proportion of the total physical volume*, in which case an increase in volume should be aimed at, but as we see in the example in Fig. 9.21, if the increase in volume is to be achieved in response to a decrease in price, then the result may be a decline both in revenue and profit.

If we turn to the dual concern of management (commonly found in practice) both with a change in revenue and a change in profit, then four possible outcomes need to be considered, as shown in Fig. 9.22:

quadrant 1—is where both revenue and profit increase and is often regarded as the most desirable outcome
quadrant 2— is where profit increases but revenue declines

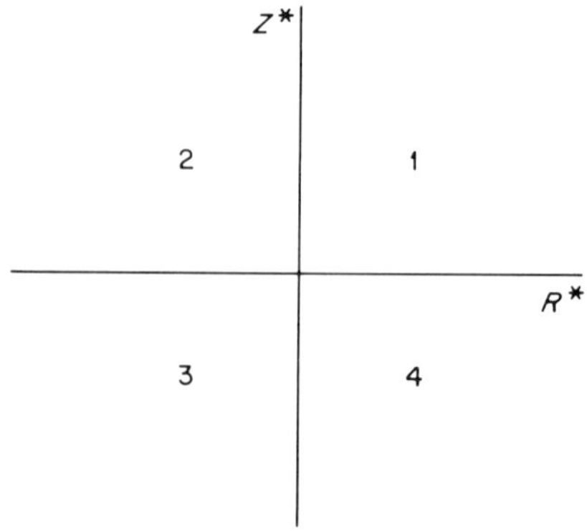

FIG. 9.22. Four domains for combinations of R^* and Z^*.

quadrant 3— is where both decline and this is the least desirable result
quadrant 4— is where revenue increases but profit declines

If the objective of management is to achieve an increase in both revenue and profit, then any result other than in quadrant 1 would not be satisfactory, but circumstances do arise when an increase in Z or in R is considered to be paramount, so that any outcome in quadrants 2 or 4 would be preferred to one in quadrant 1, if the latter involves only a relatively small improvement in the corresponding variable.

This point is illustrated in Fig. 9.23, where seven possible outcomes are

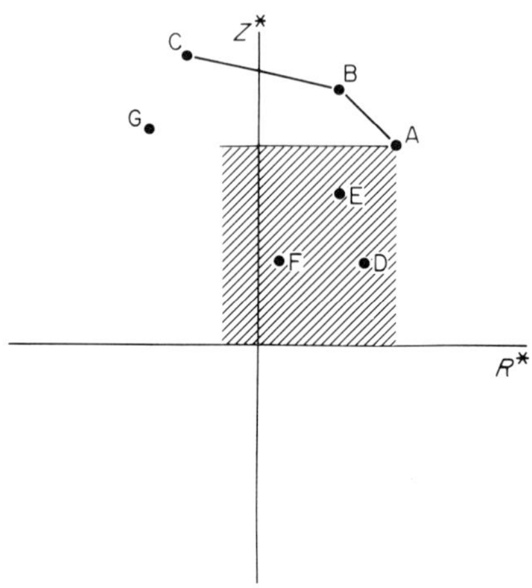

FIG. 9.23. How to rank several possible outcomes.

shown, and some are clearly preferable to others. If the dual objective is to achieve high values for both Z^* and R^*, then a comparison of outcomes A and E reveals that the former is preferable on both counts. Thus, superimposing Fig. 9.22 with point A as the origin, the third quadrant is shown by the shaded area and any point in that area is worse than A with respect to both Z^* and R^*. We thus borrow the *concept of dominance* from investment portfolio theory (where the dual objective is the maximization of the expected return and the minimization of risk), so that outcome A is said to dominate D, E and F. Similarly, both B and C dominate G. We are left with the points A, B and C, none of which is dominated by any of the others. If a mixture of A and B is possible (for example, when each is the outcome of a specific pricing strategy and when a mixed strategy can be designed to cover several markets), then the outcome would lie on the line AB, so that ABC would represent the *dominant frontier* that overshadows any possible outcome below it. If a mixed strategy is not possible, then the discrete outcomes A, B and C represent the *dominant set*, whose members are the only points that need to be considered.

The concept of dominance allows partial ranking to be performed, as shown below:

Outcome	Dominated outcome
A	D, E, F
B	E, F, G
C	G
D	—
E	F
F	—
G	—

A complete ranking cannot, however, be carried out, unless a specific value relationship is stated for the trade-off between Z^* and R^*. It is only then that preference between members of the dominant set A, B and C can be determined.

A further illustration of the concept of dominance can be obtained from the example in Fig. 9.21. If for any given value of p^* the expected result is marked on a diagram with the coordinates Z^* and R^*, the pattern shown in Fig. 9.24 is derived. Starting with $p^* = 0$ at the origin, we move down the curve when the price is reduced (i.e. $p^* < 0$); when the price is increased ($p^* > 0$) we move upwards until the maximum value of Z^* is reached, beyond which both Z^* and R^* decline. The corresponding values of p^* are noted along the curve.

The presentation in Fig. 9.24 helps to highlight the fact that R^* is relatively insensitive to price changes within a fairly wide range, whereas Z^* is very sensitive even to a small price change. The dominant section of the curve is

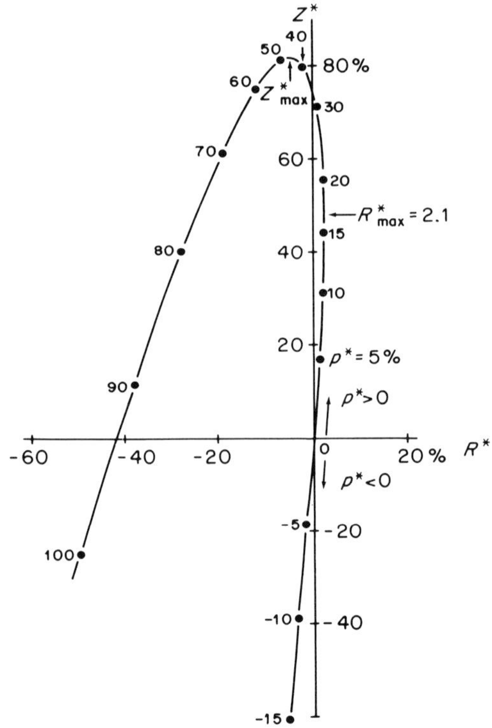

FIG. 9.24. Z^* vs R^* for the example in Fig. 9.19.

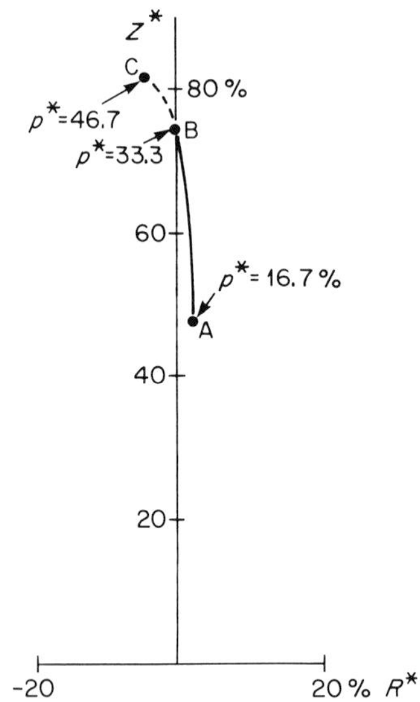

FIG. 9.25. The dominant section in Fig. 9.22.

TABLE 9.14
Three points on the dominant section in Fig. 9.25
(all figures in %).

Point	p^*	R^*	Z^*
A	16.7	$R^*_{max} = 2.1$	47.9
B	33.3	0	75.0
C	46.7	−4.7	$Z^*_{max} = 81.7$

between R^*_{max} and Z^*_{max} and is shown by the curve ABC in Fig. 9.25. The characteristics of three noteworthy points are given in Table 9.14. The dominant section in Fig. 9.25 is divided into two segments, the first between A and B corresponds to $R^* \geqslant 0$, while the second (shown by the broken line BC) promises a higher profit but involves $R^* < 0$. If management concludes that the solution domain must be constrained to ensure that revenue does not decline, then only the segment AB needs to be considered. In view of the very substantial increase in profit at the expense of foregoing a very minor improvement in revenue, as we move from point A to point B, management may well decide that B represents the most attractive outcome in this case.

EFFECT OF ADDITIONAL PROMOTION †

As we have seen from equation (9.5), an increase in profit can be achieved through one (or a combination) of the following courses of action:

(1) *A reduction in the unit cost*—as a result of a reduction of one or several of the cost components (namely, the variable unit cost or the fixed cost or both). The effect of such reductions were examined in earlier chapters dealing with cost models.
(2) *A unit price increase*—provided the price elasticity of demand is such that the consequent loss in volume is not excessively detrimental, as we have seen earlier in this chapter.
(3) *An increase in demand*—which is expected to take place either as a consequence of a price reduction, provided the effect of the price elasticity of demand is not too detrimental to revenue, or by stimulating demand through an increase in the expenditure on promotional activity (such as advertising), provided the resultant unit cost is kept under control. We now turn to the analysis of this alternative.

† This section relies heavily on the work of Eilon and Cosmetatos in [3].

The cost function

The total cost C and the unit cost c as given in Chapter 7 are

$$\left.\begin{aligned} C &= sV+F+J+D \\ &\text{and} \\ c &= s+\frac{F}{V}+\frac{J}{V}+\frac{D}{V} \end{aligned}\right\} \tag{9.38}$$

where the cost proportions are defined as

$$\begin{aligned} f_1 &= s/c \\ f_2 &= F/C \\ f_3 &= J/C \\ f_4 &= D/C \end{aligned}$$

We shall assume here that $f_3 = 0$ (though it is not difficult to extend the analysis for $f_3 \neq 0$), from which it follows that

$$\left.\begin{aligned} C^* &= f_1(sV)^* + f_2F^* + f_4D^* \\ &\text{and} \\ c^* &= f_1s^* + \frac{1}{1+V^*}(f_2F^* + f_4D^* - fV^*) \end{aligned}\right\} \tag{9.39}$$

and

$$f = f_2 + f_3 + f_4 = 1 - f_1$$

If the fixed cost and the unit variable cost do not change, i.e. if

$$s^* = F^* = 0$$

then

$$\left.\begin{aligned} C^* &= f_1V^* + f_4D^* \\ &\text{and} \\ c^* &= \frac{1}{1+V^*}(f_4D^* - fV^*) \end{aligned}\right\} \tag{9.40}$$

These equations show how the combined effect of changes in advertising expenditure and in volume affect the unit cost.

The relationship between the relative increment in advertising expenditure and the resultant relative incremental volume is conveniently described by the introduction of the promotion elasticity of demand σ as indicated in Chapter 7

$$V^* = \sigma D^* \tag{9.41}$$

where we assume that $\sigma > 0$. Substituting this relationship into (9.40) we obtain

$$C^* = (f_1\sigma + f_4)D^*$$

so that the total cost obviously rises with increased advertising, and

$$c^* = \frac{\gamma V^*}{1+V^*} = \frac{\gamma\sigma D^*}{1+\sigma D^*} \tag{9.42}$$

where, as defined in Chapter 7,

$$\gamma = \frac{f_4}{\sigma} - f \tag{9.43}$$

Thus, the effect of increasing D on the unit cost is

$$\left.\begin{aligned} c^* \gtrless 0 \qquad & \text{if } f_4 - f\sigma \gtrless 0 \\ \text{i.e. if} \qquad & \frac{f_4}{f} \gtrless \sigma \end{aligned}\right\} \tag{9.44}$$

As indicated in Chapter 7, f_4/f is a measure of the internal distribution of the total fixed cost, i.e. the proportion of the total fixed cost $F + J + D$ devoted to the promotion expenditure D. By definition, this ratio is smaller than 1, which leads to the sufficient condition that if $\sigma \geqslant 1$ then the unit cost is expected to decline. It is possible, however, for this result to obtain even when $\sigma < 1$, provided the initial value of f_4 is sufficiently small; the larger the value of f_4 and the smaller the promotion elasticity σ the more difficult it is for the condition to be met.

The effect of D^* on profit when $p^* = 0$

For the case of no change in unit price we recall from Equations (9.15) and (9.17) that the effect on profit may be determined by the condition (for $\alpha > 0$)

$$Z^* \gtrless 0 \qquad \text{if } c^* \lesseqgtr \frac{\alpha}{1-\alpha}\frac{V^*}{1+V^*} \tag{9.45}$$

In conjunction with Equation (9.42) this means that (for $V^* > 0$)

$$Z^* \gtrless 0 \qquad \text{if } \gamma \lesseqgtr \frac{\alpha}{1-\alpha} \tag{9.46}$$

The ratio $\alpha/(1-\alpha)$ is positive (for $\alpha > 0$), hence when γ is positive it must lead to an increase in unit cost by the condition (9.44) and yet, if its value is sufficiently small, it may lead to an increase in profit, following the condition (9.46). This result is shown graphically in Fig. 9.26, where unit cost and total profit both increase in the range

$$0 < \gamma < \frac{\alpha}{1-\alpha}$$

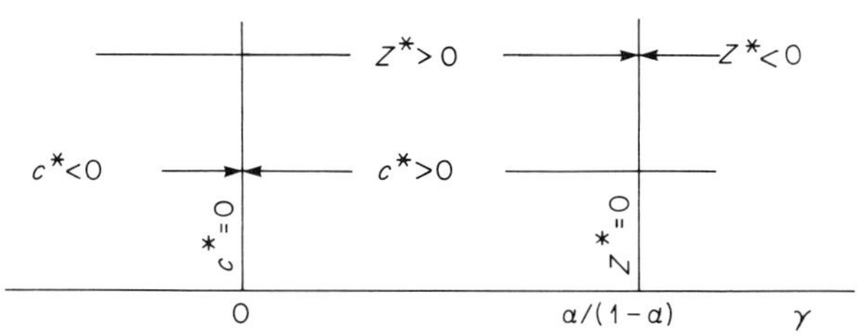

FIG. 9.26. Correspondence between c^* and Z^* [3] ($p^* = 0$).

An alternative expression for the condition in (9.46) is obtained by using the definition of β given in Equation (9.22), namely

$$\beta = \alpha + f(1-\alpha)$$

from which the cost proportion of the total fixed cost is derived as

$$f = \frac{\beta-\alpha}{1-\alpha}$$

Substituting this value into the definition of γ in Equation (9.43), we have

$$\gamma = \frac{f_4}{\sigma} - \frac{\beta-\alpha}{1-\alpha}$$

so that the condition in (9.46) becomes

$$\frac{f_4}{\sigma} - \frac{\beta-\alpha}{1-\alpha} \lesseqgtr \frac{a}{1-\alpha}$$

or

$$Z^* \gtrless 0 \qquad \text{if } \beta \gtrless \frac{(1-\alpha)f_4}{\sigma} = \beta_0 \tag{9.47}$$

β_0 is the limiting value for β at which profit does not change, and if $\beta > \beta_0$ then profit rises.

There are many possible relationships between the promotion elasticity σ and D^*, and three examples are shown in Fig. 9.27:

(a) σ is constant and does not depend on the value of D^*
(b) σ increases with D^*, i.e. large increases in promotion expenditure are proportionately more effective than small increases, but σ eventually levels off
(c) very small and very large relative increases in D have a limited effect, so that σ first rises and then declines with D^* [3]

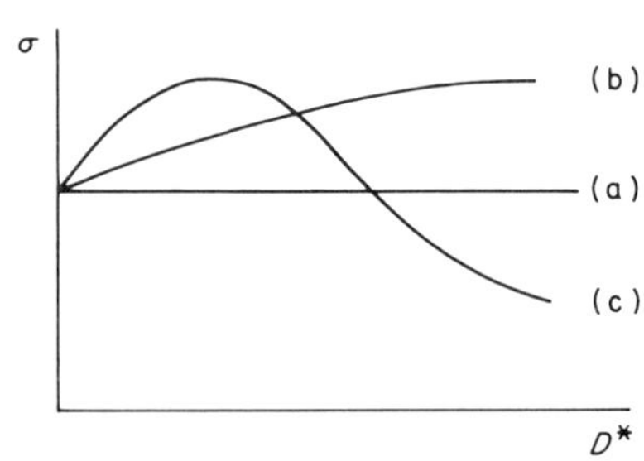

FIG. 9.27. Promotion elasticity of demand as a function of D^*.

The effect of these cases on β_0 is shown in Fig. 9.28, and in each case profit increases above the curve and declines below the curve.

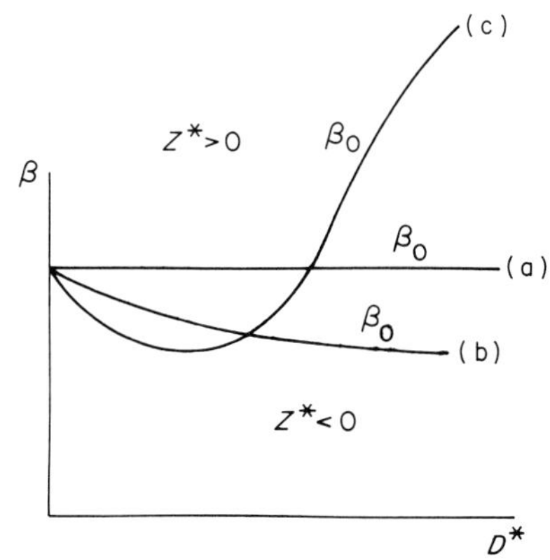

FIG. 9.28. Effect of an increase in promotion expenditure on profit.

The effect of D^* when $p^* \neq 0$

So far we have considered the consequences of an increase in promotion expenditure when the unit price remains unchanged. The more general case when $p^* \neq 0$ is shown in Fig. 9.29, where the volume is affected in two ways:

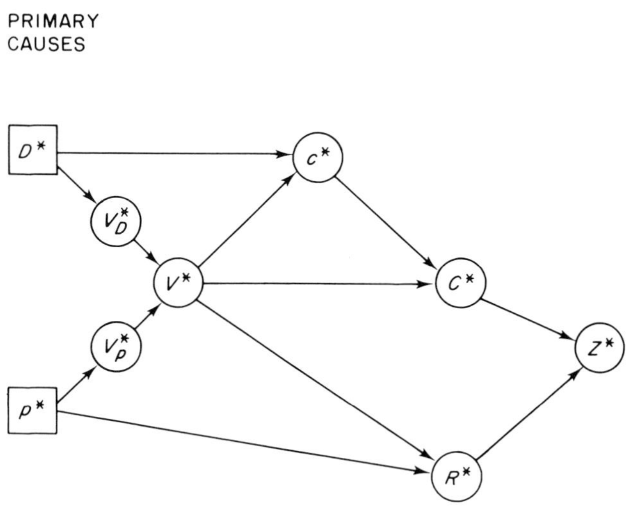

FIG. 9.29. The network of relationships with D^* and p^* as prime causes.

(1) an incremental volume corresponding to $V_D^* = \sigma D^*$ due to the stimulus of added promotion;
(2) a relative incremental volume change of $V_p^* = -\varepsilon p^*$ due to the price change p^*.

Assuming that the two effects are independent of each other, the total result for the change in volume may be expressed as

$$V^* = V_D^* + V_p^* = \sigma D^* - \varepsilon p^* \tag{9.48}$$

As shown in Fig. 9.29, the following sequential changes take place:

- D^* and V^* affect the unit cost c
- c^* and V^* affect the total cost C
- p^* and V^* affect the revenue R
- C^* and R^* affect the profit Z

The effect on the total cost and the unit cost is given in Equation (9.40). When V^* is substituted it can be shown that

$$c^* = \frac{\gamma V_D^* - f V_p^*}{1 + V^*} \tag{9.49}$$

Thus, for the special case $p^* = 0$ Equation (9.49) is reduced to (9.42).

The effect on profit may be derived after substituting Equation (9.49) into (9.14):

$$Z^* = \frac{1}{\alpha}[p^* + (p^* + \alpha)V^* + (1-\alpha)f V_p^* - \gamma(1-\alpha)V_D^*]$$

The first term in the square brackets indicates the contribution to increased profit due to the price change p^*, the second term expresses the combined effect of p^* and the total incremental volume V^* (which may be disaggregated into $V_D^* + V_p^*$), the third term shows the effect of V_p^* due to a price change and the last term is the effect of V_D^* due to a change in promotion expenditures. After disaggregating V^* and introducing β and β_0 from their respective definitions in (9.22) and (9.47) we obtain

$$Z^* = \frac{1}{\alpha}\left[(p^* + \beta - \beta_0)V_D^* + \left(p^* + \beta - \frac{1}{\varepsilon}\right)V_p^*\right] \tag{9.50}$$

We shall assume throughout (unless the contrary is stated) that the original profit margin α is positive, so that profit rises if the sum of the two terms in the square brackets is positive. The following comments may be added (assuming D^* and p^* are non-negative):

(1) If $D^* > 0$ and $p^* = 0$ then $V_D^* > 0$ and $V_p^* = 0$, reducing the result in (9.50) to

$$Z^* = \frac{1}{\alpha}(\beta - \beta_0) V_D^*$$

so that $Z^* > 0$ if $\beta > \beta_0$, as found in the expression (9.47).

(2) If $D^* = 0$ and $p^* > 0$ then $V_D^* = 0$ and $V_p^* < 0$, reducing Z^* to

$$Z^* = \frac{1}{\alpha}\left(p^* + \beta - \frac{1}{\varepsilon}\right) V_p^*$$

which is the result given in Equation (9.33), so that $Z^* > 0$ when the condition (9.34) is satisfied.

(3) If both D^* and p^* are positive, then V_D^* is positive and V_p^* negative. A sufficient condition for profit to increase is when both terms in Equation (9.50) are positive, or

$$Z^* > 0 \qquad \text{if } \beta_0 < p^* + \beta < \frac{1}{\varepsilon} \tag{9.51}$$

and profit decreases if the inequalities read the other way. For the cases where neither (9.51) nor its converse hold, the result for Z^* depends on the values D^* and p^*, as shown diagrammatically in Fig. 9.30.

(4) From Equation (9.49), the unit cost rises if $\gamma > 0$, and it is not difficult to show that a sufficient condition for that to obtain is

$$\beta_0 > \beta - \alpha$$

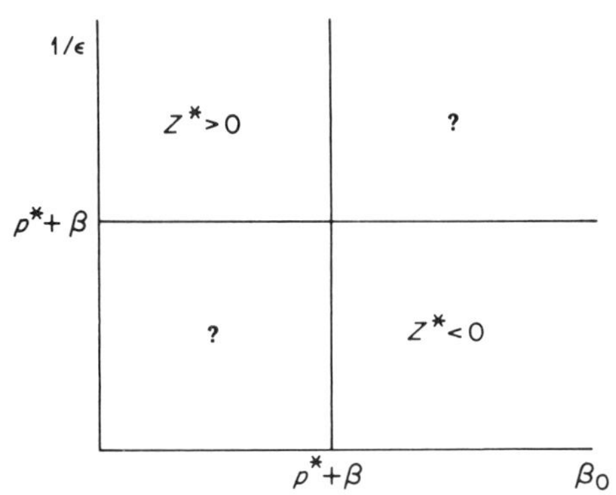

Fig. 9.30. The effect of p^*, $D^* > 0$ on Z^* [3].

Thus, Fig. 9.30 may be extended to show this result, as indicated in Fig. 9.31. The domains for which the various conditions referred to earlier are not sufficient are shown by question marks.

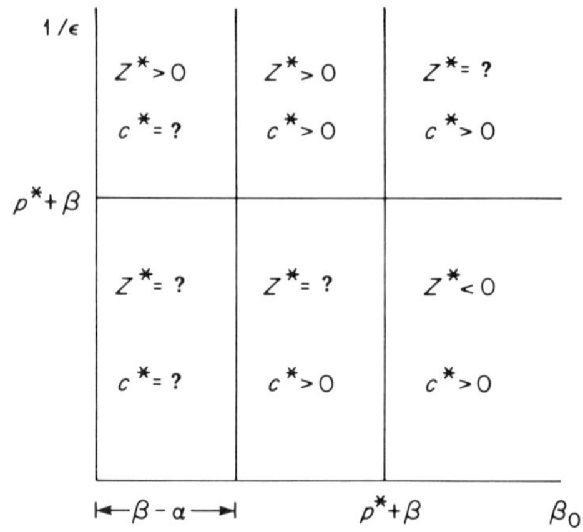

FIG. 9.31. The effect of p^*, $D^* > 0$ on Z^* and c^* [3].

(5) In the special case where the combined effect of D^* and p^* results in the volume remaining unchanged ($V^* = 0$), i.e. when the promotional effort offsets the adverse effect of the price increase, then $V_D^* = -V_p^*$, or $\sigma D^* = \varepsilon p^*$, and Equation (9.50) becomes

$$Z^* = \frac{\sigma D^*}{\alpha}\left(\frac{1}{\varepsilon} - \beta_0\right)$$

so that profit increases if $D^* > 0$ and

$$\frac{1}{\varepsilon} > \beta_0$$

or, after substituting the value of β_0,

$$\frac{\sigma}{\varepsilon} > (1 - \alpha) f_4$$

If this condition is met, then the incremental profit Z^* rises linearly with D^*, although the incremental unit cost also rises at the same time, since for $V^* = 0$ we get $c^* = f_4 D^*$ from Equation (9.40).

Example [3]

Find the effect of p^* and D^* (when both are non-negative) on

$$V^*, c^*, R^* \text{ and } Z^*$$

for the following data:

$$f_1 = 0.5,\ f_2 = 0.4,\ f_3 = 0,\ f_4 = 0.1 \text{ (i.e. } f = 0.5)$$
$$\alpha = 20\%,\ \varepsilon = 1,\ \sigma = 0.3$$

Hence,

$$\beta = \alpha + f(1-\alpha) = 0.2 + 0.5 \times 0.8 = 0.6$$

$$\beta_0 = \frac{1-\alpha}{\sigma} f_4 = \frac{0.8}{0.3} 0.1 = 0.267$$

$$\gamma = \frac{f_4}{\sigma} - f = \frac{0.1}{0.3} - 0.5 = -0.167$$
$$\gamma V_D^* - f V_p^* = -0.05\, D^* + 0.5\, p^*$$

From these figures it is clear that:

(1) $V^* > 0$ if from Equation (9.48)

$$V^* = 0.3D^* - p^* > 0$$

or if $p^* < 0.3\, D^*$

(2) $c^* < 0$ if $-0.05\, D^* + 0.5\, p^* < 0$ from Equation (9.49), i.e. if $p^* < 0.1\, D^*$; also,

$$c^* = \frac{1}{1+V^*}(0.5\, p^* - 0.05\, D^*) \simeq 0.5\, p^* - 0.05\, D^*$$

(3) $R^* > 0$ if $V^* > \dfrac{-p^*}{1+p^*}$.

(4) From Equation (9.50),

$$\begin{aligned} Z^* &= \frac{1}{0.2}[(p^* + 0.6 - 0.267)\, 0.3\, D^* - (p^* + 0.6 - 1)\, p^*] \\ &= 0.5\, D^* + 1.5 p^*\, D^* + 2\, p^* - 5 p^{*2} \\ &\simeq 0.5\, D^* + 2 p^* \end{aligned}$$

The approximations given above only hold for relatively small values for the starred variables.

The results can be presented in several ways, three of which are shown in Figs. 9.32–9.34:

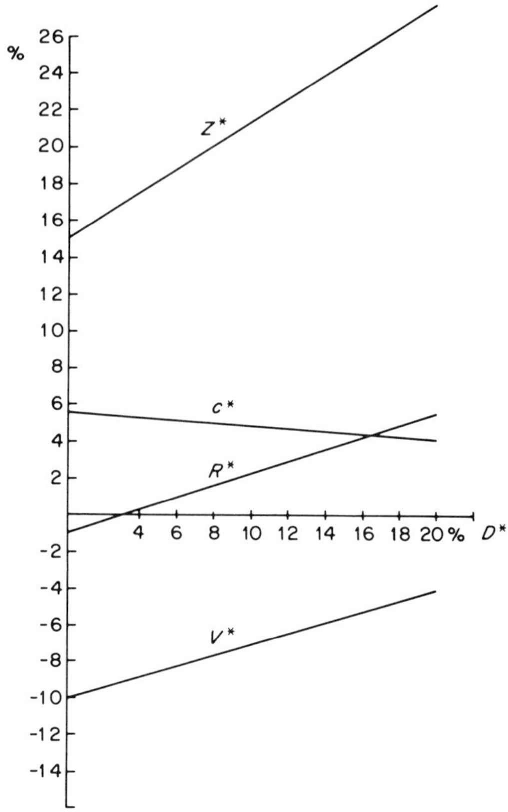

FIG. 9.32. An example for the effect of D^* (when $p^* = 10\%$).

Figure 9.32. This shows the outcomes for V^*, c^*, R^* and Z^* when D^* is varied and when the unit price is increased by 10%, all the curves here being linear. Volume and revenue obviously increase with an increase in promotion expenditure, and the rise in profit is even steeper, while the increase in unit cost declines slowly with D^*.

Figure 9.33. This shows the results for p^* as a variable, when $D^* = 10\%$. Here the relative incremental decrease in volume (excluding the effect of promotion) is equal to the relative incremental price (since $\varepsilon = 1$), the increase in revenue declines with p^* while the unit cost rises (owing to the fall in volume); profit increases sharply within the range of p^* shown in the diagram, but then tails off as p^* increases, and eventually starts to fall (see Table 9.16). Some of these results are also summarized in Table 9.15.

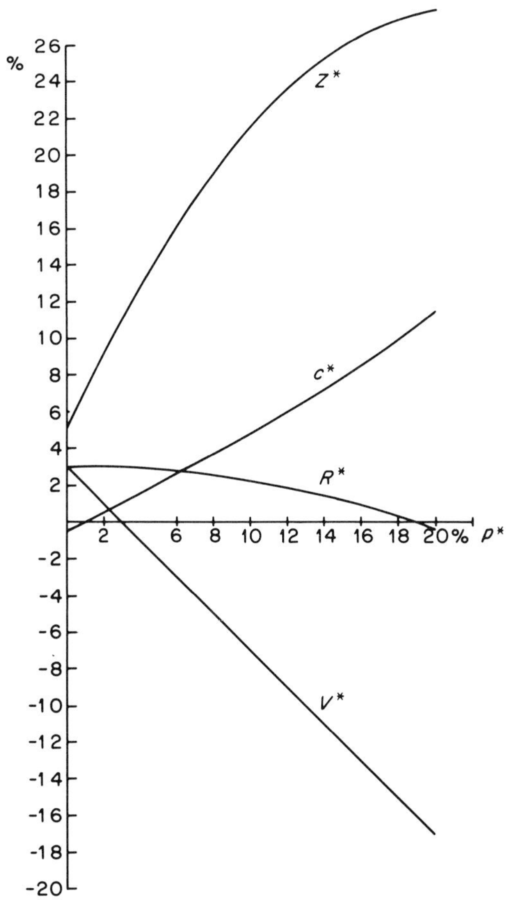

FIG. 9.33. An example for the effect of p^* (when $D^* = 10\%$).

Figure 9.34. This shows iso-lines for V^*, c^* and Z^* when both p and D are subject to change, so that for any combination of values of p^* and D^* the outcomes can be estimated. For example, for $p^* = D^* = 8\%$ we find from Fig. 9.34 that the volume is expected to rise by just under 6%, the unit cost by just under 4% and profit by about 18%. Results for Z^* for pairs of values of p^* and D^* are also given in Table 9.16, and similar tables can be constructed for the consequences for V^*, c^* and R^*. It should be stressed here that the results are based on the assumptions that the elasticity coefficients ε and σ remain constant throughout and are unaffected by the values of p^* and D^*. In reality such assumptions are rarely warranted, and consequently the models have to be modified when a wide range of values for p^* and D^* is considered.

TABLE 9.15
Effect of a price change for given D^* (for the example: $f_1 = 0.5$, $f_2 = 0.4$, $f_4 = 0.1$, $\alpha = 0.2$, $\varepsilon = 1$, $\sigma = 0.3$, $D^* = 0.1$).

p^* (%)	V^* (%)	c^* (%)	R^* (%)	Z^* (%)
0	3.0	−0.5	3.0	5.0
2	1.0	0.5	3.0	9.1
4	−1.0	1.5	3.0	12.8
6	−3.0	2.6	2.8	16.1
8	−5.0	3.7	2.6	19.0
10	−7.0	4.8	2.3	21.5
12	−9.0	6.0	1.9	23.6
14	−11.0	7.3	1.5	25.3
16	−13.0	8.6	0.9	26.6
18	−15.0	10.0	0.3	27.5
20	−17.0	11.4	−0.4	28.0

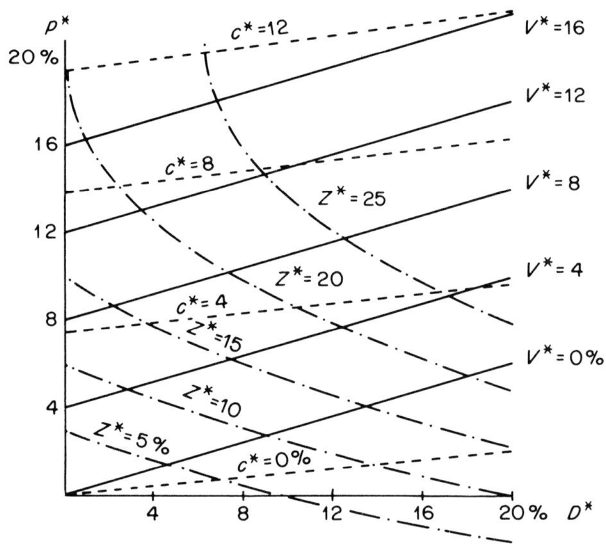

FIG. 9.34. An example of iso-curves for V^*, c^* and Z^*.

EFFECT OF OTHER PRIMARY CHANGES

The general case of a combination of primary causes of change is shown in Fig. 9.35, which depicts the following primary changes, all assumed to be independent of each other [2, 4]

TABLE 9.16
Z^* for given values of p^* and D^* (all in %) (for the example: $f_1 = 0.5$, $f_2 = 0.4$, $f_4 = 0.1$, $\alpha = 0.2$, $\varepsilon = 1$, $\sigma = 0.3$).

D^* =	0	5	10	15	20	25	30
p^* = 0	0	2.5	5.0	7.5	10.0	12.5	15.0
5	8.8	11.6	14.5	17.4	20.3	23.1	26.0
10	15.0	18.3	21.5	24.8	28.0	31.3	34.5
15	18.8	22.4	26.0	29.6	33.3	36.9	40.5
20	20.0	24.0	28.0	32.0	36.0	40.0	44.0
25	18.8	23.1	27.5	31.9	36.3	40.6	45.0
30	15.0	19.8	24.5	29.3	34.0	38.8	43.5

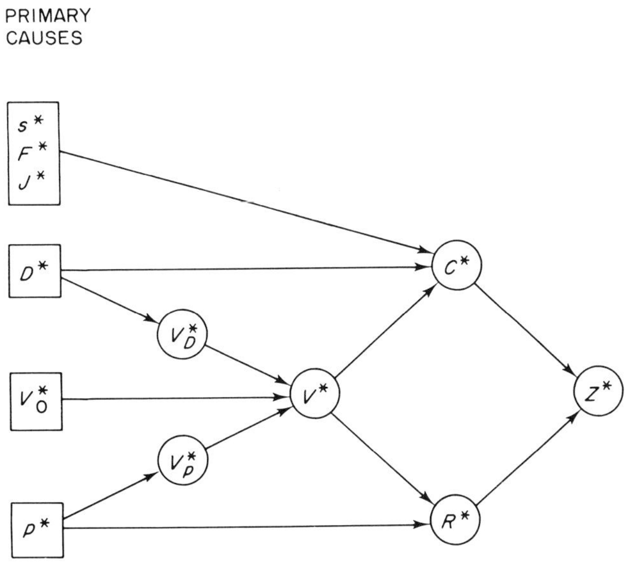

FIG. 9.35. The general case of primary changes.

(1) Changes in the cost components s and/or F (and/or J, if the interest charge J is separated from the remaining fixed cost F), which directly affect the total cost C and thereby the profit Z.
(2) A change in demand owing to external factors in the market place, the effect of competition and the general economic environment; this change is indicated by V_0^*.
(3) A change in the promotion budget, denoted by D^*, which apart from affecting the volume also affects the total cost.

(4) A change in the unit price, thereby affecting the revenue, but also affecting the total volume and hence total cost, as discussed earlier in this Chapter.

The incremental total volume change V^* thus consists of three elements:

V_0^* denoting the external primary change

$V_D^* = \sigma D^*$ denoting the effect of the promotion elasticity

$V_p^* = -\varepsilon p^*$ denoting the effect of the price elasticity

Assuming the three factors are independent of each other, then

$$V^* = V_O^* + V_D^* + V_p^*$$

The analytical methodology presented in this chapter applies to the general case, except that the composite value V^* needs to be substituted into the incremental cost functions in (9.39) and then into the incremental profit function in (9.14). In this way, various combinations of the primary causes in Fig. 9.35, and particularly of the control variables that executives are able to manipulate, can be explored and their effects on cost, revenue and profit can be ascertained.

CONCLUSIONS

Some of the main results derived in this chapter are summarized below:

(1) The relative incremental profit Z^* is sensitive to changes in revenue or cost, particularly when the initial profit margin α is low.
(2) It is possible for profit margin to decline and yet for profit to rise.
(3) A 1% change in unit price on its own (assuming it does not affect c and V) has a greater effect on profit than a 1% change in unit cost on its own, while a 1% change in volume has a markedly lower effect. Consequently, a relatively large increase in volume is needed to compensate for even a moderate increase in unit cost, if price and profit are to remain unchanged. This effect is particularly marked when the initial profit margin is low.
(4) It is possible for unit cost to rise, or for revenue to decline, and yet for profit to rise.
(5) A price increase to maximize profit is greater than the price increase that maximizes revenue.
(6) If $\varepsilon \geqslant 1$ then a price rise must lead to a fall in revenue.
(7) A price rise can lead to a profit increase, provided the necessary condition $\beta < 1/\varepsilon$ is met.
(8) If it is desirable to improve both profit and revenue, a dominant set of alternative outcomes need to be determined, and the most preferred

outcome can then be ascertained on the basis of managerial trade-off between profit and revenue.

(9) Increased advertising expenditure results in increased volume, but unit cost declines only when $f_4/f < \sigma$, for which $\sigma \geqslant 1$ is a sufficient condition.

(10) In the case of p^*, $D^* > 0$, a sufficient condition for profit to increase is $p^* + \beta < 1/\varepsilon$.

The general case is described in Fig. 9.35, from which all the more restricted or special cases discussed in this chapter may be derived. A guide to the various cases covered is given in the following table.

Problem characteristics	*References*
Given global values of C^*, R^*	Eqs (9.9), (9.10); Figs 9.2, 9.5; Tables 9.4, 9.7
Effect of α	Figs 9.6, 9.7, 9.8
Effect on α	Eq. (9.11)
Given increments δC, δR	Figs 9.3, 9.4
Effect of p^*	Fig. 9.12
Given c^*, V^*	Figs 9.8, 9.10; Table 9.9
Given p^*, c^*	Fig. 9.9
Given p^*, V^*	Fig. 9.11
Effect of p^*, ε and β	Eqs (9.27), (9.33), (9.34); Figs 9.13–9.20; Table 9.12
Given p^*, c^*, V^*	Eq. (9.14); Fig. 9.10; Table 9.8
Given p^*, V^*, β	Eqs (9.23), (9.24)
Dominance for two objectives	Figs 9.22, 9.23
R^* vs Z^*	Figs 9.15, 9.24
Effect of promotion D^*	Eqs (9.39), (9.40), (9.42), (9.45)
Combined effect of D^* and p^*	Eq. (9.50); Figs 9.29–9.31
Iso-curves	Figs 9.5, 9.14, 9.20, 9.34

A summary of various expressions for the profit Z and the incremental change Z^* is given below:

$$\begin{aligned} Z &= R - C \\ &= \alpha R \\ &= zV \\ &= (p - c)V \\ Z^* &= \frac{1}{\alpha}[R^* - (1-\alpha)C^*] \\ &= \alpha^* + R^* + \alpha^* R^* \\ &= z^* + V^* + z^* V^* \end{aligned}$$

$$= \frac{1}{\alpha}[p^* - (1-\alpha)c^*](1+V^*) + V^*$$

$$= \frac{1}{\alpha}[p^*(1+V^*) + \beta V^*] \qquad \text{note (1)}$$

$$= \frac{1}{\alpha}[R^* - V^*(1-\beta)] \qquad \text{note (1)}$$

$$= \frac{p^*}{\alpha}(1 - \varepsilon\beta - \varepsilon p^*) \qquad \text{note (1), (2)}$$

Notes: (1) $s^* = F^* = 0$
(2) $V^* = V^*_p = -\varepsilon p^*$

REFERENCES

[1] Eilon, S. (1975). Changes in profitability components. *Omega* **3**, 353–354.
[2] Eilon, S. and Cosmetatos, G. P. (1977). A profitability model for tactical planning. *Omega* **6**, 673–688.
[3] Eilon, S. and Cosmetatos, G. P. (1979). Effect of price and allied changes on profit. *Appl. Econ.* **11**, 117–131.
[4] Eilon, S. and Cosmetatos, G. P. (1980). Models for collective decision making in industry. *Europ. J. Operational Res.* **4**, 374–379.

EXERCISES

1. Figure 9.2 shows the condition for $Z^* \geqslant 0$ when $C^* > 0$. Draw a similar chart for the case $C^* < 0$.
2. Derive the general conditions as relationships between p^*, c^* and V^* for

   (a) breakeven
   (b) profit remaining unchanged

   when $\alpha = 10\%$ and 20% respectively.
3. If $p^* = 0$ and $c^* = 10\%$, by how much would volume have to increase to achieve an increase in profit? When is the breakeven point reached?
4. If the unit cost function takes the form

$$c = s + F/V$$

   and if the unit price increases by 10% and the volume falls by 20%, find the expected percentage change in total cost, revenue and profit, when $f_2 = 0.5$ and $\alpha = 10\%$.

5. If

$$(1-\alpha)c^* = p^* + \alpha$$

find the value of V^* that results in the profit remaining unchanged.

6. Investigate the relationship between the relative incremental profit and incremental revenue when the profit margin

   (a) remains unchanged
   (b) decreases

7. Consider the following two statements:
   (i) Profit will be maintained if revenue increases by 20% and total cost by 25%
   (ii) If revenue increases by 40% and cost by 60% then a breakeven point will be reached

   Are these compatible statements?

8. Is the condition

$$p^* > c^* > 0$$

sufficient to ensure an increase in profit when $\alpha > 0$?

9. What conditions must hold for profit to be doubled when $\alpha = 0.2$ while

   (a) $V^* = 0$ and is unaffected by the elasticity of demand?
   (b) volume is affected by price elasticity, and it is estimated that $\varepsilon = 1$?

10. If the unit cost is given by

$$c = s + F/V$$

and if $p^* = s^* = F^* = 0$, while $V^* \neq 0$, answer the following:

   (a) What conditions must prevail for profit to remain unchanged?
   (b) Is it possible for profit to decline while revenue increases?
   (c) Is it possible for profit to increase by a higher percentage than that of volume (assume $\alpha > 0$) and, if so, under what conditions?
   (d) If to start with $s = F/V$ and $\alpha = 20\%$, find an expression for the relative change in profit.

11. Show how R^*_{max} and Z^*_{max} are affected by ε in the range $0.5 \leqslant \varepsilon \leqslant 1.5$ for the case where $f = 0.2$ and $\beta = 0.3$ and $\beta = 0.5$ respectively.
12. Show that $p^*_{opt}(Z^*)$ in Equation (9.35) yields maximum Z^*
13. The following data are given:

$$f_1 = 0.7,\ f_2 = 0.2,\ f_4 = 0.1$$
$$\alpha = 15\%$$

The promotion elasticity is estimated below:

$$0 < D^* < 20\% \qquad \sigma = 0.1$$
$$20\% \leqslant D^* \leqslant 40\% \qquad \sigma = 0.2$$

Find the effect of increasing the promotion expenditure as well as the unit price by 0, 10% and 20% respectively on the

unit cost
total cost
revenue
profit

when the price elasticity is $\varepsilon = 0.9$.
Do you agree with the view that an increase in promotion expenditure is expected to be more effective than an increase in price?

14. For the data in the previous question, explore the effect of $0 \leqslant p^* \leqslant 40\%$ for the three values of $D^* = 0$, 20% and 40% respectively.
15. An analysis of past operations of a plant revealed the following information: initially, $f_1 = 0.5$ and $\alpha = 0.20$. No increase in advertising took place, and although price was increased by 20%, volume increased by 4%, owing to a general increase in demand in the marketplace, while revenue increased by 24.8%. The price elasticity of demand prevailing at the time is estimated at 0.8, but it is suggested that it would have risen sharply if the price increase had gone beyond 30%.

    It is now suggested that a higher profit increase could have been achieved. Analyse the situation and find what p^* within the range 0–30% would have maximized Z^*.
16. The increase in market demand due to external factors is estimated to lead to $V_0^* \simeq 5\%$. Management is interested in maximizing revenue as well as profit. Investigate the pricing policy that should be considered when the price elasticity is estimated as follows:

$p^*(\%) =$	−20	−10	−5	5	10	15	20	25
$\varepsilon =$	1.10	1.05	1.025	0.85	1.0	1.05	1.10	1.15

    Comment on the effect of V_0^* on the result (for example, what would happen if V_0^* is only 2% or turns out to be zero?). Other given data: $\alpha = 0.16$, $f = 0.3$. Assume no change in advertising expenditure.
17. What conditions should prevail to ensure that a combination of changes in price and advertising expenditure leaves the unit cost unchanged? Derive the result for the case where $f = 0.5$, $\varepsilon = 1$.
18. Consider the following two statements:

    (a) Since

    $$R = Z + C$$

then

$$R^* = \alpha Z^* + (1-\alpha)C^*$$

Assume that the total cost does not change, hence $C^* = 0$, leading to

$$R^* = \alpha Z^*$$

(b) Since

$$Z = \alpha R$$

and α is assumed constant in the base period, it follows from the rules of incremental calculus that

$$Z^* = R^*$$

How do you explain the difference between the two results?

19. Prove the following result:

$$Z^* = \frac{1}{\alpha}[R^* - (1-\beta)V^*]$$

20. Analyse the circumstances for which the following proposition holds: "If

$$p^* + V^* > c^*$$

then this is a sufficient condition for profit to increase".

21. Prove that if revenue remains constant, profit will rise provided α and β are positive.
22. Prove that $\sigma > \varepsilon$ is not a sufficient condition for profit to rise when α, p^* and D^* are positive.
23. If $p^* = 0$ and $D^* > 0$, prove that

$$\beta_0 > \beta - \alpha$$

is a sufficient condition to result in $c^* > 0$.

24. A company is operating at a profit margin of $\alpha = 20\%$. Questions have arisen about the possible effect of a price change on total cost, revenue, and profit. Price changes in the range of -20% to $+20\%$ are being considered and it is estimated that throughout this range the price elasticity of demand is $\varepsilon = 1$. An analysis of the total cost function in this company suggests the relationship

$$C = k_1 R + k_2 V$$

where R and V are revenue and physical volume respectively. At present, the two terms of this function constitute 90% and 10% of the total cost respectively.

Taking the current level of revenue, cost and profit as 100, find the effect of a relative change in price in the range specified above. Show the results with the aid of a table and/or a graph for selected values. Determine the value of p^* for which the profit Z (or the relative change Z^*) is maximized.

25. The value of p^* at which Z^* is maximized inevitably results in a decline in revenue.

    True or false?

26. (a) Comment on the result for Z^*_{max} in Equation (9.36) for the case when α is negative.
    (b) If $\alpha = -5\%$, $\varepsilon = 1.6$, $\beta = 0.7$, tabulate Z^* for various values of p^*.
27. If a combination of changes is likely to occur in fixed costs, marginal unit costs, intrinsic market demand, promotion expenditure, interest rate and unit price, draw a diagram to show the network of relationships from which the effect on unit cost and on profit can be determined.
28. For the case $\varepsilon\beta = 1$ find the condition for

    (a) profit to increase
    (b) revenue to increase

29. The current performance of a company is as follows:

Revenue		40
Cost: manpower	20	
rent and depreciation	9	
other overheads	4	
total		33
Operating profit		7
Bad debts		2
Net profit		5

On the assumption that bad debts remain at 5% of revenue, and other overheads at 10% of revenue, while rent and depreciation remain unchanged, calculate what combinations of changes in staff costs and revenue are expected to

(a) leave the net profit level at its current value;
(b) increase the profit to 7;
(c) increase the profit to 10.

30. Calculate (tabulate and/or plot) the expected relative change (in %) in volume, total cost, unit cost, revenue and profit for the following case for which current data are given (K signifies 10^3):

Fixed costs	
labor	53 400 K
materials	30 200 K
other (rent, depreciation, etc.)	47 600 K
Interest charges	5700 K

Advertising and promotion	4100 K
Marginal costs per unit	
direct labor	3.940
materials	7.262
Volume	16 400 K
Revenue	360 800 K

Number of people employed 8200

Expected changes in costs:

wage rates to be increased by 8 %

materials by 10 %

interest rate currently 12 % expected to fall to 8 %

Other expected changes:

total demand in the market will increase by 2 %

productivity of direct labor (defined as the ratio of output to direct labor) will increase by 0–4 %.

Further data: working capital is currently 20 % of revenue, 66 % of the working capital is financed by borrowings;

percentage change in working capital is proportional to the percentage change in revenue, the proportionality coefficient being 0.5 (i.e. 10 % change in revenue results in 5 % change in working capital).

Policies to consider: Should the pricing policy and/or advertizing expenditure be changed in order to increase revenue, profit and profit margin? Elasticities are estimated in the following tables.

Increase in promotion expenditure D^* (%)	2	4	6	8	10	12	⩾ 14	−1	−2	−4
Expected increase in volume V_1^* (%)	0.5	1.5	2.5	4.0	6.0	8.5	10.0	−1	−2	−4

Unit price	18	19	20	21	22	23	24	25	26	27	28	29	30
Expected volume $\times 10^6$	19.5	18.6	18.0	17.1	16.4	15.8	15.0	14.3	13.4	12.2	11.0	9.5	8.0

10
The Profit Margin

Themes in this chapter

Definitions of the net and gross profit margins α and β
Definitions of the fixed cost proportions f and g
Relationships between α, β, f and g
The α-g chart
Analysis of the effect of changes in unit price and volume
Conditions for profit margins to increase
Conditions for unit profit to increase
Compatibility of these conditions with profit increase

DEFINITIONS

A distinction is made between the *net profit margin* α and the *gross profit margin* β, as shown by the following definitions:

$$\alpha = \frac{Z}{R} = \frac{R-C}{R} = \frac{p-c}{p} \tag{10.1}$$

$$\beta = \frac{Z+F}{R} = \frac{R-S}{R} = \frac{p-s}{p} \tag{10.2}$$

Generally, whenever the profit margin is referred to in this volume it is understood to mean the net margin α, the definition of which does not depend on the chosen cost model. The range for α for a profitable operation is between 0 and 1. In the hypothetical case where no costs are involved, $C = 0$ and then α attains its maximum value, namely 1. If revenue is just sufficient to cover costs ($R = C$), α reduces to 0, and when the cost exceeds revenue, α is negative.

The gross margin β, however, relies on the fixed and variable unit and total cost model in its definition, namely on

$$c = s + \frac{F}{V}$$

$$C = S + F = sV + F$$

The gross margin β measures the degree to which the operation manages to exceed the mere recovery of the variable cost S. If all the costs incurred may be regarded as fixed, then $S = 0$ leading to $\beta = 1$. If all the costs are variable and $F = 0$, then the value for β coincides with that of α. Thus, the range $0 \leqslant \beta \leqslant 1$ signifies that at least the variable costs are recovered (otherwise, β becomes negative), but the condition $\beta > 0$ is not sufficient to yield a positive profit, namely to ensure that α is also positive.

Nevertheless, many trading companies not involved in the manufacture of goods (including retail and wholesale business) find β a convenient measure of their profitability and use it for pricing purposes. From Equation (10.2)

$$\beta = 1 - \frac{s}{p}$$

hence

$$p = \frac{s}{1-\beta}$$

where the multiplier $1/(1-\beta)$ signifies the "markup" applied to the marginal cost s. A trading company purchasing goods at s per unit and operating at a gross margin of, say, $\beta = 40\%$, will apply a markup of $1/0.6 = 1.667$, namely it will add 66.7% to s to arrive at the unit price p. In many trading operations s represents the cost of materials, in which case $p - s$ denotes the added value per unit, and the gross margin β then becomes the ratio of added value to revenue.

As we know, the cost proportions in this case are

$$\left.\begin{aligned} f_1 &= \frac{S}{C} = \frac{s}{c} \\ f_2 &= f = \frac{F}{C} \end{aligned}\right\} \tag{10.3}$$

where f_2 (or f for short) is the fixed cost proportion. Similarly, we can deduce the revenue proportions from

$$R = sV + F + Z$$

namely:

$$\left.\begin{aligned} &\text{variable cost as proportion of revenue} \quad & g_1 &= \frac{sV}{R} \\ &\text{fixed cost as proportion of revenue} \quad & g_2 &= g = \frac{F}{R} \\ &\text{and profit as proportion of revenue} \quad & \alpha &= \frac{Z}{R} \end{aligned}\right\} \tag{10.4}$$

f and g relate to the fixed cost element F in both definitions and are used here for short to signify f_2 and g_2 respectively, so that

$$f_1 + f = 1 \tag{10.5}$$

$$g_1 + g + \alpha = 1 \tag{10.6}$$

In the case of a more elaborate cost function, for example

$$C = sV + F + J + D$$

where the definition of the fixed cost does not include such elements as the interest charge J and the promotion expenditure D, the cost proportions are

$$f_1 = \frac{s}{c}$$

$$f_2 = \frac{F}{C}$$

$$f_3 = \frac{J}{C}$$

$$f_4 = \frac{D}{C}$$

so that here

$$f = 1 - f_1 = f_2 + f_3 + f_4$$

Similar definitions for revenue proportions can be derived, leading to

$$g = g_2 + g_3 + g_4$$

and hence the general results in Equations (10.5) and (10.6) hold.

BASIC RELATIONSHIPS

From the definitions of the two profit margins,

$$\frac{s}{p} = 1 - \beta$$

$$\frac{c}{p} = 1 - \alpha$$

we get

$$\frac{s}{c} = \frac{1 - \beta}{1 - \alpha}$$

Substitute

$$\frac{s}{c} = f_1 = 1 - f$$

to obtain

$$1 - f = \frac{1-\beta}{1-\alpha}$$

or

$$\beta = \alpha + f - f\alpha$$

Also, by definition,

$$\beta = \frac{Z}{R} + \frac{F}{R} = \alpha + g$$

hence

$$\beta = \alpha + g = \alpha + f - f\alpha \tag{10.7}$$

and

$$g = \beta - \alpha = f(1-\alpha) \tag{10.8}$$

from which we conclude that:

(1) Generally, $\beta > \alpha$ (since $g > 0$).
(2) If $\alpha > 0$ then $g < f$, otherwise the inequality reads the other way (in the special case of $\alpha = 0$ then $g = f = \beta$).
(3) If $f = 1$ then $\beta = 1$, but if $f = 0$ (i.e. $F = 0$) then $g = 0$ and $\beta = \alpha$, hence the value of β falls in the range $\alpha \leqslant \beta \leqslant 1$.

Some of the relationships are summarized in Table 10.1, where the ratios of various variables to cost and to revenue respectively are given. Thus, if the

TABLE 10.1
Summary of ratios.

Ratio of/to	Cost C	Revenue R
Cost C	1	$1-\alpha$
Fixed cost F	f	$g = f(1-\alpha)$
Variable cost S	$1-f$	$1-\beta$
Profit Z	$\frac{\alpha}{1-\alpha}$	α
Added value A (when $S = M$)	$\frac{\beta}{1-\alpha}$	β
Revenue R	$\frac{1}{1-\alpha}$	1

variable in question is x, its ratio to cost is x/C and its ratio to revenue is x/R, and since

$$\frac{x}{R} = (1-\alpha)\frac{x}{C}$$

it follows that any ratio in the revenue column is readily obtained from the ratio in the cost column after multiplying the latter by $1-\alpha$.

It is clear from Equation (10.7) or (10.8) that, given two values of the quartet α, β, f and g, the remaining two can be readily determined. Figure 10.1 shows the relationship between α and β with f as a parameter. As shown above, the case of $f = 0$ leads to $\beta = \alpha$, and this is represented by the bottom line, drawn at 45° to the horizontal axis. As f increases, β becomes increasingly insensitive to a change in α, and the extreme case of $f = 1$ results in $\beta = 1$ irrespective of the value of α.

Lines for equal values of g can be superimposed on Fig. 10.1, as shown in

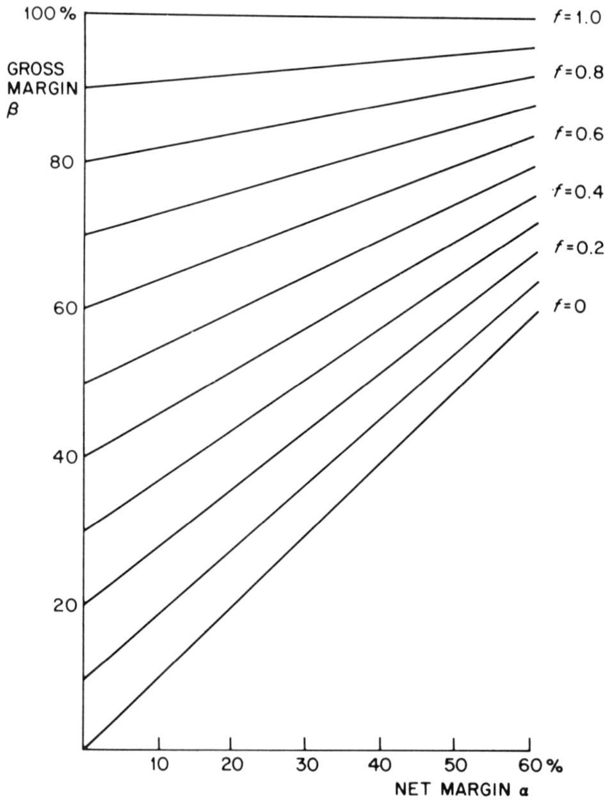

FIG. 10.1. The profit margins α and β.

Fig. 10.2, from which any pair of values of α, β, f and g can be read for any given values of the other pair. Following the conclusions derived above, the domain below the line $f = g = 0$ is infeasible, since it violates the requirement $\beta > \alpha$.

There are several other ways in which the relationships between α, β, f and g can be presented graphically. A simple diagram, called the $\alpha - \beta$ chart, is shown in Fig. 10.3. On the assumption that we are dealing with a profitable operation (i.e. $\alpha > 0$), the chart consists of a square confined to the positive quadrant (i.e. where both α and β are positive). The diagonal 0A divides the square into two parts, with the infeasible domain (where $\beta < \alpha$) below the diagonal. Any point above the diagonal signifies a theoretically feasible operating point. For example, the point X has the coordinates α and β, the values of which may be read on the horizontal and vertical axes respectively, which for the example shown correspond to $\alpha = 0.25$ and $\beta = 0.475$. The extension of the line AX cuts the vertical axis at a point that yields the value of f; in our example $f = 0.3$.

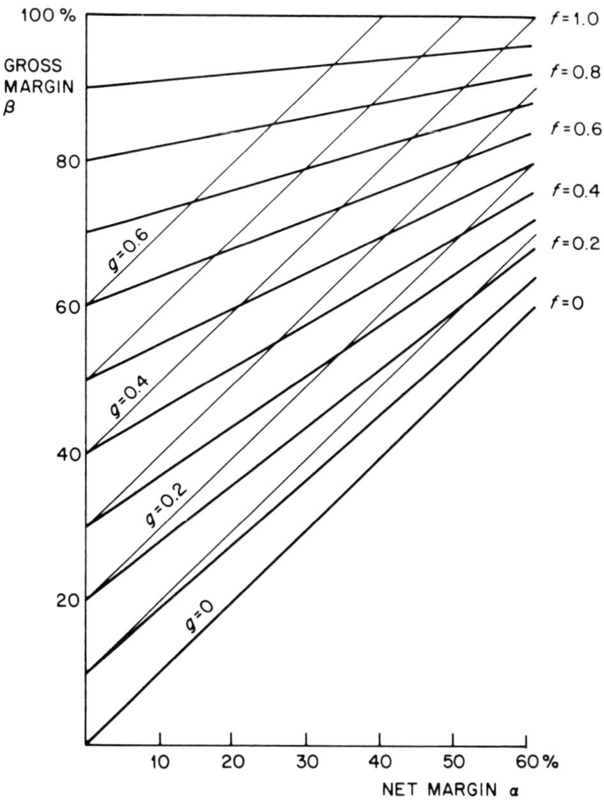

FIG. 10.2. Iso-g lines superimposed on Fig. 10.1.

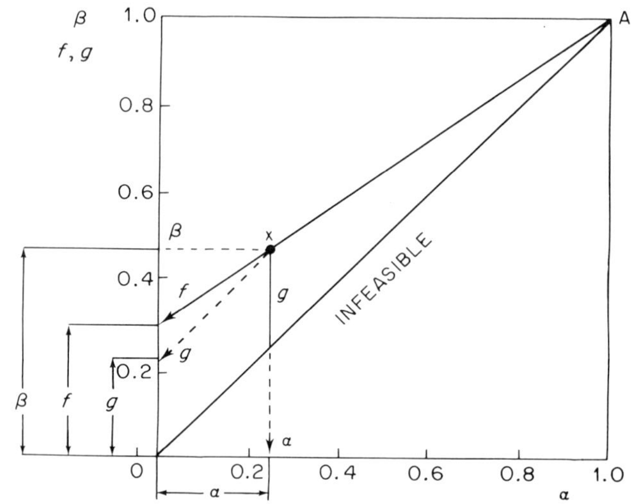

FIG. 10.3. The α-β chart.

The vertical distance between X and the diagonal A0 gives the value of g. Alternatively, g may be obtained by drawing a line through X parallel to the diagonal, and where this line intersects the vertical axis, the value of g may be read on the vertical scale.

THE $\alpha - g$ CHART

An alternative presentation, that makes better use of the space in the positive quadrant, is shown in the $\alpha - g$ chart in Fig. 10.4. The β lines are drawn at an angle of 45° to the horizontal axis, thus ensuring that $\beta > \alpha$ in this quadrant for $g > 0$, and the lines $g = f(1 - \alpha)$ ensure that $g < f$.

Any point on this chart can be seen as an intersection of an f-line (on which f is constant) and a β-line (on which β is constant), as shown in Fig. 10.5. The intersection of these lines yields four zones, with their properties summarized in Table 10.2. If we move from the point of intersection to any point in zone 1 then f increases while β declines, while in zone 3 these results are reversed. In zone 2 both f and β increase, while in zone 4 both decline. As for α and g it is interesting to observe that they move in opposite directions in zone 1 (where g increases while α declines) and zone 3 (where the reverse occurs), but the results are inconclusive in zones 2 and 4.

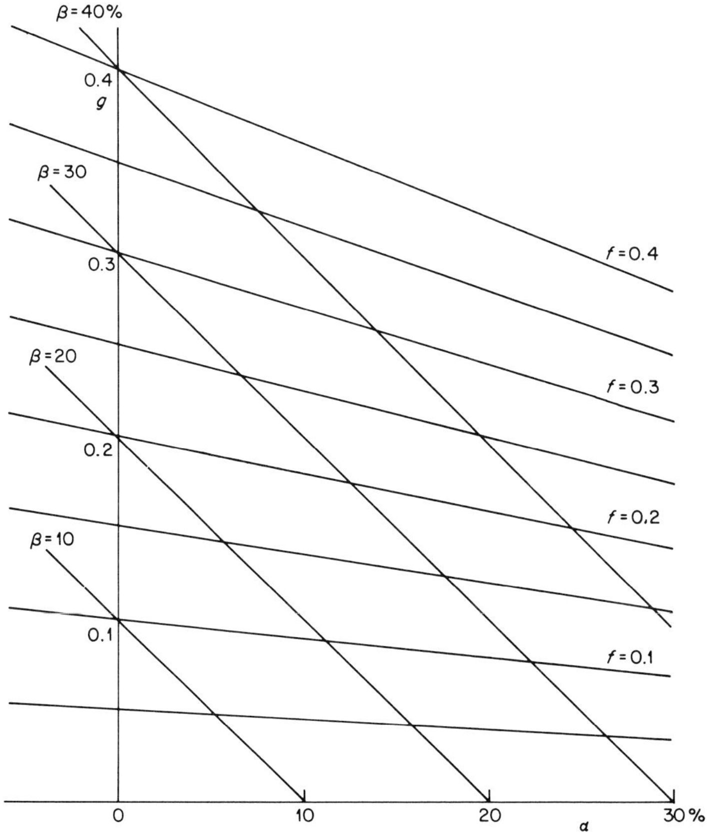

FIG. 10.4. The α-g chart.

Zone 2 may, in fact, be subdivided into three subzones, as shown in Fig. 10.6, where

in subzone 2A: α declines while g increases
in subzone 2B: both increase
in subzone 2C: α increases while g declines

TABLE 10.2
Four zones in the α-g chart.

Zone	α^*	β^*	f^*	g^*
1	<0	<0	>0	>0
2	?	>0	>0	?
3	>0	>0	<0	<0
4	?	<0	<0	?

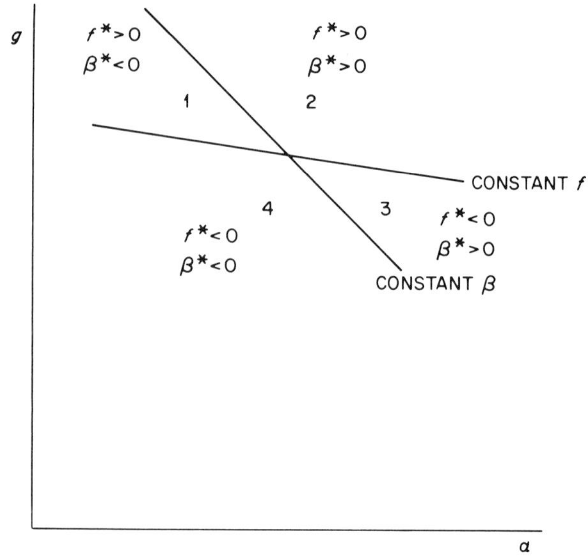

FIG. 10.5. Four zones in the α-g chart.

Similar observations can be made on the subdivisions of zone 4.

A change in the net profit margin α_1 to α_2 from point A to C in Fig. 10.7 may be interpreted as a combination of two movements, the first along the f-line from A to an interim point B (where the profit margin is α_1'), followed by a movement along the β-line from B to C (or a movement along the β-line from A to D followed by a movement along the f-line from D to C). The diagram is reminiscent of the parallelogram of forces in mechanics, except that the geometrical figure is not quite a parallelogram (while the β-lines are parallel to each other, the f-lines are not).

Thus, an increase in the net profit margin from α_1 to α_2 can be achieved by an increase in β and/or a reduction in f. The combination of such actions in Fig. 10.7 is to reduce g, though it is possible for circumstances to arise where both α and g increase, as illustrated in Fig. 10.8. Here the change from A to C may be interpreted as a movement from A to an interim point B along the f-line, thereby increasing the net margin from α_1 to α_1' (and causing β to increase in the process), followed by a movement from B to C along the β-line, thereby *decreasing* the net margin from α_1' to α_2. The net result in the case depicted in Fig. 10.8 is for the net margin α to increase while g may increase as well.

The change shown in Fig. 10.7 represents a movement into zone 3 in the

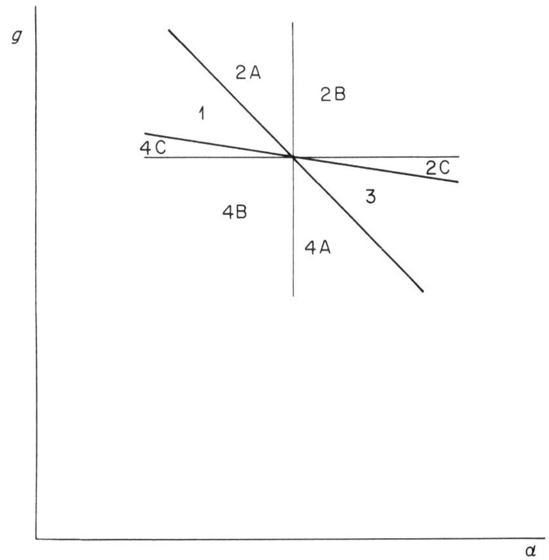

FIG. 10.6. Subdividing zones 2 and 4 in Fig. 10.5.

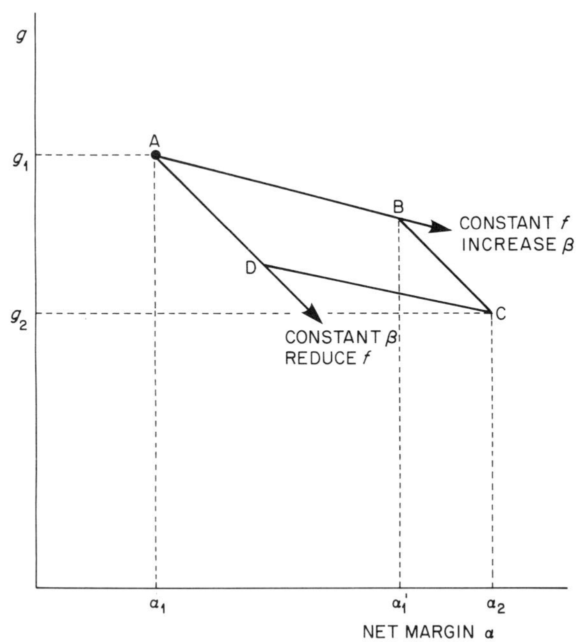

FIG. 10.7. A change from α_1 to α_2 (zone 3).

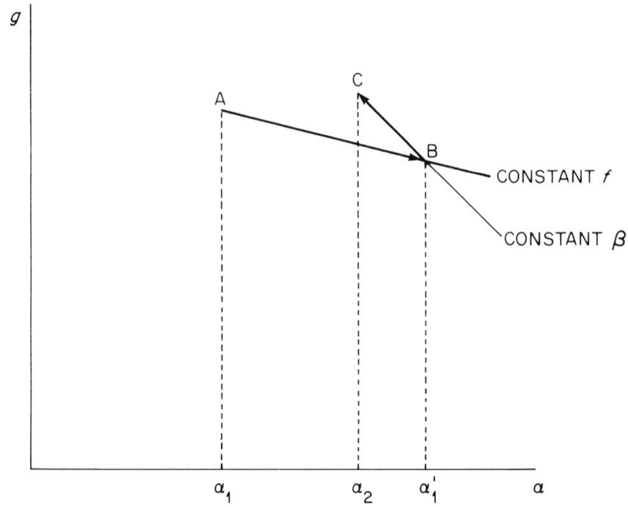

FIG. 10.8. A case where both α and g increase (zone 2).

$\alpha - g$ chart in Fig. 10.5, whereas the change depicted in Fig. 10.8 is a movement into zone 2. As indicated in Table 10.2, it is possible for α to increase in zone 4 as well, and an example is shown in Fig. 10.9, where the change from α_1 to α_2 may be regarded as a change along the β-line from point A to B, with both f and g declining, then along the f-line to point C, with g increasing and β falling. The

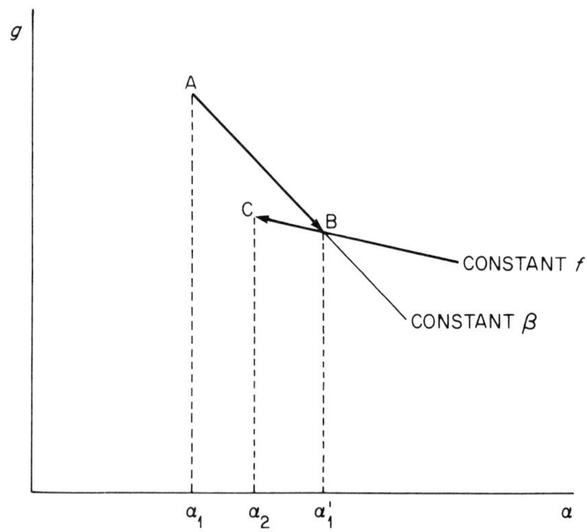

FIG. 10.9. A case where α increases in spite of a fall in β (zone 4).

overall effect is a fall in β, and clearly the reduction in f must be sufficient to offset the fall in β, if α is to rise.

To summarize the results for the net profit margin α in the four zones in Fig. 10.5:

Zone 1. α declines
Zone 2. It is possible for α to increase if β increases sufficiently to compensate for the increase in f
Zone 3. Both the reduction in f and the increase in β contribute to an improvement in α
Zone 4. The reduction in f must be significant enough to compensate for a decline in β and then it is possible for α to improve.

RELATIVE CHANGES

The net profit margin

When a change in the operating conditions takes place, the net profit margin changes from its original value α to $\alpha+\delta\alpha$, and the new value is stated in Chapter 9 in Equation (9.11). This derivation of α^* follows from the definition

$$\alpha = \frac{Z}{R}$$

hence

$$\alpha^* = \frac{Z^* - R^*}{1+R^*} \tag{10.9}$$

Alternatively, this result may be written as

$$1+\alpha^* = \frac{1+Z^*}{1+R^*}$$

From Equation (9.9)

$$Z^* = \frac{1}{\alpha}[R^* - (1-\alpha)C^*]$$

If this value of Z^* is substituted in Equation (10.9), we obtain

$$\alpha^* = \frac{1}{1+R^*}\left(\frac{R^*}{\alpha} - \frac{1-\alpha}{\alpha}C^* - R^*\right)$$

or

$$\alpha^* = \frac{1-\alpha}{\alpha}\;\frac{R^* - C^*}{1+R^*} \tag{10.10}$$

where for the fixed and variable cost model

$$C^* = f_1 S^* + f F^*$$

In the case where $s^* = F^* = 0$

$$C^* = f_1 (sV)^* = f_1 V^*$$

which may then be substituted into Equation (10.10).

It is clear from Equations (10.9) and (10.10) that the net profit margin remains unchanged ($\alpha^* = 0$) only if $Z^* = R^* = C^*$ and more generally

$$\alpha^* \gtrless 0 \qquad \text{if } Z^* \gtrless R/ \text{ or } R^* \gtrless C^* \tag{10.11}$$

The first condition is shown in Fig. 10.10, where the line drawn at a 45° angle stands for $\alpha^* = 0$ and the domain above the line signifies an increase in α. Thus, it is possible for profit to increase and yet for the net profit margin to decline, if the relative increase in revenue exceeds that of profit. Similarly, it is possible for profit to decline and yet for the net profit margin to increase, if the relative decline in profit is less severe than that of revenue.

This result can be further demonstrated when we consider the incremental changes δR and δC in revenue and cost respectively. From Equation (9.9) we recall that the condition for profit to increase (provided $\alpha > 0$) was derived as

$$Z^* > 0 \qquad \text{if } R^* > (1-\alpha)C^* \tag{10.12}$$

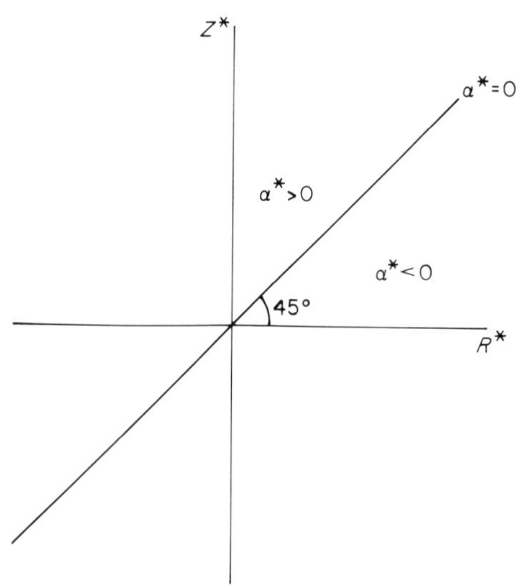

FIG. 10.10. The condition $\alpha^* = 0$.

whereas from (10.11)

$$\alpha^* > 0 \qquad \text{if } R^* > C^* \tag{10.13}$$

Rewriting (10.13) we get

$$\frac{\delta R}{R} > \frac{\delta C}{C}$$

and in general

$$\alpha^* \gtrless 0 \qquad \text{if } \delta R \gtrless \frac{\delta C}{1-\alpha} \tag{10.14}$$

In Fig. 9.3 the condition for $Z^* = 0$ is presented by a straight line drawn at 45°. If the condition with the equality sign in (10.14) is superimposed on Fig. 9.3 then Fig. 10.11 is obtained to incorporate the results for Z^* and for α^*. The line for $\alpha^* = 0$ has the slope $1/(1-\alpha)$, which (for $\alpha > 0$) has a value between 1 and ∞ and therefore involves an angle that exceeds 45° (but less than 90°) to the horizontal axis. This line goes through the regions 1 and 4 in Fig. 9.3, yielding (see Fig. 10.11):

Region 1A—where profit increases but the net profit margin decreases
Region 1B—where both increase (as they do in Regions 2 and 3)
Region 4A—where the results are the reverse of those in 1A
Region 4B—where both α and Z decline (as they do in Regions 5 and 6)

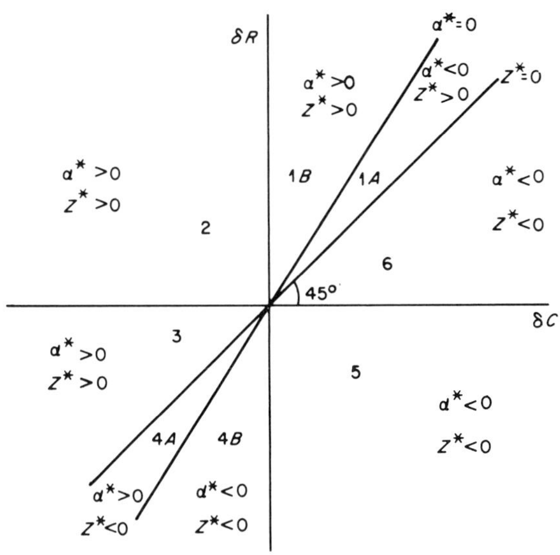

FIG. 10.11. Effect of δR and δC on the profit margin.

Thus, *profit and the net profit margin generally move in the same directions, but there are circumstances where they do not*, namely when regions 1A and 4A apply. It follows that the generally held view that α can *always* be regarded as a surrogate objective, which automatically ensures an improvement in profit, is erroneous.

If the starting point for deriving α^* is the definition

$$\alpha = \frac{p-c}{p}$$

then

$$\alpha^* = \left(\frac{p-c}{p}\right)^* = \frac{(p-c)^* - p^*}{1+p^*}$$

Substituting

$$(p-c)^* = \frac{1}{\alpha}p^* - \frac{1-\alpha}{\alpha}c^*$$

we obtain

$$\alpha^* = \frac{1-\alpha}{\alpha}\,\frac{p^*-c^*}{1+p^*} \tag{10.15}$$

so that (assuming $\alpha > 0$)

$$\alpha^* \geqslant 0 \qquad \text{if } p^* \geqslant c^* \tag{10.16}$$

and this result may be compared with that derived in the condition (9.18), namely

$$Z^* \geqslant 0 \qquad \text{if } p^* + \frac{\alpha V^*}{1+V^*} \geqslant (1-\alpha)c^* \tag{10.17}$$

Again, it is clear that the two conditions do not coincide. While it is true that if the relative increment of unit price exceeds that of unit cost then the net profit margin must rise, such a categorical statement cannot be made with respect to a change in profit. Notice also that while a change in volume does not directly have a role in the condition (10.16), except indirectly by affecting c^*, it does have a distinct role in the condition (10.17) for $Z^* \geqslant 0$. Even in the case of $V^* = 0$, when the unit cost is affected by other reasons than a change in volume, the condition in (10.16) is more demanding when $p^* > 0$ than that in (10.17), so that when the former is not met and the latter is, then the result lies in the region 1A in Fig. 10.11.

A further illustration is given in Fig. 10.12, which relates to the example of $\alpha = 20\%$ and $p^* = 5\%$. The indifference line $Z^* = 0$ is obtained (in a similar fashion to Fig. 9.10) by substituting these values into (10.17) to yield

$$0.05 + \frac{0.2\,V^*}{1+V^*} = 0.8\,c^*$$

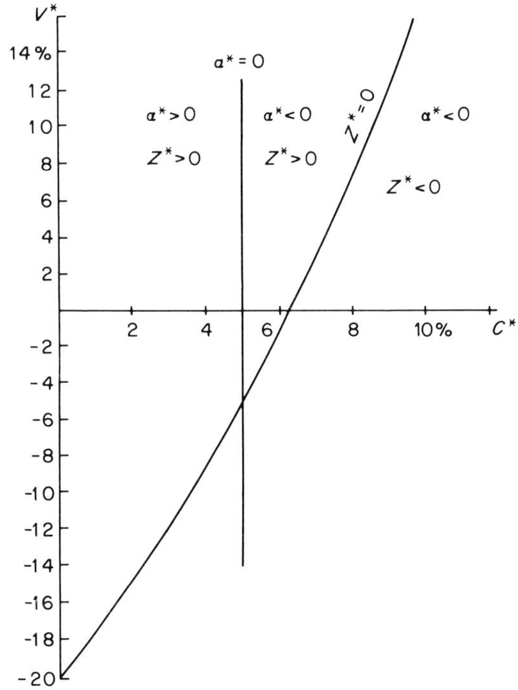

FIG. 10.12. An example for $\alpha = 20\%$, $p^* = 5\%$.

The curve in Fig. 10.12 shows this relationship between c^* and V^*. The domain above the curve corresponds to $Z^* > 0$ and that below the curve to $Z^* < 0$. Now, the condition for $\alpha^* = 0$ is given in (10.16) by $p^* = c^*$, i.e. in this example by the vertical straight line at $c^* = 5\%$. The domain to the left of this line satisfies the requirement $p^* > c^*$, hence it corresponds to $\alpha^* > 0$, while the domain to the right signifies $\alpha^* < 0$. Between this vertical line and the curve there is an area in which profit increases but the net profit margin decreases.

The unit profit

Since the unit profit $z = p - c$, it follows that

$$\alpha^* = \left(\frac{z}{p}\right)^* = \frac{z^* - p^*}{1 + p^*}$$

where, as shown earlier,

$$z^* = (p - c)^* = \frac{1}{\alpha}[p^* - (1 - \alpha)c^*]$$

The following conclusions may, therefore, be drawn:

(1) Irrespective of whether p rises or falls, $1 + p^* > 0$; hence

$$\alpha^* > 0 \qquad \text{if } z^* > p^*$$

which means that the net profit margin can rise even if the unit profit falls, but that can only occur when the unit price also falls, and by a larger amount than the fall in the unit profit. In the case of a price rise ($p^* > 0$), a necessary (but not sufficient) condition for the margin α to rise is that the unit profit z rises as well.

(2) In order to achieve an increase in the unit profit, we must have (for $\alpha > 0$) the following condition

$$z^* > 0 \qquad \text{if } p^* > (1-\alpha)c^*$$

(3) Thus, if the price is increased and if

$$p^* > c^*$$

then the condition for $z^* > 0$ is met, and from (10.16) we find that $\alpha^* > 0$ as well

(4) However, if

$$p^* > (1-\alpha)c^*$$

then $z^* > 0$, but the condition in (10.16) is not necessarily met, and it is possible then for the net margin to fall.

The fixed and variable cost model

In the case of the fixed and variable cost model

$$c^* = f_1 s^* + f\frac{F^* - V^*}{1 + V^*}$$

which needs to be substituted into (10.16) and (10.17). Of particular interest are the following special cases:

(1) $s^* = 0$, so that

$$c^* = f\frac{F^* - V^*}{1 + V^*}$$

and (10.16), (10.17) then read

$$\alpha^* \geqslant 0 \qquad \text{if } p^* \geqslant \frac{f(F^* - V^*)}{1 + V^*}$$

$$Z^* \geqslant 0 \qquad \text{if } p^* \geqslant \frac{1}{1 + V^*}[(1-\alpha)fF^* - \beta V^*]$$

(2) In the special case where it is assumed that both $s^* = F^* = 0$ then the sole cause of a change in unit cost is a change in volume, namely

$$c^* = -\frac{fV^*}{1+V^*} \tag{10.18}$$

so that the condition (10.16) becomes

$$\alpha^* \geqslant 0 \qquad \text{if } p^* \geqslant -\frac{fV^*}{1+V^*} \tag{10.19}$$

whereas the condition in (10.17) is then as stated in (9.25), namely

$$Z^* \geqslant 0 \qquad \text{if } p^* \geqslant -\frac{\beta V^*}{1+V^*} \tag{10.20}$$

and the difference between the two conditions then rests on the difference between the values of f and β. If $\alpha > 0$ then from Equation (10.7): $\beta > f$. Hence, when the volume decreases the bound in condition (10.20) is more demanding than that in (10.19), so that an increase in profit must mean that the profit margin increases as well, but it is possible for the latter to increase without the former. When volume increases, however, condition (10.19) dominates (10.20), so that an increase in profit margin is accompanied by an increase in profit, but the reverse may not necessarily hold.

The gross profit margin

Starting from

$$\beta = \frac{p-s}{p}$$

we derive, similar to Equation (10.15)

$$\beta^* \geqslant \frac{1-\beta}{\beta}\frac{p^*-s^*}{1+p^*} \tag{10.21}$$

so that

$$\beta^* \geqslant 0 \qquad \text{if } p^* \geqslant s^* \tag{10.22}$$

and for the case where $s^* = 0$

$$\beta^* = \frac{1-\beta}{\beta}\frac{p^*}{1+p^*} \tag{10.23}$$

the condition becomes

$$\beta^* \geqslant 0 \qquad \text{if } p^* \geqslant 0 \tag{10.24}$$

Hence, in Figs 10.4–10.9 the β-lines signify not only that β is constant but that the unit price p is constant as well (assuming $s^* = 0$). This means that a movement from left to right along an f-line, which increases the value of β, is obtained by increasing the unit price.

Alternatively, it can be shown that

$$\beta^* = -\frac{1-\beta}{\beta}\left(\frac{S}{R}\right)^* = -\frac{1-\beta}{\beta}g_1^*$$

and this expression is a convenient way of determining the change in β. Since management accounts usually provide data on S and R for the current as well as for the base period, g_1 and hence g_1^* can be readily computed, and it is evident from this expression that

$$\beta^* > 0 \qquad \text{when } g_1^* < 0$$

A comparison of the conditions (10.19) and (10.24) reveals that for the fixed and variable cost model both profit margins increase with a price increase, in the event that there is no change or an increase in volume ($V^* \geqslant 0$). If volume declines, then it is possible for β to increase while α declines when the unit price is increased.

Example

If $\alpha = 20\%$ and $\beta = 40\%$, find the relationships between the relative increments of price and volume that determine whether profit and the profit margins rise or fall.

Assuming $s^* = F^* = 0$, the conditions in (10.19), (10.20) and (10.24) apply. From Equation (10.7)

$$f = \frac{\beta - \alpha}{1 - \alpha} = \frac{0.2}{0.8} = 0.25$$

hence the results are:

$$Z^* \geqslant 0 \qquad \text{if } p^* \geqslant -0.4\frac{V^*}{1+V^*}$$

$$\alpha^* \geqslant 0 \qquad \text{if } p^* \geqslant -0.25\frac{V^*}{1+V^*}$$

$$\beta^* \geqslant 0 \qquad \text{if } p^* \geqslant 0$$

If volume increases, the third of these conditions is the most demanding, and when it is met then not only β increases but α and Z as well. However, if volume declines, then the first of the three conditions is the most demanding and, if met, causes Z, α and β to rise. These results are shown graphically in Fig. 10.13.

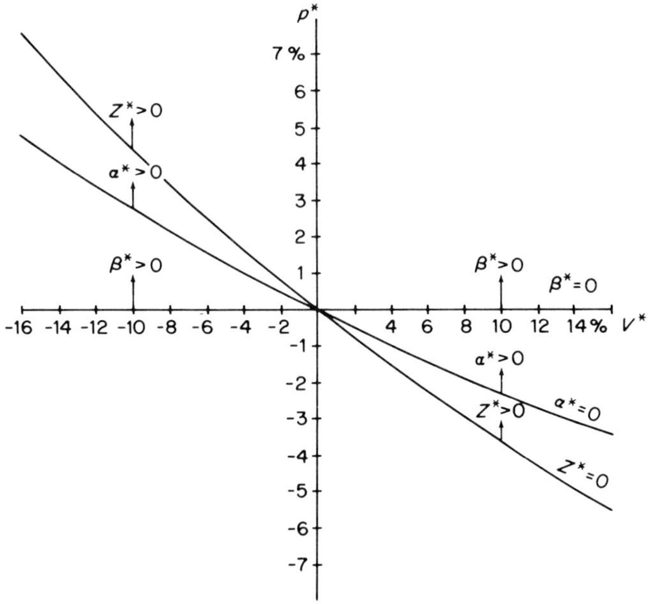

FIG. 10.13. The case of $\alpha = 20\%$, $\beta = 40\%$.

Other expressions for β^*

Another derivation for β^* can be obtained from the relationship

$$\beta = \alpha + f - \alpha f$$

hence

$$\beta^* = \frac{1}{\beta}[\alpha\alpha^* + ff^* - \alpha f(\alpha^* + f^* + \alpha^* f^*)]$$

or

$$\beta^* = \frac{\alpha f}{\beta}\left(\frac{1-f}{f}\alpha^* + \frac{1-\alpha}{\alpha}f^* - \alpha^* f^*\right) \tag{10.25}$$

Alternatively, a simpler expression for β^* is obtained from

$$\beta = \alpha + g$$

yielding

$$\beta^* = \frac{\alpha}{\beta}\alpha^* + \frac{g}{\beta}g^* \tag{10.26}$$

Special cases

(1) A movement along the f-line in the α-g chart means that $f^* = 0$, in which case Equation (10.25) is reduced to

$$\beta^* = \frac{1-f}{\beta}\alpha\alpha^*$$

If α and β are positive, then both change in the same direction, thus corroborating the conclusion drawn earlier from Fig. 10.4, except that now we have a quantitative result for β^*. If we substitute $\delta\alpha = \alpha\alpha^*$ and $\delta\beta = \beta\beta^*$ this result may be rewritten as

$$\delta\beta = (1-f)\delta\alpha \qquad \text{when } f^* = 0 \tag{10.27}$$

which means that in absolute terms the effect of change on α is greater than it is on β (since $1-f < 1$).

(2) A movement along the β-line in the α-g chart implies $\beta^* = 0$, for which Equation (10.25) becomes

$$\frac{1-f}{f}\alpha^* + \frac{1-\alpha}{\alpha}f^* - \alpha^* f^* = 0$$

i.e.

$$\alpha^* = \frac{-\dfrac{1-\alpha}{\alpha}f^*}{\dfrac{1-f}{f} - f^*}$$

or

$$\delta\alpha = -\frac{(1-\alpha)\delta f}{1-f-\delta f} \qquad \text{when } \beta^* = 0$$

which means that on the β-line α and f change in opposite directions. A simpler expression is derived from Equation (10.26), which reduces to

$$\alpha\alpha^* + gg^* = 0$$

or

$$\delta\alpha = -\delta g \qquad \text{when } \beta^* = 0 \tag{10.28}$$

(3) A change in the α-g chart in the horizontal direction requires that $g^* = 0$. In this case Equation (10.26) becomes

$$\delta\alpha = \delta\beta \qquad \text{when } g^* = 0 \tag{10.29}$$

which, again, is easily derived from Fig. 10.4.

(4) A change such that $\alpha^* = 0$ (i.e. along a vertical line in the $\alpha - g$ chart), when

substituted into Equation (10.26)

$$\delta g = \delta\beta \qquad \text{when } \alpha^* = 0 \tag{10.30}$$

The cost proportions

For the fixed and variable cost model, the definitions of f_1 and f lead to

$$f_1^* = \left(\frac{S}{C}\right)^* = \frac{S^* - C^*}{1 + C^*}$$

and, similarly,

$$f^* = f_2^* = \frac{F^* - C^*}{1 + C^*}$$

but

$$C^* = f_1 S^* + f F^*$$

hence

$$f_1^* = \frac{S^* - f_1 S^* - f F^*}{1 + C^*} = \frac{f(S^* - F^*)}{1 + C^*}$$

and similarly

$$f^* = \frac{f_1(F^* - S^*)}{1 + C^*}$$

leading to the conclusion that

$$f_1 f_1^* + f f^* = 0 \tag{10.31}$$

or simply

$$\delta f_1 + \delta f = 0$$

This result is only to be expected, since f_1 and f must change in opposite directions by equal absolute amounts in order to preserve the general condition that $f_1 + f = 1$ at all times.

If the marginal unit cost and the fixed cost do not change ($s^* = F^* = 0$), then

$$C^* = f_1 S^* = f_1 (sV)^* = f_1 V^*$$

so that f_1^* and f^* are reduced to

$$f_1^* = \frac{fV^*}{1 + f_1 V^*} \tag{10.32}$$

$$f^* = -\frac{f_1 V^*}{1 + f_1 V^*} \tag{10.33}$$

Combining the result in (10.32) with that in (10.18) it is interesting to note that f_1^* can be rewritten as

$$f_1^* = -\frac{1 + V^*}{1 + f_1 V^*} c^*$$

TABLE 10.3

Relative changes in c, f_1 and f (all changes in %).

V^*		= −30	−25	−20	−15	−10	−5	0	5	10	15	20	25	30
$f = 0.75$	c^* =	32.1	25.0	18.8	13.2	8.3	3.9	0	−3.6	−6.8	−9.8	−12.5	−15.0	−17.3
	f_1^* =	−24.3	−20.0	−15.8	−11.7	−7.7	−3.8	0	3.7	7.3	10.8	14.3	17.6	20.9
	f^* =	8.1	6.7	5.3	3.9	2.6	1.3	0	−1.2	−2.4	−3.6	−4.8	−5.9	−7.0
$f = 0.5$	c^* =	21.4	16.7	12.5	8.8	5.6	2.6	0	−2.4	−4.5	−6.5	−8.3	−10.0	−11.5
	f_1^* =	−17.6	−14.3	−11.1	−8.1	−5.3	−2.6	0	2.4	4.8	7.0	9.1	11.1	13.0
	f^* =	17.6	14.3	11.1	8.1	5.3	2.6	0	−2.4	−4.8	−7.0	−9.1	−11.1	−13.0
$f = 0.25$	c^* =	10.7	8.3	6.3	4.4	2.8	1.3	0	−1.2	−2.3	−3.3	−4.2	−5.0	−5.8
	f_1^* =	−9.7	−7.7	−5.9	−4.2	−2.7	−1.3	0	1.2	3.2	3.4	4.3	5.3	6.1
	f^* =	29.0	23.1	17.6	12.7	8.1	3.9	0	−3.6	−7.0	−10.1	−13.0	−15.8	−18.4

which shows the relative change in f_1^* compared with the relative change in unit cost. Some results for $c^*, f_1^* +$ and f^* are shown in Table 10.3 for various initial values of f; the case of $f = 0.25$ is shown graphically in Fig. 10.14.

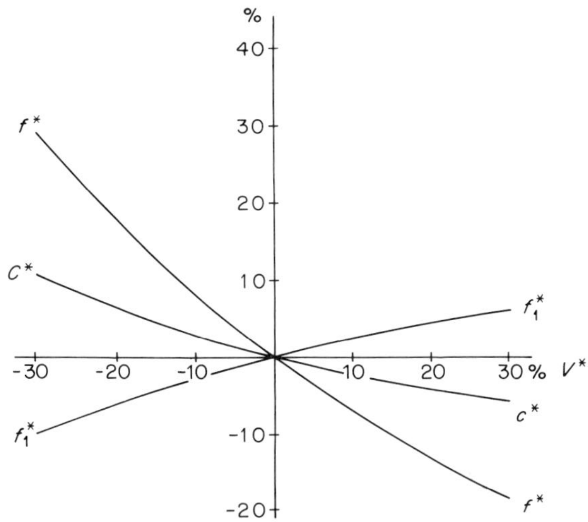

FIG. 10.14. Changes in c, f_1 and f for $f = 0.25$.

The revenue proportions

As we recall from Equations (10.4) and (10.6), the revenue proportions for the fixed and variable cost model are g_1, g and α. The relative change in g_1 is

$$g_1^* = \left(\frac{S}{R}\right)^* = \frac{S^* - R^*}{1 + R^*}$$

For the case $s^* = 0$ substitute $S^* = V^*$ and $R^* = V^* + p^*(1 + V^*)$ in the numerator

$$g_1^* = -\frac{1 + V^*}{1 + R^*}\,p^* \tag{10.34}$$

For g^* the result in Equation (10.26) applies; alternatively,

$$g^* = \left(\frac{F}{R}\right)^* = \frac{F^* - R^*}{1 + R^*}$$

and if $F^* = 0$, then

$$g^* = -\frac{R^*}{1 + R^*} \tag{10.35}$$

The relative change in α^* is given in Equations (10.9) and (10.15). If we substitute

$$Z^* = \frac{1}{\alpha}[R^* - (1-\alpha)C^*]$$

and (for $s^* = F^* = 0$)

$$C^* = f_1 V^*$$

in Equation (10.9), then the result for α^* becomes

$$\alpha^* = \frac{1-\alpha}{\alpha} \quad \frac{R^* - f_1 V^*}{1 + R^*} \tag{10.36}$$

Incorporating the price elasticity of demand

Consider the case where $s^* = F^* = 0$ and where the volume changes only in response to a price change, namely

$$V^* = -\varepsilon p^*$$

The condition for $\alpha^* > 0$ in (10.19) then beomes

$$p^* > \frac{f\varepsilon p^*}{1 - \varepsilon p^*}$$

or

$$\left.\begin{aligned} \alpha^* > 0 \qquad & \text{if } 0 < p^* < \frac{1}{\varepsilon} - f \qquad \text{when } \frac{1}{\varepsilon} > f \\ & \text{or } \frac{1}{\varepsilon} - f < p^* < 0 \qquad \text{when } \frac{1}{\varepsilon} < f \end{aligned}\right\} \tag{10.37}$$

The conditions for $R^* > 0$ and $Z^* > 0$ were derived in the previous chapter in expressions (9.29) and (9.34). Similarly, for the unit profit we obtain

$$\begin{aligned} z^* &= \frac{1}{\alpha}[p^* - (1-\alpha)c^*] \\ &= \frac{p^*}{\alpha}\left[1 - \frac{f\varepsilon(1-\alpha)}{1-\varepsilon p^*}\right] \end{aligned} \tag{10.38}$$

Hence,

$$z^* > 0 \qquad \text{if } p^* > (1-\alpha)\frac{f\varepsilon p^*}{1-\varepsilon p^*}$$

leading to

$$\left.\begin{array}{lll} z^* > 0 & \text{if } 0 < p^* < \dfrac{1}{\varepsilon} - \beta + \alpha & \text{when } \dfrac{1}{\varepsilon} > \beta - \alpha \\ & \text{or } \dfrac{1}{\varepsilon} - \beta + \alpha < p^* < 0 & \text{when } \dfrac{1}{\varepsilon} < \beta - \alpha \end{array}\right\} \quad (10.39)$$

SUMMARY OF RESULTS FOR $s^* = F^* = 0$

The results derived above, and various others, are summarized in Table 10.4, where the upper bound for p^* is given in each case. The assumption throughout is that $s^* = F^* = 0$, that $V^* = -\varepsilon p^*$ and that the net profit margin α is positive to start with. The following comments are noteworthy:

TABLE 10.4
Bounds for p^* for the case $s^* = F^* = 0$ and $\alpha > 0$.

Desired result		Condition for p^*	Optimum p^*
$z^* \geqslant 0$		$0 \leqslant p^* \leqslant \frac{1}{\varepsilon} - \beta + \alpha$	$\frac{1}{\varepsilon}[1 - (g\varepsilon)^{1/2}]$
	or	$\frac{1}{\varepsilon} - \beta + \alpha \leqslant p^* \leqslant 0$	
$\alpha^* \geqslant 0$		$0 \leqslant p^* \leqslant \frac{1}{\varepsilon} - f$	$\frac{1}{\varepsilon}[1 - (f\varepsilon)^{1/2}]$ approx
	or	$\frac{1}{\varepsilon} - f \leqslant p^* \leqslant 0$	
$Z^* \geqslant 0$		$0 \leqslant p^* \leqslant \frac{1}{\varepsilon} - \beta$	$\frac{1}{2\varepsilon}(1 - \beta\varepsilon)$
	or	$\frac{1}{\varepsilon} - \beta \leqslant p^* \leqslant 0$	
$R^* \geqslant 0$		$0 \leqslant p^* \leqslant \frac{1}{\varepsilon} - 1$	$\frac{1}{2\varepsilon}(1 - \varepsilon)$
	or	$\frac{1}{\varepsilon} - 1 \leqslant p^* \leqslant 0$	
$\beta^* \geqslant 0$		$p^* \geqslant 0$	
$c^* \leqslant 0$		$p^* \leqslant 0$	
$C^* \leqslant 0$		$p^* \leqslant 0$	
$f^* \leqslant 0$		$p^* \leqslant 0$	
$g^* \leqslant 0$		$R^* \geqslant 0$	

(1) There is a striking similarity between the functions of z^*, α^*, Z^* and R^*. They are all quadratic functions of p^* and in all cases $z^* = \alpha^* = Z^* = R^* = 0$ at $p^* = 0$. To ensure that z, α, Z and R increase in value, p^* must lie within a specified range. If the lower bound of this range is 0, the upper bound is

$$\frac{1}{\varepsilon} - f(1-\alpha) = \frac{1}{\varepsilon} - g \qquad \text{for } z^* > 0$$

$$\frac{1}{\varepsilon} - f \qquad \text{for } \alpha^* > 0$$

$$\frac{1}{\varepsilon} - \beta \qquad \text{for } Z^* > 0$$

$$\frac{1}{\varepsilon} - 1 \qquad \text{for } R^* > 0$$

(2) From Equation (10.7)

$$\beta = \alpha + f(1-\alpha)$$

so that if $0 < \alpha < 1$, then

$$\beta > f > f(1-\alpha)$$

and it follows that

$$\frac{1}{\varepsilon} - \beta < \frac{1}{\varepsilon} - f < \frac{1}{\varepsilon} - f(1-\alpha)$$

The three values in this statement are the three bounds listed above for p^* to ensure that Z, α and z increase. If the three bounds are positive, then the first (namely $(1/\varepsilon) - \beta$) is the most demanding. Hence, if the condition for $Z^* > 0$ is met, then the one for $\alpha^* > 0$ is met as well, and if $\alpha^* > 0$ occurs then the condition for $z^* > 0$ is also satisfied.

(3) If $\varepsilon < 1$, then all three upper bounds for p^* are positive (since $(1/\varepsilon) > 1$ while $f < \beta < 1$), and so is the upper bound $(1/\varepsilon) - 1$ that ensures that $R^* > 0$. As already shown in Chapter 9, the range of p^* that results in $R^* > 0$ is narrower than that corresponding to $Z^* > 0$, so that if $R^* > 0$ then Z^*, α^* and z^* are all positive. This result is summarized in Fig. 10.15, which shows the domains for p^* for which increases in the various measures take place and indicates how these domains overlap.

(4) The quadratic nature of the functions for z^*, α^*, Z^* and R^* means that there is an optimum value for p^* in each case. The optimum points for Z^* and R^* were derived in Chapter 9. For z^* this point is found, using Equation (10.38), at $(\mathrm{d}z^*/\mathrm{d}p^*) = 0$, or

$$(1-\varepsilon p^*)[1 - 2\varepsilon p^* - f\varepsilon(1-\alpha)] + \varepsilon p^*[1 - \varepsilon p^* - f\varepsilon(1-\alpha)] = 0$$

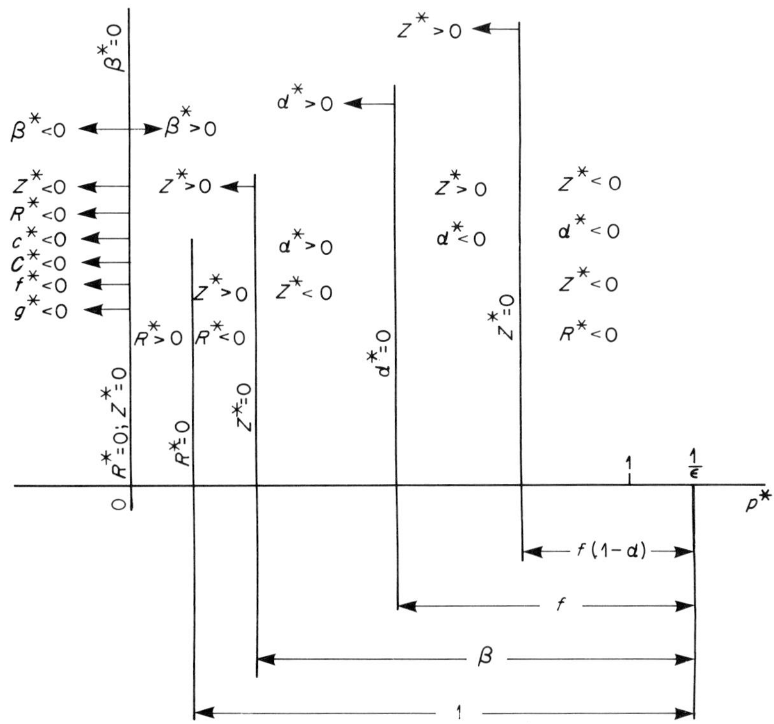

FIG. 10.15. Effect of $p^* > 0$ (for $\varepsilon < 1$ and $\alpha > 0$).

After simple manipulation this becomes

$$(1 - \varepsilon p^*)^2 = f\varepsilon(1 - \alpha)$$

or

$$p^*_{\text{opt}}(z^*) = \frac{1}{\varepsilon}[1 - (f\varepsilon(1 - \alpha))^{1/2}] \tag{10.40}$$

and this result is included in Table 10.4. Similarly, from Equations (10.15) and (10.18) for α^*,

$$\alpha^* = \frac{1 - \alpha}{\alpha} \frac{p^*}{1 + p^*} \frac{1 - f\varepsilon - \varepsilon p^*}{1 - \varepsilon p^*} \tag{10.41}$$

and it can be shown that the optimum value of α^* at $(\mathrm{d}\alpha^*/\mathrm{d}p^*) = 0$ occurs at the solution of the quadratic equation

$$(1 - \varepsilon p^*)^2 - f\varepsilon(1 + \varepsilon p^{*2}) = 0$$

When $p^{*2} \ll 1$ this equation becomes

$$(1-\varepsilon p^*)^2 \simeq f\varepsilon$$

and the approximate solution is

$$p^*_{\text{opt}}(\alpha^*) \simeq \frac{1}{\varepsilon}[1-(f\varepsilon)^{1/2}] \tag{10.42}$$

(5) The results for p^*_{opt} are summarized in Table 10.4. Comparing the solutions in Equations (10.40) and (10.42) it is simple to show that

$$p^*_{\text{opt}}(z^*) > p^*_{\text{opt}}(\alpha^*)$$

It can further be shown that

$$p^*_{\text{opt}}(\alpha^*) > p^*_{\text{opt}}(Z^*)$$

which, in conjunction with the result in (9.37), leads to the general conclusion that

$$p^*_{\text{opt}}(z^*) > p^*_{\text{opt}}(\alpha^*) > p^*_{\text{opt}}(Z^*) > p^*_{\text{opt}}(R^*) \tag{10.43}$$

Example

Consider the case where

$$\varepsilon = 0.9,\ \alpha = 10\%,\ f = 0.35$$

Hence

$$\beta = \alpha + f(1-\alpha) = 0.1 + 0.35 \times 0.9 = 0.415$$

Since $\varepsilon < 1$, the conclusions in Fig. 10.15 apply. The results for the various performance and operating measures are as follows:

Unit profit. From Equation (10.38)

$$\begin{aligned} z^* &= \frac{p^*}{\alpha}\left[1-\frac{f\varepsilon(1-\alpha)}{1-\varepsilon p^*}\right] \\ &= \frac{10p^*}{1-0.9p^*}(1-0.9p^*-0.284) \\ &= \frac{9p^*(0.796-p^*)}{1-0.9p^*} \end{aligned}$$

$$z^* > 0 \text{ when} \qquad 0 < p^* < 0.796\ (79.6\%)$$

optimum value of z^* at

$$p^*_{\text{opt}}(z^*) = \frac{1}{0.9}[1-(0.35\times 0.9\times 0.9)^{1/2}] = 0.520\ (52.0\%)$$

and at that point

$$z^*_{max} = \frac{0.9 \times 0.520(0.796 - 0.520)}{1 - 0.9 \times 0.520} = 0.243 \ (24.3\,\%)$$

Net profit margin. From Equation (10.41)

$$\alpha^* = \frac{p^*(6.165 - 8.10\,p^*)}{(1 + p^*)(1 - 0.9\,p^*)}$$

$$\alpha^* > 0 \text{ when } \quad 0 < p^* < 0.761$$

$$p^*_{opt}(\alpha^*) \simeq \frac{1}{0.9}[1 - (0.315)^{1/2}] = 0.488$$

and

$$\alpha^*_{max} \simeq 1.294$$

The quality of this approximate solution can be judged by comparison with the actual optimum, which can be found to be at

$$p^*_{opt}(\alpha^*) = 0.436$$
$$\alpha^*_{max} = 1.316$$

Gross profit margin.

$$\beta^* = \frac{1-\beta}{\beta}\,\frac{p^*}{1+p^*}$$
$$= \frac{0.585}{0.415}\,\frac{p^*}{1+p^*} = \frac{1.410p^*}{1+p^*}$$

Total profit.

$$Z^* = \frac{p^*}{\alpha}(1 - \varepsilon p^* - \varepsilon\beta)$$
$$= p^*(6.265 - 9p^*)$$

$$Z^* > 0 \text{ when } \quad 0 < p^* < 0.696$$

$$p^*_{opt}(Z^*) = \frac{1}{2 \times 0.9}(1 - 0.415 \times 0.9) = 0.348$$

and

$$Z^*_{max} = 1.090$$

Revenue.

$$R^* = p^*(1 - \varepsilon - \varepsilon p^*)$$
$$= p^*(0.1 - 0.9p^*)$$

$$R^* > 0 \text{ when } \quad 0 < p^* < 0.111$$

$$p^*_{opt}(R^*) = \frac{1}{1.8}(1 - 0.9) = 0.056$$

and

$$R^*_{max} = 0.003$$

TABLE 10.5

Results for the case $\varepsilon = 0.9$, $\alpha = 10\%$, $f = 0.35$ (all in %).

p^*	−20	−10	0	10	20	30	40	50	60	70
V^*	18.0	9.0	0	−9.0	−18.0	−27.0	−36.0	−45.0	−54.0	−63.0
c^*	−5.3	−2.9	0	3.5	7.7	12.9	19.7	28.6	41.1	59.6
C^*	−11.7	−5.9	0	5.9	11.7	17.6	23.4	29.3	35.1	41.0
R^*	−5.6	−1.9	0	0.1	−1.6	−5.1	−10.4	−17.5	−26.4	−37.1
Z^*	−161.3	−71.7	0	53.7	89.4	107.0	106.6	88.3	51.9	−2.5
z^*	−15.2	−7.4	0	6.9	13.1	18.3	22.3	24.2	23.0	16.4
α^*	−164.9	−71.1	0	53.5	92.4	118.1	130.6	128.2	106.4	55.1
β^*	−35.2	−15.6	0	12.8	23.5	32.5	40.3	47.0	52.9	58.0
f^*	−10.5	−5.5	0	6.2	13.3	21.3	30.5	41.3	54.1	69.3
g^*	5.9	1.9	0	−0.1	1.6	5.4	11.6	21.2	27.8	46.0

Cost and volume.

Unit cost

$$c^* = -\frac{fV^*}{1+V^*} = \frac{f\varepsilon p^*}{1-\varepsilon p}$$
$$= \frac{0.315p^*}{1-0.9p^*}$$

Total cost

$$C^* = f_1 V^* = -(1-f)\varepsilon p^*$$
$$= -0.585\, p^*$$

Volume

$$V^* = -\varepsilon p^* = -0.9\, p^*$$

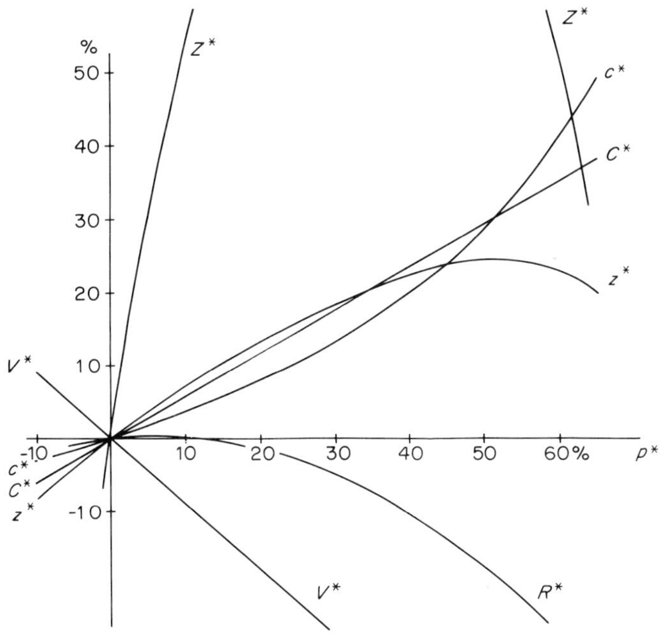

FIG. 10.16. Effect of a price change for the example in Table 10.5.

The proportions f and g

$$f^* = -\frac{f_1 V^*}{1+f_1 V^*} = \frac{f_1 \varepsilon p^*}{1-f_1 \varepsilon p^*}$$

$$= \frac{0.585p^*}{1-0.585p^*}$$

$$g^* = -\frac{R^*}{1+R^*}$$

$$= \frac{-p^*(0.1-0.9p^*)}{1+p^*(0.1-0.9p^*)}$$

Some results for the relative changes are tabulated in Table 10.5 and shown in Fig. 10.16. The actual values of α, β, f and g are shown in Fig. 10.17 and the path of change is drawn on the $\alpha - g$ chart in Fig. 10.18. This example illustrates how β and f increase with p^* while in this case g remains static for a relatively wide range of p^* before it begins to increase rapidly. The net profit margin first rises with p^*, until it reaches a maximum, where its value is more than double its original magnitude, before it begins to fall.

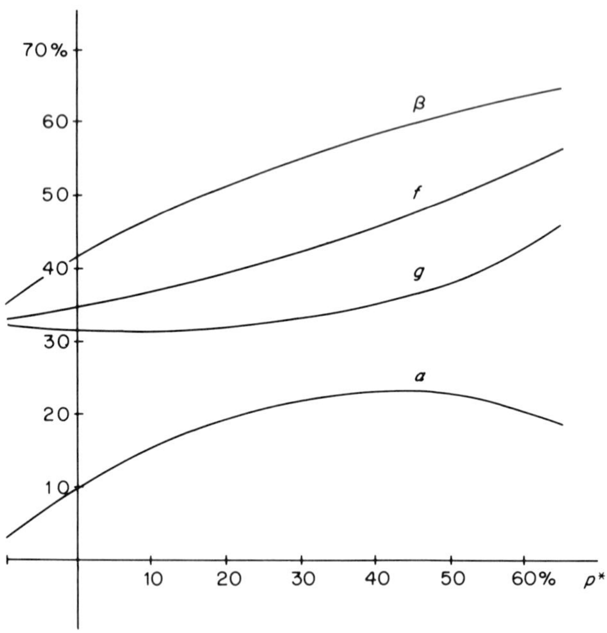

FIG. 10.17. α, β, f and g for the example in Table 10.5.

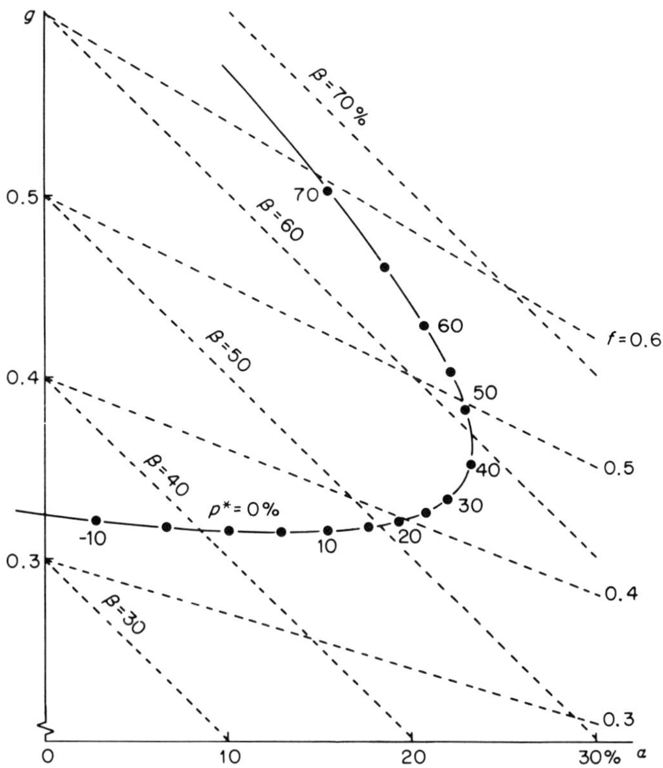

FIG. 10.18. The p^* path on the α-g chart for the example in Table 10.5.

CONCLUSION

The net profit margin α is the ratio of profit to revenue while the gross profit margin β is the ratio of gross operating profit (i.e. revenue minus variable costs) to revenue. The gross margin β is often used by trading companies, both as a measure of profitability and as a means for determining the mark-up for pricing purposes. Also, a useful distinction needs to be made between the fixed cost proportion f and the fixed cost to revenue proportion g: the former is the ratio of fixed cost to total cost, while the latter is the ratio of the fixed cost to revenue. All these definitions are summarized in Table 10.1. The relationships between the four measures is simply

$$\beta = \alpha + g = \alpha + f - f\alpha$$

A convenient way to illustrate these relationships graphically is shown in the

$\alpha - g$ chart in Fig. 10.4 and an example of how they are affected by a change in unit price is presented in Fig. 10.18. Of particular interest are the following movements in the $\alpha - g$ chart:

along an f-line, i.e. $f^* = 0$, leading to $\delta\alpha = \delta\beta/f_1$
along a β-line, i.e. $\beta^* = 0$, leading to $\delta\alpha = -\delta g$
horizontally, i.e. $g^* = 0$, leading to $\delta\alpha = \delta\beta$
vertically, i.e. $\alpha^* = 0$, leading to $\delta g = \delta\beta$

The following five changes in performance, which are discussed in this chapter, are often sought by management:

an increase in the net profit margin ($\alpha^* > 0$)
an increase in the gross profit margin ($\beta^* > 0$)
an increase in unit profit ($z^* > 0$)
an increase in profit ($Z^* > 0$)
an increase in revenue ($R^* > 0$)

Another desirable objective is an increase in the return on investment ($r^* > 0$), which is discussed in the next chapter. It is clear from the analysis that these increases are not always compatible with each other. If the unit price is the control variable used by management to affect performance, then it is possible for a range of values of p^* to exist for which all the first five performance measures listed above improve, as shown in Fig. 10.15, but there is a wide range of p^* for which some measures improve and others deteriorate. There is a belief often held by managers in industry and commerce that α and β are infallible surrogate criteria in that

(1) they are positively correlated with each other (i.e. if one rises or falls, the other must follow in the same direction), and
(2) each of the margins is always positively correlated with profit.

In view of the analysis in this chapter such a belief is erroneous. The same conclusion may be extended to the notion that a rise in the unit profit is invariably beneficial to profit, and as shown in the previous chapter, profit and revenue do not necessarily change in the same direction either. These conclusions are summarized in Table 10.4.

Of particular interest in this analysis is the case where the fixed and variable cost function is used and where $s^* = F^* = 0$. Results for the relative changes in α, β, f, g and z are given below:

$$\alpha^* = \frac{1-\alpha}{\alpha}\,\frac{p^*-c^*}{1+p^*} = \frac{1-\alpha}{\alpha}\,\frac{R^*-f_1V^*}{1+R^*}$$

$$\beta^* = \frac{1-\beta}{\beta}\frac{p^*-s^*}{1+p^*} = \frac{\alpha f}{\beta}\left(\frac{1-f}{f}\alpha^* + \frac{1-\alpha}{\alpha}f^* - \alpha^* f^*\right) = -\frac{1-\beta}{\beta}g_1^*$$

$$f^* = -\frac{f_1 V^*}{1+f_1 V^*}$$

$$g^* = -\frac{R^*}{1+R^*}$$

$$z^* = \frac{1}{\alpha}\left[p^* + \frac{(1-\alpha)f V^*}{1+V^*}\right]$$

EXERCISES

1. Show that

$$\frac{\delta\beta}{\delta\alpha} \simeq \frac{1-\beta}{\alpha}\,\frac{p^*}{1+fV^*}$$

and examine how good this approximation is for the case where $p^* = 10\%$ and $V^* = 10\%$.

2. The profit margin in the base period was $\alpha = 30\%$. In the subsequent period revenue and total cost increased by 10% and 13% respectively. In the next period again increases of 10% and 13% were recorded compared with the previous period. This pattern continued for five periods in succession. Calculate the effect on the profit and the net profit margin for the five periods.
3. Prove that f and g as shown in Fig. 10.3 are correct.
4. (a) Find the effect of a price increase on the gross profit margin, when the value of the latter in the base period is $\beta = 60\%$. Consider the range of p^* between -20% and $+30\%$.
   (b) What is the effect of the initial value of β on β^*?
5. Consider the effect on α and β of changes in the marginal unit cost and the fixed cost when $V^* = 0$ and

   (a) $s^* = F^*$
   (b) $s^* = -F^*$

6. Find the relationship between s^* and F^* for f to remain unchanged when:

   (a) $V^* \neq 0$
   (b) $V^* = 0$

7. (a) Find the effect of a change in the unit price on the two profit margins when the relative change in volume is V^* and when

$$V^* = V_0^* + V_p^*$$

where $V_p^* = -\varepsilon p^*$
and $\varepsilon = 1$, $\alpha = 12\%$, $f = 0.25$, $V_0^* = 4\%$

(b) Find the effect of V_0^* when the unit price is to be increased by 10%.

8. Is it possible for the relative change in the net profit margin to exceed that of profit, and if so when?
9. In Fig. 10.11 there is a segment where $Z^* > 0$ and $\alpha^* < 0$ whereas in Fig. 10.15 there is a domain where $Z^* < 0$ and $\alpha^* > 0$. How do you explain this discrepancy?
10. A sales director has decided to introduce an incentive scheme for his salesmen and it has been suggested that they be given a choice between the following alternatives:

    (1) 1% of turnover
    (2) 10% of profit before tax
    (3) 0.5% of turnover plus 5% of profit

    What advice would you offer the salesmen as to which alternative they should prefer? Explore the sensitivity of the solution.
11. Using incremental analysis, show how the relative change in the ratio of the gross to the net profit margins is affected by changes in total cost and revenue.
12. A trading company has a pricing policy expressed as

$$p = ys$$

where s = purchase price (i.e. the marginal cost) per unit
y = the mark-up, set by the chief executive at 2.0.

Find the effect of a change in y on the net and gross profit margins and on profit, when:

(a) the company can replenish stock at short notice and purchases only the amount needed to meet market demand;

(b) because of long lead-times in supply, the company purchases a stock V, all of which is expected to be sold at the price p, but a price increase of δp will reduce the demand by $|\delta V|$, which will have to be sold at a 70% discount.

The price elasticity of demand in both cases is $\varepsilon = 1$ and currently the fixed cost is 35% of revenue.

13. In the case (b) in the previous exercise, where the residual $|\delta V|$ is offered for sale at a discount (i.e. at a reduced price which is a fraction x of the full price), it is postulated that the amount sold at this discount is $(1-x)|\delta V|$,

leaving $x|\delta V|$ to be disposed of as scrap (at nil value). What level of discount would maximize profit and the net profit margin?

14. (a) Prove that

$$\beta^* = -\frac{1-\beta}{\beta} g_1^*$$

(b) The following data are extracted from a company's financial report (figures in thousands):

Year	Sales	Fixed cost	Total cost	Profit
1	5108	694	4628	480
2	6900	1060	5920	980
3	8200	1380	6910	1290
4	7520	1200	6500	1020
5	8000	1838	6936	1064

Find the percentage change in the profit margins and show the five year trend on an $\alpha-\beta-f-g$ chart.

(c) Of the following variables:

$$R, C, F, S, Z, \alpha, \beta, f, g,$$

select a minimum set from which all the others are easily derived.

15. The breakeven point V_{bep} is defined as the volume at which revenue is just adequate to cover total cost (i.e. $Z = 0$). In general (see Chapter 5) profit may be expressed as

$$Z = \phi V - F$$

where $\phi = p - s$ is the marginal profit per unit. It is argued that for any given level of volume V profit is improved if the breakeven point is decreased.

(a) Determine the conditions for p^* that will result in reducing V_{bep}.
(b) Show that if volume is unchanged then V_{bep} declines if $\beta^* > g^*$.
(c) If $V^* = 0$ and $\alpha, \beta > 0$ when is $V^*_{\text{bep}} < 0$ a necessary but not sufficient condition for profit to increase?

16. If $V^* = 0$, prove the result

$$Z^* = \frac{1}{\alpha}[p^* - (1-\beta)s^* - gF^*]$$

Examine the conditions that lead to an increase in profit and compare them with the conditions for a fall in V_{bep}.

11
Return on Investment and the Effect of Working Capital

Themes in this chapter

The general expression for r^* (the change in return on investment)
Conditions for r to increase
The relationship between profit and return on investment
Working capital and its effect on
total cost and unit cost
profit
return on investment
The effect of a change in volume
The effect of a price change and the elasticity of demand
Conditions for $Z^* \gtrless R^*$ and $r^* \gtrless Z^*$

THE GENERAL EXPRESSION FOR r^*

The return on capital (the terms "capital" and "total investment" are used interchangeably) is defined (see Chapter 2) as

$$r = \frac{Z}{I} \tag{11.1}$$

from which the relative incremental change r^* is given by

$$r^* = \frac{Z^* - I^*}{1 + I^*} \tag{11.2}$$

The following conclusions are immediately evident:

(1) If the total investment I is constant, i.e. if $I^* = 0$, then

$$r^* = Z^*$$

and the results for Z^* derived in Chapter 9 apply to r^* as well.

(2) Since $1+I^* > 0$ (even when the total investment falls) it follows that

$$r^* \gtrless 0 \qquad \text{if } Z^* \gtrless I^* \tag{11.3}$$

Thus, it is possible for profit to increase and yet for return on capital to decline.

(3) If $I^* \ll 1$ then the following approximation can be used:

$$r^* \simeq Z^* - I^* \tag{11.4}$$

(4) The relationship in Equation (11.2) may be presented graphically as shown in Fig. 4.5, in Chapter 4, except that the vertical axis becomes r^*, the horizontal axis is I^* and the curves are iso-Z^* lines.

(5) A simple tabulation of the value of r^* for various combinations of Z^* and I^* is given in Table 11.1 and some of the results are also presented in Fig. 11.1, where the relationship between r^* and Z^* is linear for any given value of I^*. Notice that $I^* = 0$ corresponds to a line drawn at 45° to the horizontal axis, resulting in $r^* = Z^*$.

(6) In the special case where the relative changes in profit and in total investment are proportional to a relative change in revenue, i.e.

$$Z^* = k_1 R^*$$
$$I^* = k_2 R^*$$

where k_1 and k_2 are proportionality coefficients, then

$$r^* = (k_1 - k_2)\frac{R^*}{1+k_2 R^*}$$

so that for $R^* > 0$

$$r^* \gtrless 0 \qquad \text{if } k_1 \gtrless k_2$$

(7) If Z^* is substituted from

$$Z^* = \frac{1}{\alpha}[R^* - (1-\alpha)C^*] \tag{11.5}$$

or from

$$Z^* = \frac{1+V^*}{\alpha}[p^* - (1-\alpha)c^*] + V^* \tag{11.6}$$

we get the general result for r^*, namely

$$\left.\begin{aligned} r^* &= \frac{1}{1+I^*}\left\{\frac{1}{\alpha}[R^* - (1-\alpha)C^*] - I^*\right\} \\ &= \frac{1}{1+I^*}\left\{\frac{1+V^*}{\alpha}[p^* - (1-\alpha)c^*] + V^* - I^*\right\} \end{aligned}\right\} \tag{11.7}$$

TABLE 11.1
Change in return on capital (r^* in %).

I^* (%)	−20	−15	−10	−5	0	5	10	15	20
Z^* (%)									
−40	−25.0	−29.4	−33.3	−36.8	−40.0	−42.9	−45.5	−47.8	−50.0
−35	−18.8	−23.5	−27.8	−31.6	−35.0	−38.1	−40.9	−43.5	−45.8
−30	−12.5	−17.6	−22.2	−26.3	−30.0	−33.3	−36.4	−39.1	−41.7
−25	−6.3	−11.8	−16.7	−21.1	−25.0	−28.6	−31.8	−34.8	−37.5
−20	0	−5.9	−11.1	−15.8	−20.0	−23.8	−27.3	−30.4	−33.3
−15	6.3	0	−5.6	−10.5	−15.0	−19.0	−22.7	−26.1	−29.2
−10	12.5	5.9	0	−5.3	−10.0	−14.3	−18.2	−21.7	−25.0
−5	18.8	11.8	5.6	0	−5.0	−9.5	−13.6	−17.4	−20.8
0	25.0	17.6	11.1	5.3	0	−4.8	−9.1	−13.0	−16.7
5	31.3	23.5	16.7	10.5	5.0	0	−4.6	−8.7	−12.5
10	37.5	29.4	22.2	15.8	10.0	4.8	0	−4.3	−8.3
15	43.8	35.3	27.8	21.1	15.0	9.5	4.6	0	−4.2
20	50.0	41.2	33.3	26.3	20.0	14.3	9.1	4.3	0
25	56.3	47.1	38.9	31.6	25.0	19.0	13.6	8.7	4.2
30	62.5	52.9	44.4	36.8	30.0	23.8	18.2	13.0	8.3
35	68.8	58.8	50.0	42.1	35.0	28.6	22.7	17.4	12.5
40	75.0	64.7	55.6	47.4	40.0	33.3	27.3	21.7	16.7

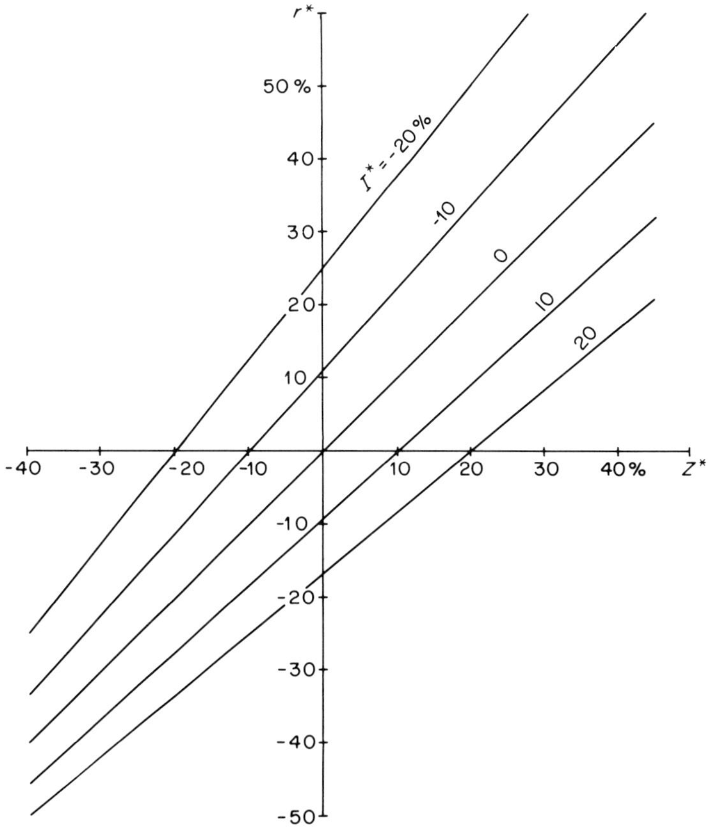

FIG. 11.1. r^* as a function of Z^* for given I^*.

EFFECTS OF VARIOUS CHANGES IN THE SYSTEM

As we have already seen, the effect of any change—whether generated by external causes or whether due to managerial decisions—can be traced through a network of relationships to determine the consequences for the unit cost, the total cost, revenue and profit. We now need to consider the possible effect on the investment I, and consequently the effect on the return on capital r.

Effect of j^* and/or s^* and/or F^*

The effect of these changes on the unit cost is discussed in some detail in Chapter 7, where the relevant network of relationships is given in Fig. 7.4. If we

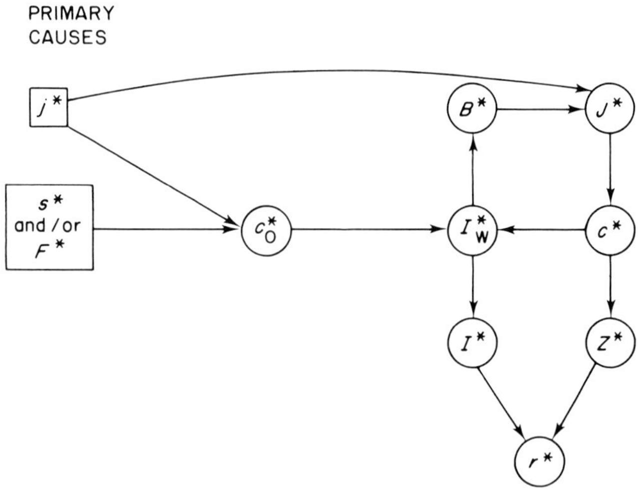

FIG. 11.2. The network of relationships affecting the return on capital when $p^* = V^* = 0$.

consider the case $p^* = V^* = 0$, the revenue is unaffected, so that Fig. 7.4 may be expanded to show the effect on I^* and r^* as shown in Fig. 11.2.

From Chapter 9, the resultant relative profit increment (for $p^* = V^* = 0$) is

$$Z^* = -\frac{1-\alpha}{\alpha}c^* \tag{11.8}$$

where c^* is given in Chapter 7 as

$$c^* = \frac{c_0^*}{1-u_1} \qquad \text{where } u_1 = (1+j^*)f_3 h_1/\lambda \tag{11.9}$$

and

$$c_0^* = f_1 s^* + f_2 F^* + f_3 j^* \tag{11.10}$$

The effect on the working capital is found from Equation (7.15) as

$$I_W^* = h_1 C^* = h_1 c^*$$

But

$$I = I_F + I_W$$

and

$$I^* = iI_F^* + (1-i)I_W^*$$

If we assume here $I_F^* = 0$, then

$$\begin{aligned} I^* &= (1-i)I_W^* \\ &= (1-i)h_1 c^* \end{aligned} \tag{11.11}$$

Substituting these relationships into Equation (11.2), we find

$$r^* = -\frac{c^*}{1+(1-i)h_1c^*}\left[\frac{1-\alpha}{\alpha}+(1-i)h_1\right] \tag{11.12}$$

so that an increase in unit cost ($c^* > 0$) leads to the expected result of a fall in the value of r.

If $I^* \ll 1$, where the approximation in (11.4) may be used, then

$$\frac{r^*}{c^*} \simeq -\frac{1-\alpha}{\alpha} - (1-i)h_1 \tag{11.13}$$

The values of the two terms, which are independent of each other, are tabulated in Table 11.2, from which it is clear that for relatively low α (say, up to $\alpha = 10\%$), the second term makes little impact on the ratio (unless h_1 is exceptionally high and i is very low). Thus, for most practical purposes:

$$r^* \simeq -\frac{1-\alpha}{\alpha}c^* \tag{11.14}$$

Example

Given data: $f_1 = 0.4,\ f_2 = 0.5,\ f_3 = 0.1$
$h_1 = 0.4,\ \lambda = 0.5,\ i = 0.6,\ \alpha = 12\%$
$p^* = V^* = j^* = 0$

If the fixed and variable cost components are likely to rise by 10%, while the interest rate is expected to remain unchanged, find the effect on unit cost, total cost, profit, investment and return on investment.

Assuming $p^* = V^* = 0$ the equations given earlier apply. Hence

$$c_0^* = 0.4 \times 0.1 + 0.5 \times 0.1 = 0.09\ (9.0\%)$$

$$u_1 = 0.1 \times 0.4/0.5 = 0.08$$

$$c^* = \frac{0.09}{1-0.08},\ \text{i.e. } 9.8\%$$

$$C^* = (cV)^* = c^*$$

$$I^* = 0.4 \times 0.4\, c^*,\ \text{i.e. } 1.6\%$$

$$Z^* = -\frac{0.88}{0.12}c^* \text{ or } -71.7\%$$

And the resultant r^*

$$r^* = \frac{Z^* - I^*}{1+I^*} = \frac{-0.717+0.016}{1.016} = -0.690\ (-69.0\%)$$

TABLE 11.2

The terms in Equation (11.13).

α (%)	1	2	5	10	15	20	25	30	35	40
$\frac{1-\alpha}{\alpha}$	99.0	49.0	19.0	9.0	5.67	4.00	3.00	2.33	1.86	1.50

	Value of $(1-i)h_1$									
$i =$	0	0.1	0.2	0.3	0.4	0.5	0.6	0.8	1.0
$h_1 = 0.1$	0.10	0.09	0.08	0.07	0.06	0.05	0.04	0.02	0
0.2	0.20	0.18	0.16	0.14	0.12	0.10	0.08	0.04	0
0.3	0.30	0.27	0.24	0.21	0.18	0.15	0.12	0.06	0
0.4	0.40	0.36	0.32	0.28	0.24	0.20	0.16	0.08	0
0.5	0.50	0.45	0.40	0.35	0.30	0.25	0.20	0.10	0
0.6	0.60	0.54	0.48	0.42	0.36	0.30	0.24	0.12	0
0.8	0.80	0.72	0.64	0.56	0.48	0.40	0.32	0.16	0
1.0	1.00	0.90	0.80	0.70	0.60	0.50	0.40	0.20	0
1.2	1.20	1.08	0.96	0.84	0.72	0.60	0.48	0.24	0

Note that the approximation (11.4) yields

$$r^* \simeq Z^* - I^* = -70.1\,\%$$

whereas the approximation (11.14) leads to

$$r^* \simeq -\frac{1-\alpha}{\alpha}c^* = -71.7\,\%$$

This example serves to illustrate several points made earlier (in this Chapter and elsewhere), namely:

- A change in working capital requirements affects the unit cost (as shown by the difference between c^* and c_0^*).
- When $\alpha < 50\,\%$ the effect of changes on profit is much more significant than the effect on unit cost, and the lower the value of α the more sensitive Z^* becomes.
- When I^* is relatively small, the effect on r^* and on Z^* is approximately the same and this explains why r^* is often found in practice to be highly correlated with Z^*.

We also see how the change in unit cost c^* becomes here the basis for further computations, since Z^* and I^*, as well as r^* for practical purposes, are proportional to c^*:

$$\frac{Z^*}{c^*} = -\frac{1-\alpha}{\alpha}$$

$$\frac{I^*}{c^*} = (1-i)h_1$$

$$\frac{r^*}{c^*} \simeq \text{Equation (11.13) or (11.14)}$$

and all these values are given in Table 11.2.

Added effects of p^* and V^*

The more general case of the effects of various changes on the level of total cost, revenue and profit is described in the network in Fig. 9.35, which we can now expand to incorporate the features shown in Figs 7.5 and 7.6, as well as the effect on total investment and on the return r. The resultant network is given in Fig. 11.3 (for the sake of simplicity the nodes for c_0^* and c^* are not shown and the total cost increments C_0^* and C^* are given instead) to highlight the following sequence of changes:

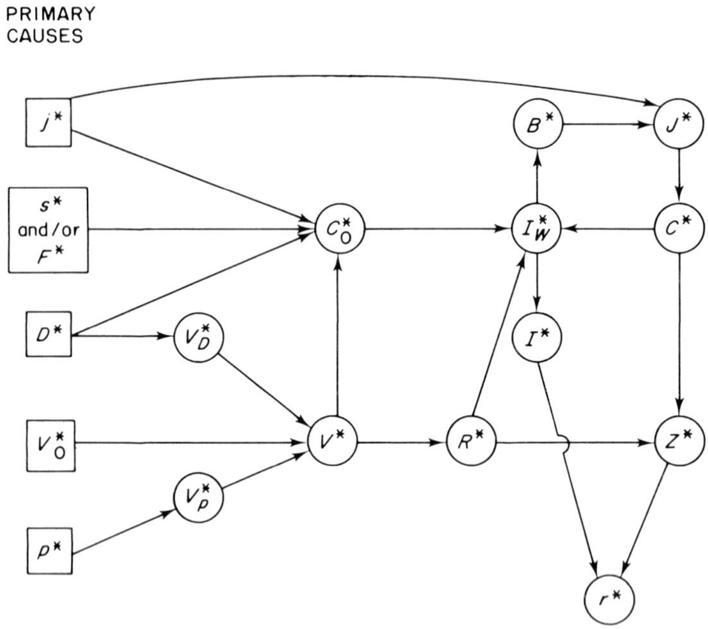

FIG. 11.3. The effect of numerous primary changes.

- The primary causes j^*, s^* and F^* (any one or a combination of them) and a change in the promotion expenditure D^* lead to an initial change in the total cost designated as C_0^*.
- D^* causes a volume change V_D^* and p^* causes V_p^*, to which there is an added volume change in the market V_0^*; the total effect on volume is assumed to be additive, i.e.

$$\begin{aligned} V^* &= V_0^* + V_D^* + V_p^* \\ &= V_0^* + \sigma D^* - \varepsilon p^* \end{aligned}$$

- V^* also affects C_0^*.
- p^* and V^* cause a change in revenue R^*.
- I_W^* is affected by C^* (initially, by C_0^*) and R^* as in Equation (7.16),

$$I_W^* = h_1 C^* + h_2 R^* \tag{11.15}$$

- The changes in the loop B^*, J^*, C^*, I_W^* are as shown in Chapter 7.
- The final result for the effect on I (assuming $I_F^* = 0$) is

$$I^* = (1 - i) I_W^*$$

- C^* and R^* affect Z^* and in conjunction with I^* affect r^*.

THE EFFECT OF WORKING CAPITAL ON COST

The first task is to determine C_0^*, which—following Equations (7.4) and (7.5)—becomes

$$C_0^* = f_1(sV)^* + f_2 F^* + f_3 j^* + f_4 D^* \tag{11.16}$$

Now, the value of C^* is derived as follows (note that the derivation of c^* in Chapter 7 is confined to the case where $p^* = 0$):

$$C^* = f_1(sV)^* + f_2 F^* + f_3 J^* + f_4 D^* \tag{11.17}$$

but

$$J^* = j^* + B^* \ (1 + j^*)$$

hence

$$C^* = C_0^* + f_3 B^* (1 + j^*) \tag{11.18}$$

where

$$B^* = \frac{I_W^*}{\lambda} = \frac{1}{\lambda}(h_1 C^* + h_2 R^*)$$

The second term in Equation (11.18) is

$$f_3 B^*(1 + j^*) = \frac{1 + j^*}{\lambda} f_3 (h_1 C^* + h_2 R^*)$$

Substituting

$$u_1 = \frac{(1 + j^*) f_3 h_1}{\lambda}$$

$$u_2 = \frac{(1 + j^*) f_3 h_1}{\lambda}$$

and we get

$$C^* = C_0^* + u_1 C^* + u_2 R^*$$

or

$$C^* = \frac{1}{1 - u_1} (C_0^* + u_2 R^*) \tag{11.19}$$

The effect on total cost of the coefficients h_1 and h_2 in the general model (11.15) for I_W^* may be summarized as follows (see Fig. 11.4):

- First, the initial value of C_0^* is computed from Equation (11.16), which determines the effect on cost prior to taking account of the effect of a change in working capital.
- The effect of h_2 is merely to add $u_2 R^*$ to C_0^*.
- The effect of h_1 is to divide the outcome by $1 - u_1$.

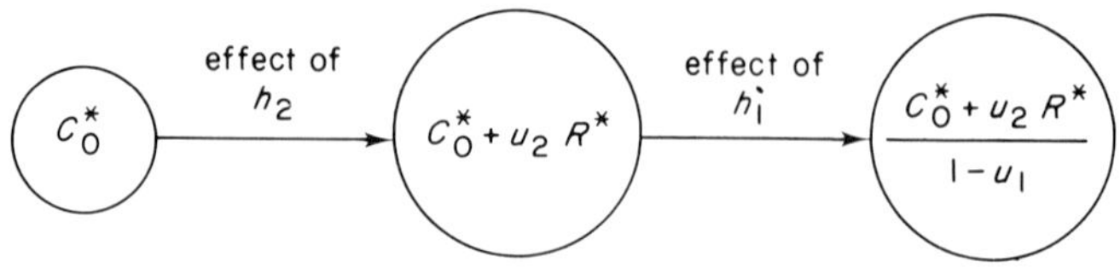

FIG. 11.4. Derivation of C^* from C_0^*.

Note the special cases

$$\text{if } h_1 = 0 \qquad \text{then } C^* = C_0^* + u_2 R^*$$

$$\text{if } h_2 = 0 \qquad \text{then } C^* = \frac{C_0^*}{1-u_1}$$

The latter result also applies when $h_1 \neq 0$, and $h_2 \neq 0$ in the case where the revenue remains unchanged ($R^* = 0$), and it is reminiscent of Equation (7.19), which holds when $p^* = V^* = 0$ (in which case $c_0^* = C_0^*$ and $c^* = C^*$).

For the general case where $p^* \neq 0$ and $V^* \neq 0$ the relative change in the unit cost is

$$c^* = \left(\frac{C}{V}\right)^* = \frac{C^* - V^*}{1 + V^*}$$

Substituting the following

C^* from Equation (11.19)
$R^* = p^*(1+V^*) + V^*$
the definition $v = (u_1 + u_2)\dfrac{V^*}{1+V^*}$

and c_0^* from $c_0^* = \left(\dfrac{C_0}{V}\right)^*$

we get

$$c^* = \frac{1}{1-u_1}(c_0^* + v + u_2 p^*) \tag{11.20}$$

Thus, the results in Chapter 7 may be derived as a special case of Equation (11.20), namely:

(1) If $p^* = V^* = v = 0$ then c^* is reduced to

$$c^* = \frac{c_0^*}{1-u_1}$$

as in Equation (7.19).

(2) If $p^* = 0$ but $V^* \neq 0$ then we have

$$c^* = \frac{c_0^* + v}{1 - u_1}$$

as in Equation (7.23).

(3) If I_W^* is affected only by a change in revenue, as in Equation (7.15), then $h_1 = 0$ and $u_1 = 0$.

(4) If I_W^* follows Equation (7.14) and is affected only by a change in cost, then $h_2 = 0$ and $u_2 = 0$.

(5) For the case where $s^* = F^* = 0$

$$C_0^* = f_1 V^* + f_3 j^* + f_4 D^*$$

$$c_0^* = \frac{1}{1 + V^*} (f_3 j^* + f_4 D^* - f V^*)$$

(6) If $s^* = F^* = 0$ and $f_3 = f_4 = 0$ then

$$C_0^* = f_1 V^*$$

$$c_0^* = -\frac{f V^*}{1 + V^*}$$

and

$$u_1 = u_2 = v = 0$$

so that $C^* = C_0^*$ and $c^* = c_0^*$, i.e. the case is arrived at where there is no change in the working capital and consequently it has no effect on the cost function.

The effect on profit and return on investment

The relative change in profit is derived from Equation (11.5). Substituting C^* from Equation (11.19),

$$Z^* = \frac{1}{\alpha(1 - u_1)} [(1 - u_3)R^* - (1 - \alpha)C_0^*] \qquad (11.21)$$

where $u_3 = u_1 + u_2(1 - \alpha)$. If the original profit margin $\alpha > 0$ and if $u_1, u_2 < 1$ then profit can increase

$$Z^* > 0 \qquad \text{if} \quad R^* > \frac{1 - \alpha}{1 - u_3} C_0^* \qquad (11.22)$$

which is a more detailed condition than the one derived from Chapter 9, based on Equation (11.5), namely

$$Z^* > 0 \qquad \text{if} \quad R^* > (1 - \alpha)C^*$$

A further expression for Z^* may be obtained by substituting c^* into (11.6) which results in

$$Z^* = -\frac{1+V^*}{\alpha(1-u_1)}[(1-u_3)p^* - (1-\alpha)(c_0^* + v)] + V^* \qquad (11.23)$$

so that if $V^* > 0$ then a sufficient condition for profit to increase (assuming $\alpha > 0$ and u_1, $u_3 < 1$) is given by

$$Z^* > 0 \qquad \text{if } p^* \geqslant \frac{1-\alpha}{1-u_3}(c_0^* + v) \qquad (11.24)$$

The same condition ensures that r increases, provided $V^* > I^*$, as indicated by Equation (11.7). Clearly, when volume declines, the unit price needs to be increased by an appropriate amount greater than that stated in (11.24) for profit to increase, and similar observations may be made about the consequences for r if $V^* < I^*$.

An alternative expression to Equation (11.7) for r^* is derived by substituting Z^* from Equation (11.21) into (11.2) to yield

$$r^* = \frac{1}{1+I^*}\left\{\frac{1}{\alpha(1-u_1)}[(1-u_3)R^* - (1-\alpha)C_0^*] - I^*\right\} \qquad (11.25)$$

and a further decomposition can take place if the expression for Z^* in Equation (11.23) is used. Thus, the condition for r to increase is given by

$$r^* > 0 \qquad \text{if} \quad R^* > \frac{1-\alpha}{1-u_3}C_0^* + \alpha\frac{1-u_1}{1-u_3}I^* \qquad (11.26)$$

which is obviously more demanding (when $I^* > 0$) than the condition in (11.22) for profit to increase.

Example

The following data are given:

$f_1 = 0.6, f_2 = 0.3, f_3 = 0.1, \alpha = 20\%$
$s^* = F^* = j^* = 10\%, p^* = 5\%$
$I_F^* = 0, i = 0.6$ and $\lambda = 1$

Investigate the effect of V^*, h_1 and h_2 on profit and the return on investment.

Here

$$u_1 = \frac{(1+j^*)f_3 h_1}{\lambda} = 0.11h_1$$

$$u_2 = 0.11h_2$$
$$u_3 = u_1 + u_2(1-\alpha) = 0.11h_1 + 0.088h_2$$

$$v = (u_1 + u_2)\frac{V^*}{1+V^*} = \frac{0.11(h_1+h_2)V^*}{1+V^*}$$

$$R^* = p^* + V^*(1+p^*) = 0.05 + 1.05\,V^*$$

$$\begin{aligned} C_0^* &= f_1[s^* + V^*(1+s^*)] + f_2F^* + f_3j^* \\ &= 0.6(0.1 + 1.1\,V^*) + 0.3 \times 0.1 + 0.1 \times 0.1 \\ &= 0.10 + 0.66\,V^* \end{aligned}$$

$$\begin{aligned} C^* &= \frac{1}{1-u_1}(C_0^* + u_2R^*) \\ &= \frac{10^{-2}}{1-0.11h_1}(10 + 66\,V^* + 0.55h_2 + 11.55h_2\,V^*) \end{aligned}$$

$$I^* = (1-i)\,I_W^* = 0.4(h_1C^* + h_2R^*)$$

Some results are summarized in Tables 11.3–11.5, from which the following observations may be made for this example:

(1) C_0^* and C^* are linear functions of V^*, as shown in Fig. 11.5, where C^* is plotted for case (5) in Table 11.5 (i.e. for $h_1 = h_2 = 1.0$). The three cases (2), (3) and (4) lie between case (1), where C^* coincides with C_0^*, and case (5).

TABLE 11.3
Results for u_1, u_2, u_3, v.

h_2	u_2	h_1 / u_1	0 / 0	0.2 / 0.022	0.4 / 0.044	0.6 / 0.066	0.8 / 0.088	1.0 / 0.11
0	0	u_3	0	0.022	0.044	0.066	0.088	0.110
		v	0	0.022	0.044	0.066	0.088	0.110
0.2	0.022	u_3	0.018	0.040	0.062	0.084	0.106	0.128
		v	0.022	0.044	0.066	0.088	0.110	0.132
0.4	0.044	u_3	0.035	0.057	0.079	0.101	0.123	0.145
		v	0.044	0.066	0.088	0.110	0.132	0.154
0.6	0.066	u_3	0.053	0.075	0.097	0.119	0.141	0.163
		v	0.066	0.088	0.110	0.132	0.154	0.176
0.8	0.088	u_3	0.070	0.092	0.114	0.136	0.158	0.180
		v	0.088	0.110	0.132	0.154	0.176	0.198
1.0	0.11	u_3	0.088	0.110	0.132	0.154	0.176	0.198
		v	0.110	0.132	0.154	0.176	0.198	0.220

Note: The entries for v need to be multiplied by $\frac{V^*}{1+V^*}$.

TABLE 11.4
Results for C_0^*, c_0^*, R^* (in %).

V^*	−20	−15	−10	−5	0	5	10	15	20	25
C_0^*	−3.2	0.1	3.4	6.7	10.0	13.3	16.6	19.9	23.2	26.5
c_0^*	21.0	17.8	14.9	12.3	10.0	7.9	6.0	4.3	2.7	1.2
R^*	−16.0	−10.8	−5.5	−0.3	5.0	10.3	15.5	20.8	26.0	31.3

The difference between the two lines reflects the effect of changing working capital, and this difference becomes increasingly marked for high values of the coefficients h_1 and h_2 and for high values of V^*.

(2) The incremental unit costs c^* (again, drawn for case (5) in Table 11.5) and c_0^* (which is identical to c^* in case (1)) decline with V^*, and again high values of h_1, h_2 and V^* lead to a significant difference between the initial effect c_0^* and the ultimate value c^*.

(3) The change in revenue R^* rises linearly with V^* (since p^* is constant here throughout) and is unaffected by h_1 and h_2.

(4) Similarly, Z^* increases linearly with V^*, as shown in Fig. 11.6, where the cases (1), (4) and (5) are plotted, the other two cases being close to case (4). This means that in this example Z^* rises linearly with R^*.

(5) The results for cases (2), (3) and (4) are often relatively close together for most of the variables tabulated in Table 11.5, thereby suggesting that in many cases either $I_W^* = h_1 C^*$ or $I_W^* = h_2 R^*$ may be an adequate model to take account of the effect of changes in the working capital, without having to resort to the more elaborate formulation in (11.15).

(6) When r^* is plotted against Z^*, as shown by cases (1), (4) and (5) in Fig. 11.7, it is found that—within the range investigated here—the function of r^* rises with Z^*. In case (1) the working capital (and hence the total investment) remains unchanged, so that $r^* = Z^*$. The results for cases (2) and (3) are close to those of case (4), and clearly the high values of h_1 and h_2 in case (5), combined with high values of V^*, cause r^* to deviate sharply from Z^*; thus, circumstances can arise (as seen from Table 11.5 and Fig. 11.7) where profit rises ($Z^* > 0$) while the return on investment falls ($r^* < 0$), thereby confirming the conclusion drawn earlier from Fig. 11.1 and Equation (11.2). These results point to the necessity of controlling the rise in working capital requirements when the output V expands, first in order to achieve the potential profit level (given by line (1) in Fig. 11.6), and secondly in order to avert a decline in the return on investment.

TABLE 11.5
Results for C^*, c^*, Z^*, I^*, r^* (in %).

	V^*	−20	−15	−10	−5	0	5	10	15	20	25
C^*	(1)	−3.2	0.1	3.4	6.7	10.0	13.3	16.6	19.9	23.2	26.5
	(2)	−5.0	−1.1	2.8	6.7	10.6	14.4	18.3	22.2	26.1	29.9
	(3)	−3.6	0.1	3.8	7.5	11.2	14.9	18.7	22.4	26.1	29.8
	(4)	−4.3	−0.5	3.3	7.1	10.9	14.7	18.5	22.3	26.1	29.9
	(5)	−5.6	−1.2	3.1	7.5	11.9	16.2	20.6	24.9	29.3	33.6
c^*	(1)	21.0	17.8	14.9	12.3	10.0	7.9	6.0	4.3	2.7	1.2
	(2)	18.8	16.4	14.2	12.3	10.6	9.0	7.6	6.2	5.1	4.0
	(3)	20.5	17.8	15.4	13.2	11.2	9.5	7.9	6.4	5.1	3.8
	(4)	19.6	17.0	14.8	12.7	10.9	9.2	7.7	6.3	5.1	3.9
	(5)	18.0	16.2	14.6	13.2	11.9	10.7	9.6	8.6	7.7	6.9
Z^*	(1)	−67.2	−54.2	−41.1	−28.1	−15.0	−2.0	11.1	24.2	37.2	50.3
	(2)	−60.2	−49.4	−38.7	−27.9	−17.2	−6.5	4.3	15.0	25.8	36.5
	(3)	−65.6	−54.2	−42.8	−31.4	−19.9	−8.5	2.9	14.3	25.7	37.1
	(4)	−62.7	−51.7	−40.6	−29.6	−18.5	−7.4	3.6	14.7	25.7	36.8
	(5)	−57.7	−48.9	−40.1	−31.2	−22.4	−13.6	−4.8	4.1	12.9	21.7
I^*	(1)	0	0	0	0	0	0	0	0	0	0
	(2)	−6.4	−4.3	−2.2	−0.1	2.0	4.1	6.2	8.3	10.4	12.5
	(3)	−1.4	0	1.5	3.0	4.5	6.0	7.5	8.9	10.4	11.9
	(4)	−4.1	−2.3	−0.4	1.4	3.2	5.0	6.8	8.6	10.4	12.2
	(5)	−8.6	−4.8	−0.9	2.9	6.7	10.6	14.4	18.3	22.1	26.0
r^*	(1)	−67.2	−54.2	−41.1	−28.1	−15.0	−2.0	11.1	24.2	37.2	50.3
	(2)	−57.4	−47.1	−37.3	−27.9	−18.8	−10.1	−1.8	6.2	13.9	21.3
	(3)	−65.1	−54.2	−43.6	−33.4	−23.4	−13.7	−4.3	4.9	13.9	22.6
	(4)	−61.2	−50.6	−40.3	−30.5	−21.0	−11.8	−3.0	5.6	13.9	21.9
	(5)	−53.7	−46.3	−39.5	−33.2	−27.3	−21.9	−16.8	−12.0	−7.6	−3.4

Note: (1) $h_1 = h_2 = 0$ (4) $h_1 = h_2 = 0.5$
(2) $h_1 = 0, h_2 = 1.0$ (5) $h_1 = h_2 = 1.0$
(3) $h_1 = 1.0, h_2 = 0$

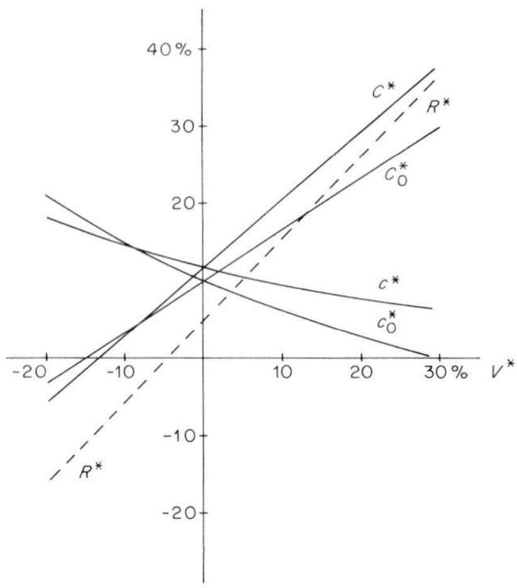

FIG. 11.5. Results for costs and revenue for a given example.

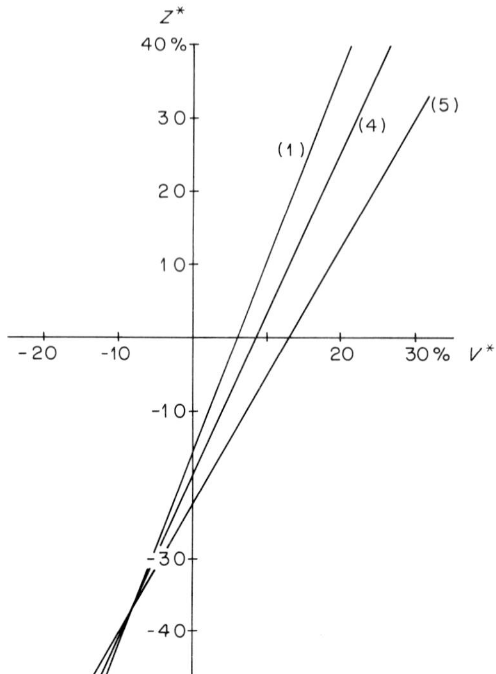

FIG. 11.6. Examples for the effect of V^* on Z^*.

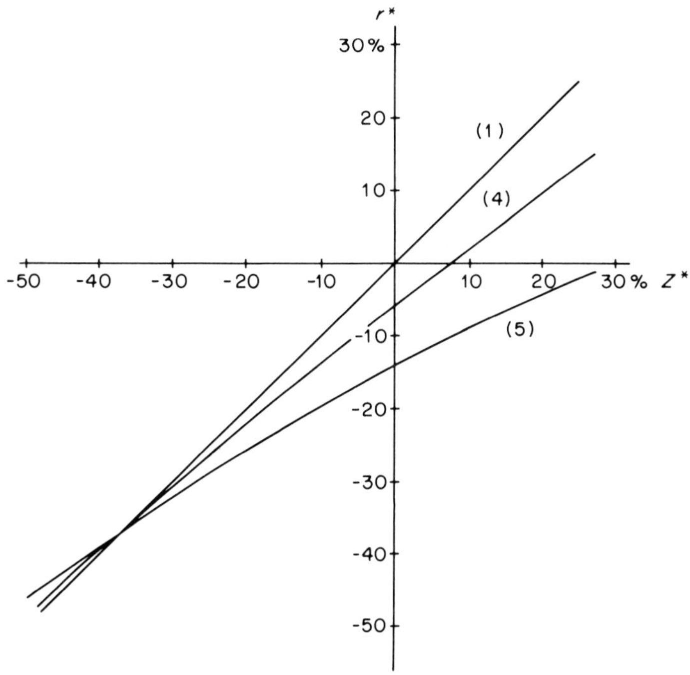

FIG. 11.7. Examples for the relationship between r^* and Z^*.

THE EFFECT OF A PRICE CHANGE

The cost function

If the change in volume is solely attributed to the effect of the price elasticity of demand, i.e.

$$V^* = -\varepsilon p^*$$

then the network of relationships may be depicted as in Fig. 11.8, which is a special case of Fig. 11.3, assuming that the only primary change is that of p^*. If the cost function is of the form

$$C = sV + F + J$$

and if $s^* = F^* = j^* = 0$, then

$$C^* = f_1 V^* + f_3 B^*$$

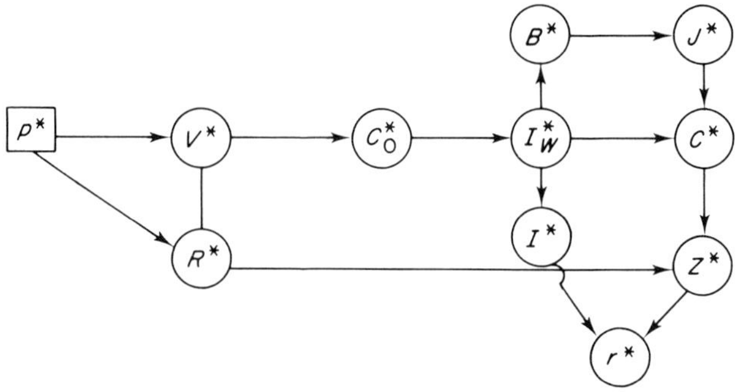

FIG. 11.8. The effect of a price change.

substituting $B^* = I_W^*/\lambda$ and V^* we get

$$C^* = -f_1 \varepsilon p^* + \frac{f_3}{\lambda} I_W^* \tag{11.27}$$

The effect on I_W

If this value of C^* is substituted into Equation (11.15), we get

$$I_W^* = h_1\left(-f_1 \varepsilon p^* + \frac{f_3}{\lambda} I_W^*\right) + h_2 R^*$$

or

$$I_W^* = \frac{1}{1-u_1}(-h_1 f_1 \varepsilon p^* + h_2 R^*) \tag{11.28}$$

Substituting

$$R^* = p^*(1-\varepsilon-\varepsilon p^*) \tag{11.29}$$

as derived in Chapter 9 and u_1 (which here becomes $u_1 = h_1 f_3/\lambda$) and we get (provided $h_2 \neq 0$)

$$I_W^* = \frac{p^* h_2}{1-u_1}\left(1-\varepsilon-\varepsilon p^* - \frac{f_1 h_1}{h_2}\varepsilon\right)$$

This means (assuming $u_1 < 1$) that the working capital remains unchanged if

$$1-\varepsilon-\varepsilon p^* - \frac{f_1 h_1}{h_2}\varepsilon = 0$$

or when $\varepsilon = \varepsilon_0$ where

$$\varepsilon_0 = \frac{1}{1+p^*+f_1 h_1/h_2} \tag{11.30}$$

Thus, the expression for I_W^* may be rewritten as

$$I_W^* = \frac{p^* h_2}{1-u_1}\left(1 - \frac{\varepsilon}{\varepsilon_0}\right) \tag{11.31}$$

from which the consequences for I_W^* may be simply summarized as follows (for $u_1 < 1$):

	$\varepsilon > \varepsilon_0$	$\varepsilon = \varepsilon_0$	$\varepsilon < \varepsilon_0$
$p^* > 0$	$I_W^* < 0$	$= 0$	> 0
$p^* < 0$	> 0	$= 0$	< 0

Results for I_W^* for selected cases are given in Table 11.6 and the limiting values ε_0 in Table 11.7 (these are also presented in a somewhat different form, i.e. for constant values of ε_0, in Fig. 11.9). All these results prompt the following observations:

(1) The choice of model for I_W^* can make a difference to the expected outcome for I_W^*, particularly when a wide range of values for p^* is considered. Similarly, the cost function in Equation (11.27) is affected by the model chosen for I_W^*, as can be seen after substituting the general expression of I_W^* to obtain

$$\left.\begin{aligned} C^* &= \frac{1}{1-u_1}(u_2 R^* - \varepsilon f_1 p^*) \\ &= \frac{p^*}{1-u_1}[u_2(1-\varepsilon-\varepsilon p^*) - \varepsilon f_1] \end{aligned}\right\} \tag{11.32}$$

(2) For any given values of h_1, h_2 and f_1, the value of I_W^* is often sensitive to the elasticity ε.
(3) When $h_1 = 0$ and $\varepsilon = 1$, a change in price (positive or negative) causes I_W to decline as seen from Equation (11.28).
(4) If $p^* > 0$ then $\varepsilon < 1$ is a necessary but not sufficient condition for I_W to rise.
(5) $\varepsilon_0 = 1$ when $p^* = -f_1 h_1/h_2$.
(6) When the change in working capital depends only on a change in revenue, i.e. when $h_1 = 0$, then

$$\varepsilon_0 = \frac{1}{1+p^*}, \quad u_1 = 0$$

and

$$I_W^* = p^* h_2\left(1 - \frac{\varepsilon}{\varepsilon_0}\right) = p^* h_2(1 - \varepsilon - \varepsilon p^*)$$

TABLE 11.6

I_W^* for various circumstances (values in %).

p^*	−25	−20	−15	−10	−5	0	5	10	15	20	25
(1)	0	0	0	0	0	0	0	0	0	0	0
(2) (a)	−10.9	−8.0	−5.4	−3.3	−1.4	0	1.1	1.8	2.1	2.0	1.6
(b)	−6.3	−4.0	−2.3	−1.0	−0.3	0	−0.3	−1.0	−2.3	−4.0	−6.3
(c)	−1.6	0	0.9	1.3	0.9	0	−1.6	−3.8	−6.6	−10.0	−14.1
(3) I	0	0	0	0	0	0	0	0	0	0	0
II (a)	9.4	7.5	5.6	3.8	1.9	0	−1.9	−3.8	−5.6	−7.5	−9.4
(b)	12.5	10.0	7.5	5.0	2.5	0	−2.5	−5.0	−7.5	−10.0	−12.5
(c)	15.6	12.5	9.4	6.3	3.1	0	−3.1	−6.3	−9.4	−12.5	−15.6
III (a)	18.8	15.0	11.3	7.5	3.8	0	−3.8	−7.5	−11.3	−15.0	−18.8
(b)	25.0	20.0	15.0	10.0	5.0	0	−5.0	−10.0	−15.0	−20.0	−25.0
(c)	31.3	25.0	18.8	12.5	6.3	0	−6.3	−12.5	−18.8	−25.0	−31.3
(4) I (a)	−5.5	−4.0	−2.7	−1.6	−0.7	0	0.5	0.9	1.0	1.0	0.8
(b)	−3.1	−2.0	−1.1	−0.5	−0.1	0	−0.1	−0.5	−1.1	−2.0	−3.1
(c)	−0.8	0	0.5	0.6	0.5	0	−0.8	−1.9	−3.3	−5.0	−7.0
II (a)	−0.8	−0.3	0.1	0.3	0.2	0	−0.4	−1.0	−1.8	−2.8	−3.9
(b)	3.1	3.0	2.6	2.0	1.1	0	−1.4	−3.0	−4.9	−7.0	−9.4
(c)	7.0	6.3	5.2	3.8	2.0	0	−2.3	−5.0	−8.0	−11.3	−14.8
III (a)	3.9	3.5	2.9	2.1	1.2	0	−1.3	−2.9	−4.6	−6.5	−8.6
(b)	9.4	8.0	6.4	4.5	2.4	0	−2.6	−5.5	−8.6	−12.0	−15.6
(c)	14.8	12.5	9.8	6.9	3.6	0	−3.9	−8.1	−12.7	−17.5	−22.7

(5)	I (a)	−10.9	−8.0	−5.4	−3.3	−1.4	0	1.1	1.8	2.1	2.0	1.6
	(b)	−6.3	−4.0	−2.3	−1.0	−0.3	0	−0.3	−1.0	−2.3	−4.0	−6.3
	(c)	−1.6	0	0.9	1.3	0.9	0	−1.6	−3.8	−6.6	−10.0	−14.1
	II (a)	−1.6	−0.5	0.2	0.5	0.4	0	−0.8	−2.0	−3.6	−5.5	−7.8
	(b)	6.3	6.0	5.3	4.0	2.3	0	−2.8	−6.0	−9.8	−14.0	−18.8
	(c)	14.1	12.5	10.3	7.5	4.1	0	−4.7	−10.0	−15.9	−22.5	−29.7
	III (a)	7.8	7.0	5.8	4.3	2.3	0	−2.7	−5.8	−9.2	−13.0	−17.2
	(b)	18.8	16.0	12.8	9.0	4.8	0	−5.3	−11.0	−17.3	−24.0	−31.3
	(c)	29.7	25.0	19.7	13.8	7.2	0	−7.8	−16.3	−25.3	−35.0	−45.3

Note: (1) $h_1 = h_2 = 0$
(2) $h_1 = 0, h_2 = 1.0$
(3) $h_1 = 1.0, h_2 = 0$
(4) $h_1 = h_2 = 0.5$
(5) $h_1 = h_2 = 1.0$
I $f_1 = 0$
II $f_1 = 0.5$
III $f_1 = 1.0$
(a) $\varepsilon = 0.75$
(b) $\varepsilon = 1.0$
(c) $\varepsilon = 1.25$

TABLE 11.7
The value of ε_0 corresponding to $l^*_W = 0$.

p^*	−25	−20	−15	−10	−5	0	5	10	15	20	25
$f_1 \frac{h_1}{h_2} = 0$	1.333	1.250	1.176	1.111	1.053	1.0	0.925	0.909	0.870	0.833	0.800
0.2	1.053	1.000	0.952	0.909	0.870	0.833	0.800	0.769	0.741	0.714	0.690
0.4	0.870	0.833	0.800	0.769	0.741	0.714	0.690	0.667	0.645	0.625	0.606
0.6	0.741	0.714	0.690	0.667	0.645	0.625	0.606	0.588	0.571	0.555	0.541
0.8	0.645	0.625	0.606	0.588	0.571	0.555	0.541	0.526	0.513	0.500	0.488

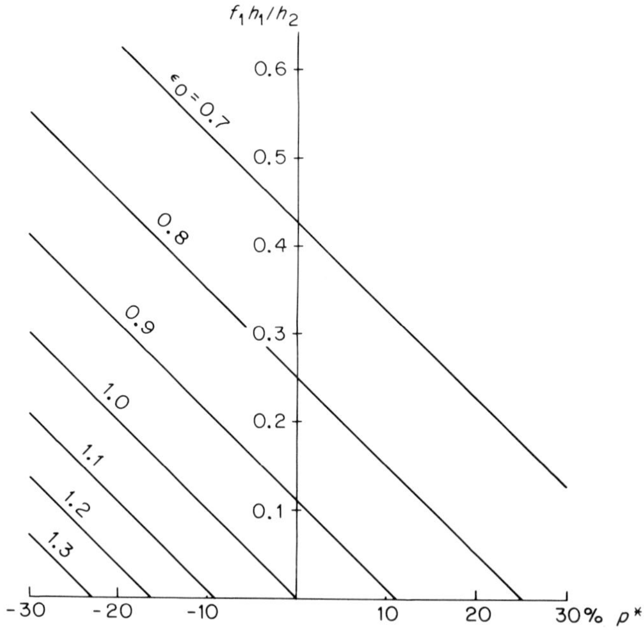

FIG. 11.9. The makeup of ε_0.

(7) If, on the other hand, $h_1 \neq 0$ and $h_2 = 0$ (i.e. the change in working capital is affected by a change in cost), then

$$I_W^* = -\frac{h_1 f_1 \varepsilon}{1 - u_1} p^*$$

The effect on profit

In Chapter 9 the simplified cost function is considered, where the interest J is incorporated into the fixed cost F, so that the assumption there that $F^* = 0$ also implies no change in working capital. Under these assumptions the value of Z^* is found as

$$Z^* = \frac{p^*}{\alpha}(1 - \varepsilon p^* - \varepsilon\beta) \tag{11.33}$$

However, when the cost function explicitly includes J, and when the change in working capital needs to be accounted for, as in Fig. 11.8, we have to start from Equation (11.5), into which C^* and R^* are substituted from (11.27) and (11.29) and I_W^* from (11.31) to yield

$$Z^* = \frac{p^*}{\alpha}\left[1 - \varepsilon p^* - \varepsilon\beta - (1-\alpha)\frac{u_2}{1-u_1}\left(1 - \frac{\varepsilon}{\varepsilon_0}\right)\right] \tag{11.34}$$

The last term in this equation represents the effect of a change in working capital on profit, and when $h_2 = 0$ then $u_2 = 0$, so that Equation (11.34) is then reduced to (11.33). Hence, Equation (11.33) may be used as an approximation when $|p^*|$ and the parameters h_1 and h_2 are not excessively high.

Example

Consider the following case:

the cost proportions $f_1 = 0.3, f_2 = 0.6, f_3 = 0.1$
the net profit margin $\alpha = 20\%$
the price elasticity $\varepsilon = 0.9$
other data $h_1 = 0.2,\ h_2 = 1.0,\ \lambda = 1,\ i = 0.5$
assumptions $s^* = F^* = j^* = 0,\ V^* = -\varepsilon p^*$

Find the effect of p^* on costs, revenue, profit, working capital, and return on investment.

Here:

$$u_1 = \frac{(1+j^*)f_3 h_1}{\lambda} = 0.02$$

$$u_2 = \frac{(1+j^*)f_3 h_2}{\lambda} = 0.10$$

$$\varepsilon_0 = \frac{1}{1 + p^* + f_1 h_1/h_2} = \frac{1}{1.06 + p^*}$$

$$\beta = \alpha + f(1-\alpha) = 0.760$$

$$I_W^* = \frac{p^* h_2}{1-u_1}\left(1 - \frac{\varepsilon}{\varepsilon_0}\right) = \frac{p^*}{0.98}(0.046 - 0.9p^*)$$

$$I^* = (1-i)I_W^* = 0.5 I_W^*$$

$$C^* = -f_1\varepsilon p^* + \frac{f_3}{\lambda} I_W^*$$

$$= -0.2653p^* - 0.0918p^{*2}$$

From Equation (11.34)

$$Z^* = 5p^*\left[1 - 0.9p^* - 0.9 \times 0.76 - \frac{0.8 \times 0.1}{0.98}(1 - 0.9 \times 1.06 - 0.9p^*)\right]$$

$$= 1.561p^* - 4.133p^{*2}$$

and

$$r^* = \frac{Z^* - I^*}{1 + I^*}$$

Some results are summarized in Table 11.8, and shown in Fig. 11.10 from which we note the following:

(1) c^* and C^* move in opposite directions; c^* rises with p^*, and although the function is not linear, we find that as a very crude approximation $c^* = 0.6p^*$ when p^* is negative in the range 0– –20 % and $c^* \simeq 0.7p^*$ when p^* is positive in the range 0– +20 %.

(2) R^* is positive only within the range of p^* of 0–11.1 %, and the increase in revenue then is very modest. Outside that range, revenue is expected to fall. This pattern of results of R^* is in line with the discussion in Chapter 9.

(3) Similarly, Z^* is positive within a given range of p^*, namely between 0 and 37.8 % and the maximum value of Z^* is 14.75 %, obtainable at $p^* = 18.9$ %. Notice that if Equation (11.33) is used as an approximation for Z^* the results would be as follows:

		Result from (11.34)	Approximation (11.33)
$Z^* > 0$	when	$0 < p^* < 37.8$ %	$0 < p^* < 35.1$ %
Z^*_{max}	at	$p^* = 18.9$	17.6
$Z^*_{max} =$		14.7	13.9

Thus, the approximation performs here reasonably well in estimating the point at which Z^*_{max} can be realized, as well as the value of Z^*_{max} itself.

(4) ε_0 attains the value of 0.9 approximately at $p^* = 5$ % (see Table 11.8), at which point, as is clear from Equation (11.31), the working capital remains unchanged ($I^*_W = 0$).

(5) In general, the curve for I^* is rather flat and its value is fairly modest for a wide range of changes in price. In this example the total investment declines irrespective of whether the change in price is positive or negative, except for the range $0 < p^* < 5$ % where I^* has a minute positive value $I^* = 0.03$ % (at $p^* = 2.5$ %).

Table 11.8

The effect of p^* for an example. (all figures in %).

p^* =	−25	−20	−15	−10	−5	0	5	10	15	20	25	30	35
V^*	22.5	18.0	13.5	9.0	4.5	0	−4.5	−9.0	−13.5	−18.0	−22.5	−27.0	−31.5
ε_0	1.235	1.163	1.099	1.042	0.990	0.943	0.901	0.862	0.826	0.794	0.763	0.735	0.709
I^*_W	−6.9	−4.6	−2.8	−1.4	−0.5	0	0	−0.4	−1.4	−2.7	−4.6	−6.9	−9.6
I^*	−3.5	−2.3	−1.4	−0.7	−0.2	0	0	−0.2	−0.7	−1.4	−2.3	−3.4	−4.8
C^*	6.1	4.9	3.8	2.6	1.3	0	−1.3	−2.7	−4.2	−5.7	−7.2	−8.8	−10.4
c^*	−13.4	−11.1	−8.6	−5.9	−3.1	0	3.3	6.9	10.8	15.0	19.7	25.0	30.8
R^*	−8.1	−5.6	−3.5	−1.9	−0.7	0	0.3	0.1	−0.5	−1.6	−3.1	−5.1	−7.5
Z^*	−64.9	−47.8	−32.7	−19.7	−8.8	0	6.8	11.5	14.1	14.7	13.2	9.6	4.0
appr. Z^* by Eq. (11.33)	−67.6	−49.6	−33.8	−20.3	−9.0	0	6.8	11.3	13.6	13.6	11.4	6.9	0.2
r^*	−63.6	−46.5	−31.8	−19.2	−8.6	0	6.8	11.7	14.9	16.3	15.8	13.5	9.3

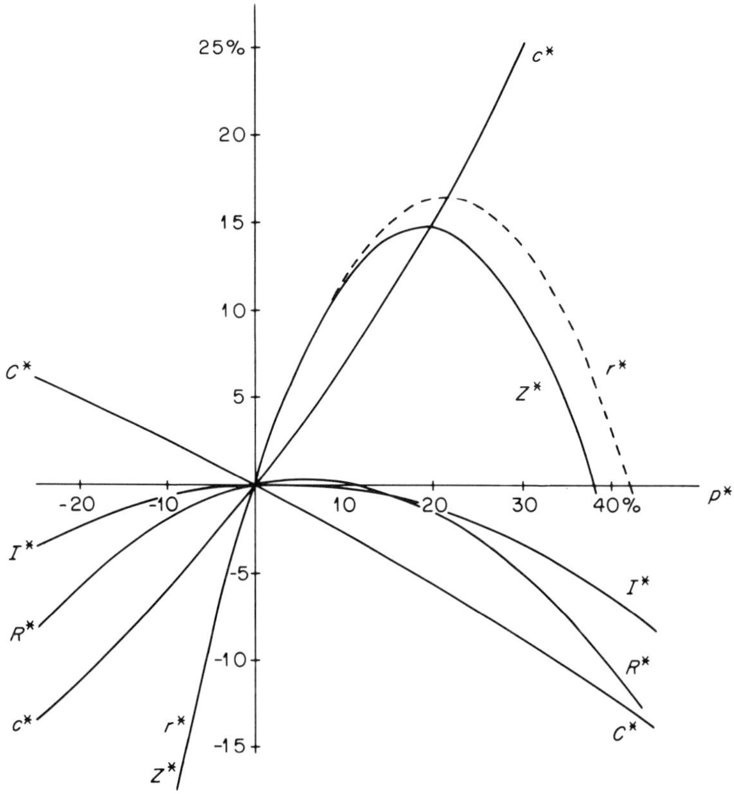

FIG. 11.10. Results for an example.

(6) This behavior of I^* affects the outcome for r^* and, as we can see from the results, r^* and Z^* are almost identical up to about 10% increase in price. Throughout the range the outcome for r^* is better than that for Z^* (since I^* is negative). It is also interesting to note that while Z^* is maximized at $p^* = 18.9\%$, r^* attains a maximum at $p^* = 21.3\%$, and while $Z^* = 0$ at $p^* = 37.8\%$ (twice the value of $p^*(Z^*_{max})$, which is compatible with the results in Chapter 9), $r^* = 0$ at $p^* = 41.9\%$ (which is not twice the value of $p^*(r^*_{max})$).

(7) When Equation (11.33) is used as an approximation for the change in profit (i.e. when the effect of a change in working capital is ignored), the results are reasonably good compared with those from Equation (11.34) within the range $-20 < p^* < 20\%$, but for higher values of $|p^*|$ the effect of I^*_W is no longer negligible. It is also noteworthy that, as in this example,

the working capital is hardly expected to rise throughout the range under consideration, the approximation from (11.33) consistently underestimates the value of Z^*.

COMPARATIVE CHANGES OF R^*, Z^*, r^*

While it is acknowledged that managerial action is often aimed at increasing profit and the rate of return on investment, it is not uncommon to find revenue as a surrogate criterion. Many managers believe, as indicated in Chapter 9, that provided revenue increases, then profit will increase by at least the same percentage, and examples are cited in that chapter where profit can go up even though revenue declines. Similarly, Z^* is often regarded as a surrogate objective for r^*. The question that we need to consider is: what are the circumstances that lead to

$$Z^* \geqslant R^* \geqslant 0$$
$$r^* \geqslant Z^*$$

when the effect of changes in the working capital is accounted for? We shall confine our discussion to the cost function $C = sV + F + J$ and to the case where $\alpha > 0$, $s^* = F^* = j^* = 0$ and $V^* = -\varepsilon p^*$.

Conditions for $Z^* \geqslant R^* \geqslant 0$

In general, since

$$\alpha^* = \left(\frac{Z}{R}\right)^* = \frac{Z^* - R^*}{1 + R^*}$$

it follows that $Z^* \geqslant R^*$ corresponds to $\alpha^* \geqslant 0$.

From the general expression (11.5) for Z^* it is clear that

$$Z^* \geqslant R^* \qquad \text{when } R^* \geqslant C^*$$

Substituting C^* from (11.32) and then R^* from (11.29), we get

$$R^* \geqslant \frac{1}{1 - u_1}(u_2 R^* - \varepsilon f_1 p^*)$$

namely

$$R^* \geqslant -\frac{\varepsilon f_1 p^*}{1 - u_1 - u_2}$$

or

$$p^*\left(1 - \varepsilon - \varepsilon p^* + \frac{\varepsilon f_1}{1 - u_1 - u_2}\right) \geqslant 0$$

Notice that this condition does not depend on α. If the unit price is increased, then the condition becomes

$$0 \leqslant p^* \leqslant \frac{1}{\varepsilon} - 1 + \frac{f_1}{1 - u_1 - u_2} \tag{11.35}$$

whereas we recall from Chapter 9 that

$$R^* \geqslant 0 \qquad \text{if } 0 \leqslant p^* \leqslant \frac{1}{\varepsilon} - 1,\ \varepsilon < 1 \tag{11.36}$$

Hence, the condition in (11.36) automatically satisfies that in (11.35) (assuming that $u_1 + u_2 < 1$, which would often apply). If the unit price is reduced, then the condition becomes

$$0 \geqslant p^* \geqslant \frac{1}{\varepsilon} - 1 + \frac{f_1}{1 - u_1 - u_2} \tag{11.37}$$

Note that here

$$R^* \geqslant 0 \qquad \text{if } 0 \geqslant p^* \geqslant \frac{1}{\varepsilon} - 1,\ \varepsilon > 1 \tag{11.38}$$

so that (11.38) is not adequate to satisfy the condition in (11.37).

Conditions for $r^* \geqslant Z^*$

From Equation (11.2) it is clear that

$$r^* \geqslant Z^* \qquad \text{if } I^*(1 + Z^*) \leqslant 0$$

i.e. if $I^* \leqslant 0$. Assuming that in this case $I_F^* = 0$, then the required condition can be met if

$$I_W^* \leqslant 0$$

For the case where $h_2 = 0$ it is clear from Equation (11.28) that the working capital declines only if the unit price is increased. If $h_2 \neq 0$, then from Equation (11.31) we find that the working capital does not rise provided

$$p^* \geqslant 0 \qquad \text{and } \varepsilon \geqslant \varepsilon_0$$

or

$$p^* \leqslant 0 \qquad \text{and } \varepsilon \leqslant \varepsilon_0$$

Hence the condition becomes:

$$\text{for a price increase} \qquad 0 \leqslant p^* \geqslant \frac{1}{\varepsilon} - 1 - \frac{f_1 h_1}{h_2} \tag{11.39}$$

$$\text{for a price decrease} \qquad 0 \geqslant p^* \leqslant \frac{1}{\varepsilon} - 1 - \frac{f_1 h_1}{h_2} \tag{11.40}$$

For the case where $h_2 = 0$ the condition for working capital to decline becomes

$$I_W^* = -\frac{h_1 f_1 \varepsilon}{1-u_1} p^* < 0$$

or

$$p^* > 0 \qquad \text{(assuming throughout that } u_1 < 1\text{)}$$

The results for $h_2 \neq 0$ may therefore be summarized as follows (see Fig. 11.11):

(1) If $\varepsilon < 1$

$$\begin{aligned} R^* &\geqslant 0 && \text{with condition (11.36)} \\ Z^* &\geqslant R^* && \text{with condition (11.35)} \\ r^* &\geqslant Z^* && \text{with condition (11.39)} \end{aligned}$$

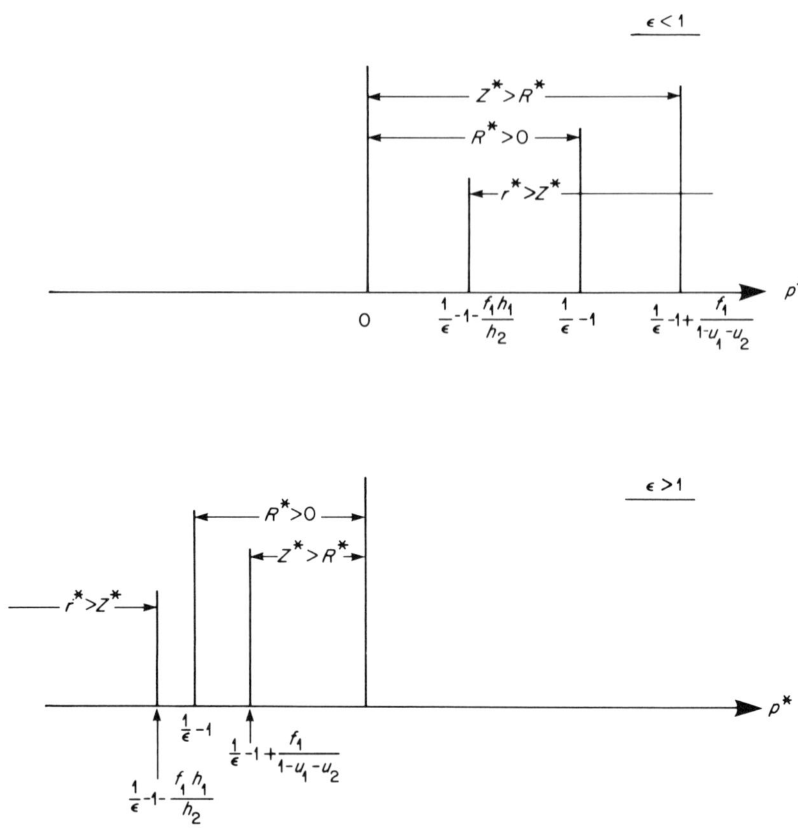

FIG. 11.11. The domains for $Z^* > R^*$ and $r^* > Z^*$ (when $h_2 \neq 0$).

Thus, in the region

$$\frac{1}{\varepsilon} - 1 - \frac{f_1 h_1}{h_2} < p^* < \frac{1}{\varepsilon} - 1 + \frac{f_1}{1 - u_1 - u_2}$$

the result $r^* > Z^* > R^*$ obtains, but $R^* > 0$ applies only to a part of this region. In Fig. 11.11 the lower bound $1/\varepsilon - 1 - f_1 h_1/h_2$ (for $r^* > Z^*$) is shown as a positive value. In order for that to hold, the value of ε must be constrained to

$$\varepsilon < \frac{h_2}{h_2 + f_1 h_1}$$

which means that:

if $h_2 \neq 0$ then $\varepsilon < 1$ is a necessary but not sufficient condition;
if $h_2 \neq 0$ and $h_1 = 0$ then the condition for ε reduces to $\varepsilon < 1$
if $h_2 = 0$, the lower bound cannot be positive, so that the condition for $r^* \geqslant Z^*$ in (11.39) is reduced to $p^* \geqslant 0$.

(2) If $\varepsilon > 1$

$R^* \geqslant 0$ with condition (11.38)
$Z^* \geqslant R^*$ with condition (11.37)
$r^* \geqslant Z^*$ with condition (11.40)

It is clear from these results, shown in Fig. 11.11, that $r^* > Z^* > R^*$ is not a feasible outcome for $\varepsilon > 1$ (for $u_1 + u_2 < 1$).

Summary

The results and their corresponding conditions are summarized in the following table:

	$R^* \geqslant 0$	$Z^* \geqslant R^*$	$r^* \geqslant Z^*$
$p^* > 0$	(11.36)	(11.35)	(11.39)
$p^* < 0$	(11.38)	(11.37)	(11.40)

Example

The following data are given:

$s^* = F^* = j^* = 0$
$h_1 = h_2 = 1$ and $u_1 = u_2 = 0.1$
$\alpha = 0.25$ and $f_1 = 0.4$
$i = 0.5$
$\varepsilon = 0.8$

Hence, from (11.29) and (11.32)

$$R^* = p^* \, (0.2 - 0.8 \, p^*)$$

$$C^* = \frac{1}{0.9} \, (0.1 \, R^* - 0.32 \, p^*)$$

Hence, from (11.5),

$$Z^* = 4 \, (R^* - 0.75 \, C^*)$$

From (11.28),

$$I_W^* = \frac{1}{0.9} \, (-0.32 \, p^* + R^*)$$

and

$$I^* = 0.5 \, I_W^*$$

while

$$r^* = \frac{Z^* - I^*}{1 + I^*}$$

The bounds for the conditions in (11.36), (11.35) and (11.39) are:

$$\frac{1}{\varepsilon} - 1 = 0.250$$

$$\frac{1}{\varepsilon} - 1 + \frac{f_1}{1 - u_1 - u_2} = 0.750$$

$$\frac{1}{\varepsilon} - 1 - \frac{f_1 \, h_1}{h_2} = -0.150$$

The last bound is negative, and as p^* must be positive in condition (11.39), it follows that $p^* \geqslant 0$ dominates. Hence, we expect that:

$$\begin{aligned} R^* &\geqslant 0 && \text{when } 0 \leqslant p^* \leqslant 0.250 \\ Z^* &\geqslant R^* && \text{when } 0 \leqslant p^* \leqslant 0.750 \\ r^* &\geqslant Z^* && \text{when } 0 \leqslant p^* \end{aligned}$$

These conclusions are highlighted by the results in Table 11.9.

CONCLUSION

The return on capital is affected by the levels of profit Z and investment I, which in turn are affected by various changes in the cost components and by changes in the working capital I_W. The general model for the sequential effects that take place is shown in Fig. 11.3. Special cases, such as $p^* = V^* = 0$ or

TABLE 11.9
Results for an example for which $\varepsilon = 0.8$ (figures in %).

p^*	0	5	10	15	20	25	30	40	50	60	70	75	80
I^*_W	0	−0.9	−2.2	−4.0	−6.2	−8.9	−12.0	−19.6	−28.9	−40.0	−52.9	−60.0	−67.6
R^*	0	0.8	1.2	1.2	0.8	0	−1.2	−4.8	−10.0	−16.8	−25.2	−30.0	−35.2
Z^*	0	8.3	15.1	20.4	24.3	26.9	27.6	25.1	16.7	2.4	−17.7	−30.0	−43.7
r^*	0	8.8	16.4	22.9	28.3	32.6	35.7	38.6	36.4	28.0	11.8	0	−15.0

$V^* = -\varepsilon p^*$, are of some interest, as shown in Figs 11.2 and 11.8 respectively, and other cases can be investigated by the same methodology.

When working capital and consequently total investment are little affected by changes in other variables, the result for r^* is similar to that of Z^* and the analysis for Z^* in Chapter 9 applies to r^* as well. But when the change in working capital is not marginal, r^* can depart significantly from the value of Z^* and the effect of I_W^* may not then be ignored.

Furthermore, while the effect of I_W^* on total cost and on unit cost is developed in Chapter 7, the results there are confined to the case when the unit price remains unchanged. That analysis is extended in this chapter to cover the case $p^* \neq 0$.

As for changes in the profit function Z^*, the discussion in Chapter 9 centers on the simplified function

$$C = sV + F$$

so that the effect of a change in working capital (which causes a change in the interest charge) is not accounted for. In this chapter the analysis is extended to the more general case

$$C = sV + F + J + D$$

where the effect of J on C, and hence on Z, is explored.

Throughout this Chapter, the general model

$$I_W^* = h_1 C^* + h_2 R^*$$

is considered, though for most practical purposes the simpler versions of $h_1 = 0$ or $h_2 = 0$ would be adequate, thereby simplifying the mathematical

TABLE 11.10
Summary of results for three cases.

Assumptions	$C = sV + F$ $s^* = F^* = I_W^* = I^* = 0$	$C = sV + F + J$ $s^* = F^* = j^* = 0$ $V^* = -\varepsilon p^*$	$C = sV + F + J + D$
c_0^*	$= c^*$	(7.25), $f_4 = 0$	(7.25)
c^*	(9.21)	(7.24), $f_4 = 0$	(7.24), (11.20)
C_0^*	$= C_0^*$	(11.16), $f_4 = 0$	(11.16)
C^*	$= f_1 V^*$ (9.2)	(11.27) (11.32)	(11.17), (11.19)
R^*	(9.4)	(11.29)	(9.4)
Z^*	(9.9), (9.23), (11.33)	(11.5), (11.6), (11.34)	(11.5), (11.6) (11.21), (11.23)
I_W^*	–	(11.15), (11.28), (11.31)	(11.15)
r^*	$= Z^*$	(11.2), (11.7)	(11.2), (11.7), (11.25)

Note: The numbers refer to the equations in the text.

formulations treated here. In the special case of volume changing solely in response to price ($V^* = -\varepsilon p^*$), the expression in Equation (11.31) is noteworthy for the relative change in working capital, so that conclusions can easily be drawn as to when I_W is expected to rise with adverse effects on the resultant return on capital.

Results for three important cases for the effect on total and unit cost, revenue, profit and working capital are summarized in Table 11.10, in which the relevant equations are listed in each case for ease of reference.

EXERCISES

1. Examine how good the approximation for I_W^* is for the case when $h_1 = h_2 = 0.5$ and $f_1 = 0.5$, when

$$I_W^* \simeq p^*[(1-\varepsilon)h_2 - f_1 h_1 \varepsilon]$$

Assume for the purpose of this exercise that $f_3 = 0$.

2. Prove that when $Z^* = 0$ then

$$u_3 = \frac{u_1 + u_2 R^*/C_0^*}{1 + u_2 R^*/C_0^*}$$

3. (a) Prove that if

$$I^* = C^* = R^*$$

then $r^* = 0$

(b) If

$$I^* = C^* = k_1 R^*$$

find an expression for r^* and state the conditions for achieving

(i) $r^* = R^*$
(ii) $r^* = Z^*$

4. Equation (11.19) reads

$$C^* = \frac{1}{1-u_1}(C_0^* + u_2 R^*)$$

Does it follow that

$$c^* = \frac{1}{1-u_1}(c_0^* + u_2 R^*)?$$

5. An analyst is faced with the choice of model for I_W^* from:

model 1: $I_W^* = h_1 C^*$
model 2: $I_W^* = h_2 R^*$

Consider the following statements:

(a) For model 1 with a given coefficient h_1 it is always possible to specify h_2 for model 2 to yield the same I_W^*.
(b) Hence, the outcome for the two alternative models that yield the same I_W^* is the same as far as C^* and Z^* are concerned.

True or false, and under what conditions?

6. (a) If $h_1 = 0$ prove that if $c_0^* > R^*$ then $c^* > R^*$.
(b) Is the result true for $h_2 = 0$?
7. For the case of no price change it has been suggested that the model

$$I_W^* = h_1 C^* + h_2 R^*$$

may be replaced by the simpler expression

$$I_W^* = h_3 V^*$$

Comment on this suggestion, and examine the case where $c_0^* = 0$.
8. The following data are given:

$f_1 = 0.5, f_2 = 0.42, f_3 = 0.08, \varepsilon = 1, \lambda = 0.5, \alpha = 12.5\%, i = 0.5$
$h_1 = 0, h_2 = 1$

and

$$V^* = -\varepsilon p^*$$

If $s^* = F^* = j^* = 0$, investigate the effect of a price change on the:

unit cost
total cost
revenue
profit
working capital
total investment
return on investment

Find the value of p^* for which Z^* and r^* is maximized respectively.
9. If $h_1 \neq 0$ and $h_2 \neq 0$ but $s^* = F^* = j^* = f_4 = 0$ and if $V^* = -\varepsilon p^*$ determine the circumstances that would result in

(a) $c^* = 0$
(b) $I_W^* = 0$
(c) $Z^* = 0$
(d) $r^* = 0$

10. For the example, which is the basis for the results in Table 11.9, calculate the values of R^*, C^*, Z^*, I_W^* and r^* for p^* between -25% and $+80\%$ in steps of 5% (i.e. $p^* = -25\%, -20\%, -15\%$, etc.).

11. Determine the conditions for $Z^* \geqslant R^* \geqslant 0$ and $r^* \geqslant Z^*$ respectively for the following assumptions:

$c^* = 0$
$\lambda = 1$
$i = 0.5$
$\alpha > 0$
$R^* > 0$
$I_F^* = 0$
$V^* = -\varepsilon p^*$
$u_1, u_2 < 1$

and (a) $h_1 = 0$
or (b) $h_2 = 0$

12. Calculate R^*, Z^* and r^* for the following case:

$s^* = F^* = j^* = 0$
$h_1 = u_1 = 0$
$h_2 = 1.2$
$u_2 = 0.1$
$\alpha = 0.2$
$f_1 = 0.5$
$i = 0.5$
$\varepsilon = 1.25$
p^* is considered in the region -25% to $+25\%$.

State the conditions for:

$R^* \geqslant 0$
$Z^* \geqslant R^*$
$r^* \geqslant Z^*$

13. Repeat the calculations for the case described above, except that now:
$h_1 = 1$, $u_1 = 0.2$
$h_2 = u_2 = 0$
(a) $\varepsilon = 1.25$
(b) $\varepsilon = 0.9$

12
The Two Product/Market Case

Themes in this chapter

Independent products versus interaction between products
Some basic relationships between performance of the total product-mix and individual product performance
Use of the α-g chart to present the effect of price changes
Strategies for price changes for the two product-case
Conditions for improvement in overall profit and profit margin
Six methods for allocating overheads
The case of a single product distributed in two markets
Direct price elasticities of demand and cross elasticities
Numerical examples to illustrate the use of the suggested models

THE PROBLEM

Analysis of performance in the preceding chapters is generally related to the uniform output of a single product. In Chapter 3 attention is drawn to the measurement of output in the multi-product case and it is pointed out that, while a simple statement for the aggregated physical output in any given period cannot be made, there are various conventions for determining the relative change in the "aggregated output volume" and its "overall price".

Since costs, volumes and prices for the individual products in a product-mix are liable to change, the effect of performance of individual products on aggregate performance needs to be considered. Such effects are relevant to a host of managerial actions that may be contemplated, for example: should unit prices be altered and, if so, for which products? Would a possible reduction in cost components or a change in factor prices be significant? Which of the changes in individual product volumes are likely to have a marked effect on overall performance, and how should advertising expenditure be adjusted in such cases?

All these questions, and many others, are considered in the preceding chapters for a single product and an extension to multi-products should not, in principle, be difficult. We need, however, to make a distinction between two possible cases:

(1) The individual products are completely independent of each other, so that each product may be regarded as a self-contained system, changes within which have no effect on the other products. This case is relatively simple to analyze and the assumption of product independence may well be justified where products are not regarded as substitutions for each other, but considered as entirely separate entities, or appeal to different sections of the market. A special example of this case is when a product is sold in different geographical markets, which are independent of each other, so that the volumes in these markets may be designated as if they involved different products (although physically they may be identical and incur the same manufacturing costs, irrespective of which market they are destined for).

(2) Interactions between products do occur, so that a change in price and/or advertising expenditure for one product may have repercussions on volumes of other products, and similarly the cost structures for the various products may be inter-related.

Although the analysis in this chapter mainly relates to the two-product case, the results are similarly extendable to any number of products.

SOME BASIC RELATIONSHIPS

We start with the assumption that the total revenue R may be disaggregated into

$$R = R_1 + R_2$$

where $R_1 = p_1 V_1 =$ revenue for product 1
$R_2 = p_2 V_2 =$ revenue for product 2
and similarly the total cost C, the fixed cost F, and the variable cost S (assuming a linear fixed and variable cost function), as well as the profit Z may be disaggregated:

$$C = C_1 + C_2$$

$$F = F_1 + F_2$$

$$S = S_1 + S_2$$

$$Z = Z_1 + Z_2$$

where

$$Z_1 = R_1 - C_1$$

and

$$Z_2 = R_2 - C_2$$

If α_1, α_2 and α denote the net profit margin for product 1, product 2 and the total operation respectively, then

$$\alpha = \frac{Z}{R} = \frac{Z_1 + Z_2}{R} = \frac{Z_1}{R_1}\frac{R_1}{R} + \frac{Z_2}{R_2}\frac{R_2}{R}$$

Substitute

$$\alpha_1 = \frac{Z_1}{R_1} \text{ and } \alpha_2 = \frac{Z_2}{R_2}$$

to derive

$$\alpha = \rho_1 \alpha_1 + \rho_2 \alpha_2 \tag{12.1}$$

where $\rho_1 = R_1/R$ and $\rho_2 = R_2/R$ are the revenue proportions of the two products (so that $\rho_1 + \rho_2 = 1$, as indicated in Chapter 4). For n products this expression expands to

$$\alpha = \rho_1 \alpha_1 + \rho_2 \alpha_2 + \ldots + \rho_n \alpha_n$$

Similarly, the gross profit margin is obtained by

$$\beta = \frac{R - S}{R} = \frac{R_1 - S_1}{R} + \frac{R_2 - S_2}{R}$$

Substituting the definition

$$\beta_1 = \frac{R_1 - S_1}{R_1}$$

and a similar definition for β_2 to obtain

$$\beta = \rho_1 \beta_1 + \rho_2 \beta_2 \tag{12.2}$$

Consider an example where the revenues of two products are in the ratio of 3 : 1 and where the net profit margins are 8% and 16% respectively. Hence

$$\rho_1 = 0.75; \qquad \rho_2 = 0.25$$
$$\alpha_1 = 8\%; \qquad \alpha_2 = 16\%$$

The total net margin by (12.1) is

$$\alpha = \rho_1 \alpha_1 + \rho_2 \alpha_2 = 6.0 + 4.0 = 10.0\%$$

Alternatively, if a desirable value for α is specified, the various combinations of α_1 and α_2 that would yield that α can be explored, as shown in Fig. 12.1, where these combinations are represented by the line

$$10 = 0.75\,\alpha_1 + 0.25\,\alpha_2$$

One point on this line corresponds to $\alpha_1 = 8$ and $\alpha_2 = 16$, while another is at $\alpha_1 = \alpha_2 = \alpha = 10$. The slope of the line depends on ρ_1 (the higher the value of

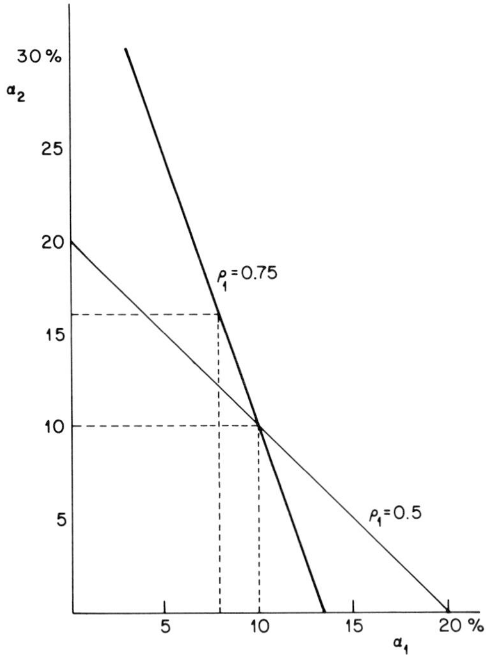

FIG. 12.1. α_1 and α_2 for $\alpha = 10\%$.

ρ_1 the steeper the slope) and it determines the tradeoff between α_1 and α_2 for a given value of α. Thus, if α_1 were to fall by 1 % from 8 % to 7 %, α_2 would have to rise by 3 % to 19 % to compensate for the fall in α_1, for $\alpha = 10\%$ to be maintained. A 1 : 1 tradeoff is obtained at $\rho_1 = 0.5$.

As for the fixed cost proportion f, it is derived as

$$f = \frac{F}{C} = \frac{C_1}{C}\frac{F_1}{C_1} + \frac{C_2}{C}\frac{F_2}{C_2}$$

hence

$$f = \eta_1 f_{(1)} + \eta_2 f_{(2)} \tag{12.3}$$

where η_1 and η_2 are the proportions of the total costs attributable to products 1 and 2 respectively, i.e.

$$\eta_1 = \frac{C_1}{C} \text{ and } \eta_2 = \frac{C_2}{C}$$

so that $\eta_1 + \eta_2 = 1$,

and

$$f_{(1)} = \frac{F_1}{C_1} \text{ and } f_{(2)} = \frac{F_2}{C_2}$$

are the respective fixed cost proportions. Note that the notation $f_{(1)}$ rather than f_1 is adopted here, in order for this ratio not to be confused with f_1 which is used in Chapter 7 to designate s/c.

The fixed cost proportion g is derived from

$$g = \frac{F}{R} = \frac{F_1}{R} + \frac{F_2}{R}$$

hence

$$g = \rho_1 g_{(1)} + \rho_2 g_{(2)} \tag{12.4}$$

where

$$g_{(1)} = \frac{F_1}{R_1}$$

and a similar expression defines $g_{(2)}$. It is not difficult to show that the results in Chapter 10 for the relationships between α, β, f and g apply here as well, namely

$$\beta = \alpha + g \tag{12.5}$$

$$g = f(1 - \alpha) \tag{12.6}$$

for the whole operation, with similar expressions for the individual products.

RELATIVE CHANGES

From the above basic relationships, the change in total cost may be expressed by

$$C^* = \eta_1 C_1^* + \eta_2 C_2^* \tag{12.7}$$

and for the case where the fixed cost and unit variable cost are unchanged (i.e. $F_1^* = F_2^* = s_1^* = s_2^* = 0$) we have

$$C_1^* = (F_1 + s_1 V_1)^* = (1 - f_{(1)}) V_1^*$$

and a similar expression is found for C_2^*, so that

$$C^* = \eta_1 (1 - f_{(1)}) V_1^* + \eta_2 (1 - f_{(2)}) V_2^* \tag{12.8}$$

The relative increment for the net profit margin is derived from Equation (12.1):

$$\alpha^* = \frac{\rho_1 \alpha_1}{\alpha} (\rho_1 \alpha_1)^* + \frac{\rho_2 \alpha_2}{\alpha} (\rho_2 \alpha_2)^* \tag{12.9}$$

where $(\rho_1 \alpha_1)/\alpha$ is easily shown to be Z_1/Z (denoted as ζ_1), namely the proportion of the profit derived from product 1, and a similar expression applies to product 2 (where $\zeta_1 + \zeta_2 = 1$).

Also

$$(\rho_1\alpha_1)^* = \left(\frac{Z_1}{R}\right)^* = \frac{Z_1^* - R^*}{1+R^*}$$

and

$$(\rho_2\alpha_2)^* = \frac{Z_2^* - R^*}{1+R^*}$$

so that

$$\alpha^* = \frac{Z^* - R^*}{1+R^*} = \frac{1-\alpha}{\alpha}\,\frac{R^* - C^*}{1+R^*} \tag{12.10}$$

which is obtainable directly from $\alpha = Z/R$ and from Equation (10.10); we note that

$$Z^* = (Z_1 + Z_2)^* = \zeta_1 Z_1^* + \zeta_2 Z_2^* \tag{12.11}$$

while

$$R^* = \rho_1 R_1^* + \rho_2 R_2^* \tag{12.12}$$

so that the new level of profit is

$$Z + \delta Z = Z(1+Z^*) = Z_1(1+Z_1^*) + Z_2(1+Z_2^*) \tag{12.13}$$

and the new level of revenue

$$R + \delta R = R(1+R^*) = R_1(1+R_1^*) + R_2(1+R_2^*) \tag{12.14}$$

In order to achieve an improvement in profit and net profit margin, the following conditions may be stated (assuming the original margin α is positive):

$Z^* > 0$ if $R^* > (1-\alpha)C^*$ (as found in Chapter 9)

or $Z_1^* > -\dfrac{1-\zeta_1}{\zeta_1} Z_2^*$

$\alpha^* > 0$ if $Z^* > R^*$ or $R^* > C^*$ (as found in Chapter 10)

or $\rho_1\alpha_1(\rho_1\alpha_1)^* + \rho_2\alpha_2(\rho_1\alpha_2)^* > 0$

or, when the distribution of revenue amongst the products is unchanged, if

$$\rho_1\delta\alpha_1 + \rho_2\delta\alpha_2 > 0$$

which may be rewritten as

$$\delta\alpha_1 > -\frac{1-\rho_1}{\rho_1}\delta\alpha_2$$

An example

Assume two independent products with the data for the base period as given in Table 12.1: explicit values for s, c and p for the total product-mix are not given, since such concepts are meaningless when a total physical output V cannot be defined, for reasons explained in Chapter 3. Notice that although the gross margin for product 2 is comparable to that of product 1, the effect of the fixed cost is significant when the net margin is computed, so that the relatively high margin of 20 % for product 1 is diluted by a low margin for product 2 to give an overall figure of $\alpha = 13\%$. Generally, it is clear from Equation (12.1) that if $\alpha_1 > \alpha_2$ then

$$\alpha_1 > \alpha > \alpha_2$$

and a similar observation applies to the gross margin β.

In this example the product proportions are:

$$\begin{array}{lll} \text{for costs:} & \eta_1 = 0.533; & \eta_2 = 0.467 \\ \text{for revenues:} & \rho_1 = 0.580; & \rho_2 = 0.420 \\ \text{for profits:} & \zeta_1 = 0.889; & \zeta_2 = 0.111 \end{array}$$

and the fixed cost proportions are:

$$\begin{array}{lll} f_{(1)} = 0.375; & f_{(2)} = 0.429; & f = 0.400 \\ g_{(1)} = 0.300; & g_{(2)} = 0.414; & g = 0.348 \end{array}$$

where f and g may be computed either direct from Table 12.1 or from Equations (12.3) and (12.4). From the foregoing analysis the following conditions are derived:

$$\begin{array}{ll} Z^* > 0 & \text{if} \quad Z_1^* > -0.13\, Z_2^* \\ \alpha^* > 0 & \text{if} \quad \delta\alpha_1 > -0.72\, \delta\alpha_2 \end{array}$$

The basic equations

To explore the effect of price changes we need data on the price elasticity of demand. If in this case $\varepsilon_1 = \varepsilon_2 = 1.20$, and assuming $s_1^* = s_2^* = F_1^* = F_2^* = 0$ and that changes in working capital may be ignored, then the expressions for cost, revenue, profit and profit margin are derived as follows:

$$R_1^* = p_1^*(1 - \varepsilon_1 - \varepsilon_1 p_1^*) = -p_1^*(0.20 + 1.20\, p_1^*)$$

$$C_1^* = (1 - f_{(1)})V_1^* = -0.750\, p_1^*$$

$$Z_1^* = \frac{1}{\alpha_1}[R_1^* - (1 - \alpha_1)C_1^*] = 5(R_1^* - 0.8\, C_1^*)$$

TABLE 12.1
An example for the two-product case.

	s	$F \times 10^3$	V	$S \times 10^3$	c	$C \times 10^3$	p	$R \times 10^3$	$Z \times 10^3$	α %	β %
Product 1	10	12	2000	20	16	32	20	40	8	20.0	50.0
Product 2	16	12	1000	16	28	28	29	29	1	3.4	44.8
Total	–	24	–	36	–	60	–	69	9	13.0	47.8

or

$$\begin{aligned} Z_1^* &= \frac{p_1^*}{\alpha_1}(1-\varepsilon_1\beta_1-\varepsilon_1 p_1^*) = 2p_1^*(1-3p_1^*) \\ R_2^* &= -p_2^*(0.20+1.20p_2^*) \\ C_2^* &= -0.686\,p_2^* \\ Z_2^* &= 29(R_2^*-0.966\,C_2^*) = p_2^*(13.4-34.8\,p_2^*) \end{aligned}$$

Results

Results for total cost, unit cost, revenue, profit and profit margins are shown in Table 12.2 for $p_1^* = p_2^* = p^*$, from which we conclude that:

- A price increase causes the total cost for each product to decline and the unit cost to increase
- Revenue declines with increasing unit price
- Profit for product 1 is maximized when its unit price increases by about 17 %, but the net profit margin reaches a maximum at a price increase of about 25 %
- For product 2 both the profit and the net profit margin are maximized at around $p^* = 20\%$
- The gross margins increase throughout the range, and so do the fixed cost to revenue proportions $g_{(1)}$, $g_{(2)}$ and g
- If the same price increase is contemplated for both products, maximum total profit is expected to rise by 29 % for a price increase of between 17.1 and 18.4 %, while the total net profit margin reaches 18.5 % when the price increases by 22 to 26 %.

INDIVIDUAL AND AGGREGATE PATHS

The paths for each product and for the product-mix in Table 12.1 are shown in the $\alpha - g$ chart in Fig. 12.2, where the points corresponding to price increases by steps of 10 % are marked. Notice how the paths curve upwards and eventually turn back, thereby demonstrating the fact that, while the gross profit margins continue to increase as prices are lifted, the net margins reach maximum levels and then drift downwards. Also, the aggregate path for the total product-mix lies between the two individual paths.

Furthermore, as shown in Fig. 12.3, any point on the aggregate path lies on a straight line that connects the corresponding points on the individual paths. For example, the base case (for which $p_1^* = p_2^* = 0$) is given by point A, for product 1 and B_1 for product 2, resulting in point C_1 for the product-mix, where C_1 lies on the straight line A_1B_1, and it is easy to verify that the position

TABLE 12.2

Effect of a price increase for an example.

$p^* =$	2	4	6	8	10	12	14	16	18	20	22	24	26	28	30
C_1^*	−1.5	−3.0	−4.5	−6.0	−7.5	−9.0	−10.5	−12.0	−13.5	−15.0	−16.5	−18.0	−19.5	−21.0	−22.5
c_1^*	0.9	1.9	2.9	4.0	5.1	6.3	7.6	8.9	10.3	11.8	13.5	15.2	17.0	19.0	21.1
R_1^*	−0.4	−1.0	−1.6	−2.4	−3.2	−4.1	−5.2	−6.3	−7.5	−8.8	−10.2	−11.7	−13.3	−15.0	−16.8
Z_1^*	3.8	7.0	9.8	12.2	14.0	15.4	16.2	16.6	16.6	16.0	15.0	13.4	11.4	9.0	6.0
α_1	20.8	21.6	22.3	23.0	23.6	24.1	24.5	24.9	25.2	25.4	25.6	25.7	25.7	25.6	25.5
β_1	51.0	51.9	52.8	53.7	54.5	55.4	56.1	56.9	57.6	58.3	59.0	59.7	60.3	60.9	61.5
$g_{(1)}$	30.1	30.3	30.5	30.7	31.0	31.3	31.6	32.0	32.4	32.9	33.4	34.0	34.6	35.3	36.1
C_2^*	−1.4	−2.7	−4.1	−5.5	−6.9	−8.2	−9.6	−11.0	−12.3	−13.7	−15.1	−16.5	−17.8	−19.2	−20.6
c_2^*	1.1	2.2	3.3	4.6	5.8	7.2	8.7	10.2	11.8	13.5	15.4	17.3	19.4	21.7	24.1
R_2^*	as R_1^*														
Z_2^*	25.4	48.0	67.9	84.9	99.2	110.7	119.4	125.3	128.4	128.8	126.4	121.2	113.2	102.4	88.8
α_2	4.3	5.2	5.9	6.5	7.1	7.6	8.0	8.3	8.5	8.7	8.7	8.6	8.5	8.2	7.8
β_2	45.9	46.9	48.0	48.9	49.8	50.7	51.6	52.4	53.2	54.0	54.8	55.5	56.2	56.9	57.6
$g_{(2)}$	41.6	41.8	42.1	42.4	42.7	43.2	43.6	44.1	44.7	45.4	46.1	46.9	47.7	48.7	49.7
C^*	−0.9	−1.7	−2.6	−3.5	−4.3	−5.2	−6.1	−6.9	−7.8	−8.7	−9.5	−10.4	−11.3	−12.1	−13.0
R^*	as R_1^*														
Z^*	6.2	11.6	16.3	20.2	23.5	26.0	27.7	28.7	29.0	28.5	27.3	25.4	22.7	19.3	15.2
α	13.9	14.7	15.4	16.1	16.6	17.1	17.6	17.9	18.2	18.4	18.5	18.5	18.5	18.3	18.1
β	48.8	49.8	50.8	51.7	52.6	53.4	54.2	55.0	55.8	56.5	57.2	57.9	58.6	59.2	59.9
g	34.9	35.1	35.4	35.6	35.9	36.3	36.7	37.1	37.6	38.1	38.7	39.4	40.1	40.9	41.8

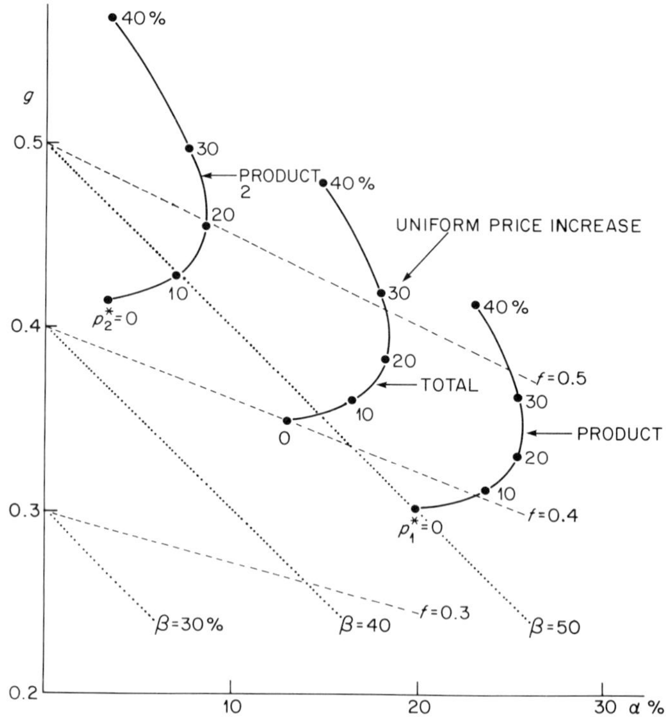

FIG. 12.2. The α-g chart for two products.

of C_1 is determined by:

$$\frac{A_1 C_1}{B_1 C_1} = \frac{\rho_2}{\rho_1} = \frac{0.42}{0.58}$$

which is the direct outcome of Equations (12.1) and (12.4).

Similarly, a price increase of 10 % for both products results in points A_2 and B_2 for product 1 and product 2 respectively and point C_2 for the total, where C_2 lies on the straight line $A_2 B_2$, and a combination of $p_1^* = p_2^* = 40\%$ corresponds to the points A_3 and B_3 while the resultant overall performance at C_3 lies on the straight line $A_3 B_3$, and so on. Thus, any point on the aggregate path is simply the outcome of a combination of two points on the individual paths.

Since the two products are assumed to be independent of each other, their individual paths on the $\alpha - g$ chart are also independent. The aggregate path, however, depends on the combination of p_1^*, and p_2^*. In Figs 12.2 and 12.3 the equality $p_1^* = p_2^*$ is maintained throughout, while in Fig. 12.4 the unit price for

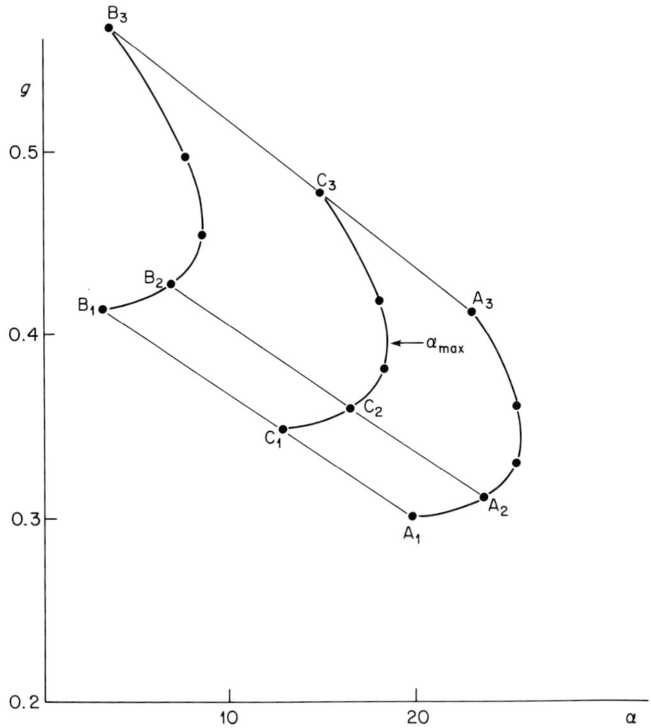

FIG. 12.3. Relationship between the aggregate and the individual α-g paths.

product 2 remains unchanged ($p_2^* = 0$) leaving p_1^* to vary from 0 to 40%, and in Fig. 12.5 the roles are reversed with $p_1^* = 0$ while p_2^* is allowed to vary. Again in every case each resultant point on the aggregate curve lies on the straight line that joins the corresponding points for p_1^* and p_2^*, but the aggregate paths in Figs 12.4 and 12.5 are different from that found in Fig. 12.3. This is further highlighted in Fig. 12.6, where the results in the previous three figures are superimposed on each other. It is interesting to note that the curve for a uniform price increase ($p_1^* = p_2^*$) dominates the other two curves here, in that for any given value of g there is a solution on the uniform-price-increase path that has a better α value than the corresponding points on the other two paths. From Fig. 12.7 it is clear that $p_1^* = p_2^*$ dominates the other two with respect to profit as well.

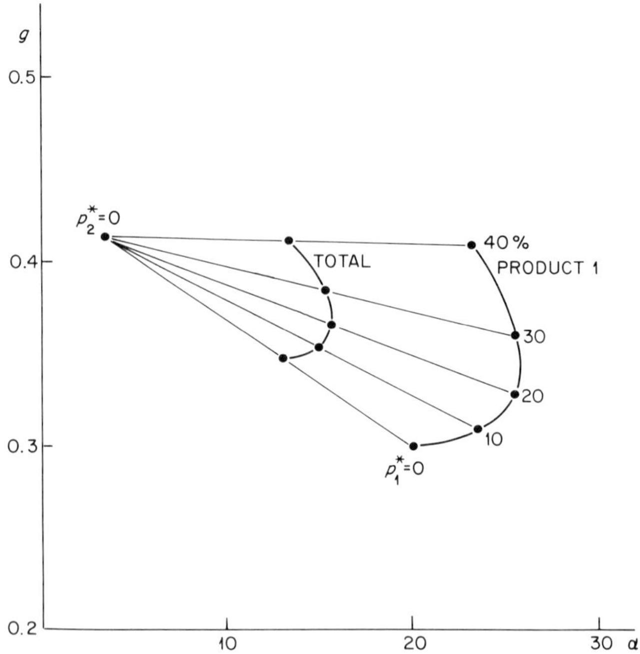

FIG. 12.4. The case of $p_2^* = 0$.

SOLUTIONS LEADING TO THE SAME α AND g

The analysis of the two-product example and the ensuing results suggest that there are many combinations of p_1^* and p_2^* that lead to the same aggregate net margin α. This conclusion is clearly demonstrated in Fig. 12.8, where four cases are presented with a constant p_1^* for each:

$$p_1^* = 0 \text{ while } p_2^* \text{ is varied from 0 to } 40\%$$
$$p_2^* = 10\% \text{ while } p_1^* \text{ is varied from 0 to } 40\%$$

and similarly for $p_2^* = 20$ and 30% respectively. The resultant overall paths are shown for the four cases. If an overall value of, say, 17.4% is sought for α we see that the first curve for $p_1^* = 0$ cannot provide any feasible combination with p_2^* to yield such a result, but the vertical line for $\alpha = 17.4$ cuts each of the other three curves twice, so that six combinations of p_1^* and p_2^* are feasible for the cases shown in Fig. 12.8, let alone the infinite number of combinations possible for other values of p_1^* not shown here.

But it is interesting to note that even a point with the same α and g coordinates can be obtained as a result of alternative pricing strategies. For

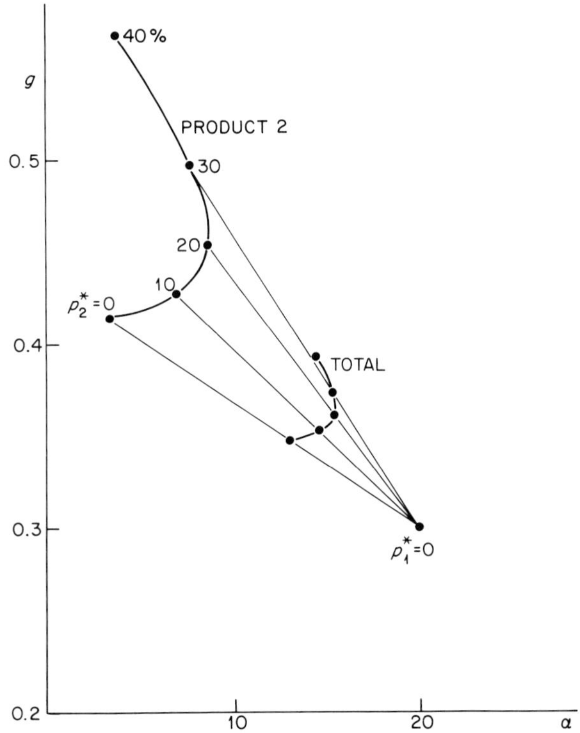

FIG. 12.5. The case of $p_1^* = 0$.

example, the three paths for $p^* = 10$, 20 and 30% intersect at three points, which can be found by stating the equations for the three paths and then solving for each pair at a time. The general $\alpha - g$ equation is simply derived from the basic definitions

$$\alpha = \frac{Z}{R} \text{ and } g = \frac{F}{R}$$

or

$$\alpha = \frac{gZ}{F} \tag{12.15}$$

while expressions for the new levels of Z and R are given in Equations (12.13) and (12.14), into which we substitute in our example $Z_1 = 8000$ and $Z_2 = 1000$ (from Table 12.1) and the expressions given earlier for Z_1^* and Z_2^* as functions of p_1^* and p_2^*. For the three cases under consideration we then have:

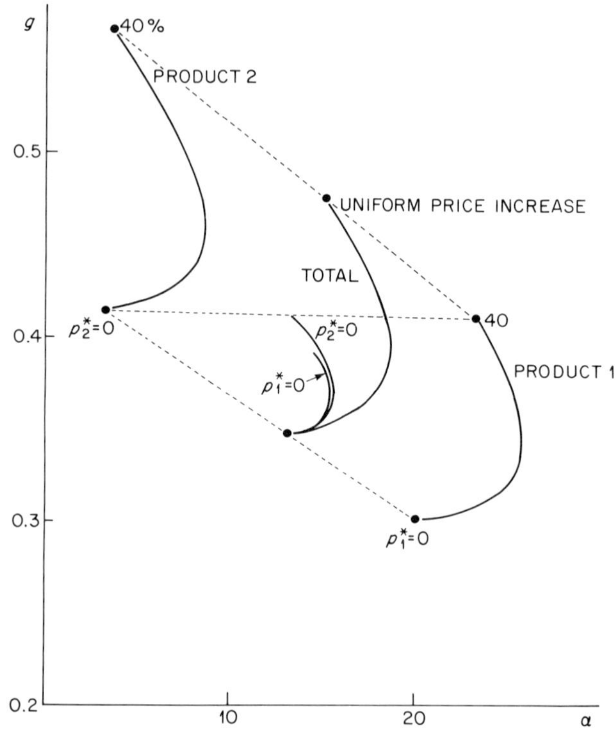

FIG. 12.6. Three price increase policies.

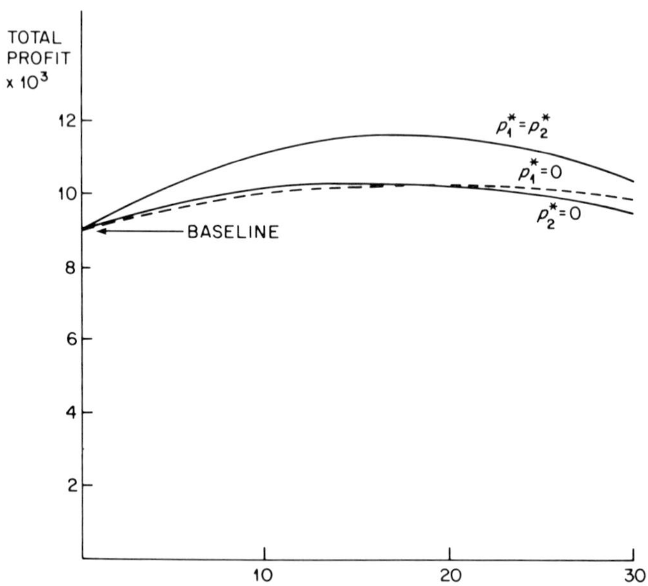

FIG. 12.7. The profit function for three pricing policies.

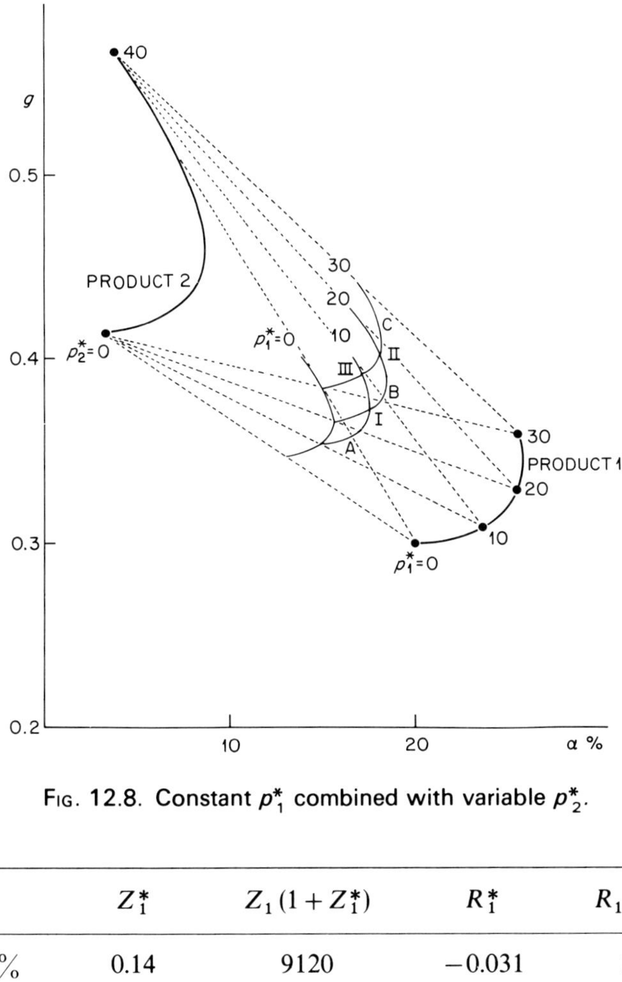

FIG. 12.8. Constant p_1^* combined with variable p_2^*.

	Z_1^*	$Z_1(1+Z_1^*)$	R_1^*	$R_1(1+R_1^*)$
$p^* = 10\%$	0.14	9120	−0.031	38 760
20%	0.16	9280	−0.088	36 480
30%	0.06	8480	−0.168	33 280

Substituting these values and F into Equation (12.13) and then into (12.15), we get the equations for the three curves in Fig. 12.8.

First, for curve A corresponding to $p_1^* = 10\%$

$$\begin{aligned} Z &= Z_1(1+Z_1^*) + Z_2(1+Z_2^*) \\ &= 9120 + 1000\,(1+Z_2^*) \end{aligned}$$

where, as given above, $Z_2^* = p_2^*\,(13.4 - 34.8\,p_2^*)$

Hence, Equation (12.15) yields

$$24\,\alpha = (10.12 + Z_2^*)g$$

Similarly, for curve B corresponding to $p_1^* = 20\%$,

$$24\,\alpha = (10.28 + Z_2^*)g$$

To find the intersection point I it would seem that we need to solve these two equations, but Z_2^* does not have the same value in both, since p_2^* at point I for the first curve is different from p_2^* at point I for the second.

It is clear, however, that as $g = F/R$ has the same value at point I for both curves, and as F is the same for both, then R must be the same. And as $\alpha = Z/R$ at point I is the same for both curves, it follows that Z too is the same. Thus, for curve A the expression for Z follows from the above, namely,

$$Z/100 = 101.2 + p_{2A}^*(134 - 348\,p_{2A}^*) \tag{12.16}$$

The expression for R is based on Equation (12.14), into which $R_1\,(1 + R_1^*)$ from the last table and R_2 are substituted to yield

$$R/100 = 677.2 - 2p_{2A}^*(1 + 6\,p_{2A}^*) \tag{12.17}$$

Similarly, for curve B, the two equations are

$$Z/100 = 102.8 + p_{2B}^*(134 - 348\,p_{2B}^*) \tag{12.18}$$

$$R/100 = 654.8 - 2\,p_{2B}^*(1 + 6\,p_{2B}^*) \tag{12.19}$$

Notice that p_2^* in Equations (12.16) and (12.17) is denoted as p_{2A}^* to indicate that it relates to curve A, while in the last two equations for curve B the notation should read p_{2B}^*, since $p_{2A}^* \neq p_{2B}^*$. From Equations (12.16) and (12.18) we get

$$p_{2A}^*(134 - 348\,p_{2A}^*) = p_{2B}^*(134 - 348\,p_{2B}^*) + 1.6$$

and from Equations (12.17) and (12.19),

$$p_{2A}^*(1 + 6\,p_{2A}^*) = p_{2B}^*(1 + 6p_{2B}^*) + 11.2$$

We thus derive two simultaneous equations from which it is possible to determine the values of the two unknowns p_{2A}^* and p_{2B}^*.

Similarly, the Z and R equations for curve C corresponding to $p_1^* = 30\%$ may be stated, which together with Equations (12.16) and (12.17) provide a solution for the intersection point II in Fig. 12.8, and if grouped with Equations (12.18) and (12.19) yield a solution for the intersection III. The results of these computations are summarized in Table 12.3.

TABLE 12.3
Solutions for the intersections in Fig. 12.8.

	p_1^*	p_2^*	α	g	$Z\times10^3$	$R\times10^3$
Intersection I						
Curve A	10.0	23.7	17.6	0.373	11.34	64.39
Curve B	20.0	11.2				
Intersection II						
Curve B	20.0	34.7	18.1	0.405	10.74	59.28
Curve C	30.0	22.2				
Intersection III						
Curve C	30.0	10.4	17.1	0.392	10.50	61.29
Curve A	10.0	35.5				

SOME POSSIBLE STRATEGIES

The examples illustrated in Fig. 12.8 and Table 12.3 suggest that there are many strategies available to management in formulating changes in price levels in relation to various possible objectives. Some of the strategies of particular interest are described below.

1. Price adjustment confined to one product

Management may have reason to keep the price of one product unchanged, for example $p_1^* = 0$, and seek to determine the value of p_2^* with the objective of:

(1) maximizing Z_2 or Z
(2) maximizing α_2
(3) maximizing α

where, for the sake of convenience, the notation Z_2, Z, α_2 and α is used here to denote the new values of these variables. The foregoing analysis in this chapter, as well as the discussion in Chapter 10, suggests that the value of p_2^* will not necessarily be the same for all these alternative objectives, which will now be examined.

For objective (1), since the new profit level for product 2 is

$$Z_2 = 1000\,(1 + Z_2^*) = 1000\,(1 + 13.4\,p_2^* - 34.8\,p_2^{*2})$$

it follows that maximum Z_2 is obtained at $\mathrm{d}Z_2^*/\mathrm{d}p_2^* = 0$ i.e.

$$13.4 - 69.6\,p_2^* = 0$$

or at

$$p_2^* = 0.1925 \text{ or about } 19.3\%$$

The total profit is

$$Z = Z_1 + Z_2$$

and as Z_1 is constant here and independent of the price for product 2, the total profit Z is maximized when Z_2 is maximized, i.e. at p_2^* found above. In general, the quadratic expressions for the new values of Z_1 and Z_2 may be written as

$$\left.\begin{aligned} Z_1 &= k_1 + a_1 p_1^* + b_1 p_1^{*2} \\ Z_2 &= k_2 + a_2 p_2^* + b_2 p_2^{*2} \end{aligned}\right\} \tag{12.20}$$

where the constants k_1, k_2, etc. are determined by the data for the base period. Hence, Z_1 is maximum at

$$p_1^* = -\frac{a_1}{2b_1}$$

and Z_2 at (12.21)

$$p_2^* = -\frac{a_2}{2b_2}$$

Turning to objective (2), that of maximizing α_2, we find that as

$$\alpha_2 = \frac{Z_2}{R_2}$$

then

$$\frac{d\alpha_2}{dp_2^*} = \frac{Z'_2 R_2 - R'_2 Z_2}{R_2^2}$$

or

$$Z'_2 R_2 - R'_2 Z_2 = 0 \tag{12.22}$$

where Z'_2 and R'_2 denote the derivatives of Z_2 and R_2 with respect to p_2^*. In our case:

$$Z'_2 = \frac{dZ_2}{dp_2^*} = 1000\,(13.4 - 69.6\,p_2^*)$$

$$R_2 = 29000\,(1 + R_2^*) = 29000\,(1 - 0.2\,p_2^* - 1.2\,p_2^{*2})$$

$$R'_2 = \frac{dR_2}{dp_2^*} = -29000\,(0.2 + 2.4\,p_2^*)$$

Substitution into Equation (10.22) leads to the cubic equation

$$(13.4 - 69.6\,p_2^*)\,(1 - 0.2\,p_2^* - 1.2\,p_2^{*2}) + (0.2 + 2.4\,p_2^*)\,(1 + 13.4\,p_2^* - 34.8\,p_2^{*2}) = 0$$

Though cumbersome, this equation can be readily solved for p_2^* (a suitable desk calculator is adequate for the purpose) to yield $p_2^* = 21.9\%$.

Objective (3) is only marginally more complicated; since

$$\alpha = \frac{Z}{R} = \frac{Z_1 + Z_2}{R_1 + R_2}$$

hence, maximum α is attained at

$$\frac{d\alpha}{dp_2^*} = \frac{Z'R - R'Z}{R^2} = 0$$

i.e.

$$Z'R - R'Z = 0$$

or

$$Z_2'R - R_2'Z = 0 \qquad (12.23)$$

which is similar to (12.22), except that we have here R and Z instead of R_2 and Z_2. Substitution of the appropriate values again reveals a cubic equation in p_2^*, which can be solved to yield $p_2^* = 24.3$.

The results of these three exercises in optimization are summarized in Table 12.4, and it is interesting to see how close they are. Maximum Z_2 and Z are achieved when $p_2^* = 19.25\%$ (rounded up in Table 12.4 to 19.3), for which α_2 and α are 8.6 and 15.5% respectively. These profit margins are relatively insensitive to p_2^* at this point, so that α_2 can marginally be increased to 8.7% (which is its maximum value) at $p_2^* = 21.9\%$ while the maximum of α is 15.6% at $p_2^* = 24.3\%$. The effect of a choice of strategy on the other results also appears to be fairly minimal here.

The example pursued in Table 12.4 relates to the case $p_1^* = 0$. Similarly, the three defined strategies may be examined for $p_2^* = 0$, except that instead of

TABLE 12.4
Solutions for four objectives when $p_1^* = 0$.

Objective	(1) Max Z_2 (or Z)	(2) Max α_2	(3) Max α
p_2^* (%)	19.3	21.9	24.3
R_2^* (%)	−8.3	−10.1	−11.9
R_2	26 594	26 061	25 536
Z_2^* (%)	129.0	126.6	120.1
Z_2	2 290	2 266	2 201
Z	10 290	10 266	10 201
Z^* (%)	14.3	14.1	13.3
α_2 (%)	8.6	8.7	8.6
α (%)	15.5	15.5	15.6
$g_{(2)}$	0.451	0.460	0.467
g	0.360	0.363	0.366
ρ_2	0.399	0.394	0.390

maximizing Z_2 and α_2 objectives (1) and (2) are concerned with Z_1 and α_1 respectively.

2. Price adjustment for one product coupled with a fixed change for the other

Market conditions, or other considerations, may lead management to decide that, say, the price for product 1 should be increased by 10% while a price level for product 2 should be sought so as to achieve one of the three objectives listed earlier. There is, in principle, no difference between this strategy and the previous one. In both cases a value for p_1^* is specified, except that in the previous strategy $p_1^* = 0$, whereas here we have the more general case of p_1^* fixed at some predetermined value.

3. A uniform price increase

There are many circumstances when management favors the same price change, in percentage terms, for all products. For the two-product case this means $p_1^* = p_2^*$ (or p^* for short), for which the aggregate path on the $\alpha - g$ chart is given in Fig. 12.2. We now need to determine the value of p^* for each of the following objectives:

(1) Maximize Z_1
(2) Maximize Z_2
(3) Maximize Z
(4) Maximize α_1
(5) Maximize α_2
(6) Maximize α

Of these, the most common objectives are the maximization of Z and α. Since the products are assumed to be independent of each other, the maximization of Z_1 and Z_2 respectively is given by the solutions in (10.21). The maximization of total profit, can be derived from substituting Equation (10.20) and $p_1^* = p_2^* = p^*$ into $Z = Z_1 + Z_2$, leading to

$$\frac{\mathrm{d}Z}{\mathrm{d}p^*} = a_1 + a_2 + 2p^*(b_1 + b_2) = 0$$

or

$$p^* = -\frac{a_1 + a_2}{2(b_1 + b_2)} \tag{12.24}$$

In our numerical example the values for the profit levels are:

$$Z_1 = 8000\,(1 + Z_1^*) = 8000\,(1 + 2p^* - 6p^{*2})$$
$$Z_2 = 1000\,(1 + Z_2^*) = 1000\,(1 + 13.4\,p^* - 34.8\,p^{*2})$$

hence

$$a_1 = 16, b_1 = -48, a_2 = 13.4, b_2 = -34.8 \text{ (all in thousands)}$$

so that Z is optimized at $p^* = 17.8\%$.

To maximize α_1 or α_2, expressions similar to (12.22) need to be formulated, namely

$$\text{maximum } \alpha_1 \text{ at} \qquad Z_1'R_1 - R_1'Z_1 = 0$$

$$\text{maximum } \alpha_2 \text{ at} \qquad Z_2'R_2 - R_2'Z_2 = 0$$

Maximization of α_2 is achieved at $p_2^* = 21.9$, derived from the cubic equation shown after Equation (12.22), and the solution for maximum α_1 is found from the solution of a similar cubic equation, which yields $p_1^* = 25.3\%$.

The overall α is given by

$$\alpha = \frac{Z}{R} = \frac{Z_1 + Z_2}{R_1 + R_2} = \frac{8\,(1 + 2p^* - 6p^{*2}) + (1 + 13.4p^* - 34.8p_2^{*2})}{69\,(1 - 0.2p^* - 1.2p^{*2})}$$

$$= \frac{9 + 29.4p^* - 82.8p^{*2}}{69\,(1 - 0.2p^* - 1.2p^{*2})}$$

α is maximized when

$$\frac{d\alpha}{dp^*} = 0$$

or when

$$Z'R - R'Z = 0$$

i.e. when

$$(29.4 - 165.6\,p^*)\,(1 - 0.2\,p^* - 1.2\,p^{*2}) + (0.2 + 2.4\,p^*)\,(9 + 29.4\,p^* - 82.8\,p^{*2}) = 0$$

for which the solution is $p^* = 23.7$. The results for the six objectives are summarized in Table 12.5, from which it is interesting to note that although the outcome for p^* varies according to which criterion is chosen, the results for

TABLE 12.5
Results for $p_1^* = p_2^* = p^*$.

Objective	p^* (%)	Z_1	Z_2	Z	α_1 (%)	α_2 (%)	α (%)
(1) Maximize Z_1	16.7	9330	2270	11 600	25.0	8.4	18.0
(2) Maximize Z_2	19.3	9300	2290	11 590	25.4	8.6	18.3
(3) Maximize Z	17.8	9330	2280	11 610	25.2	8.5	18.2
(4) Maximize α_1	25.3	8980	2160	11 140	25.7	8.5	18.5
(5) Maximize α_2	21.9	9200	2270	11 470	25.6	8.7	18.5
(6) Maximize α	23.7	9100	2220	11 320	25.7	8.7	18.5

profits and profit margins are fairly close, particularly within the first group of three objectives as well as within the last group of three. It is clear, though, that if (as often found in practice) management wishes to maximize two objectives, say both Z and α, a range of values of p^* needs to be determined where a tradeoff between the objectives can be identified. This is shown in Fig. 12.9, where Z^* is plotted against α^* (based on the results in Table 12.2) for various uniform price increases. Up to about $p^* = 18\%$ both Z and α benefit from a price increase and beyond approximately $p^* = 24\%$ both decline. The region in between is where a tradeoff between Z and α exists.

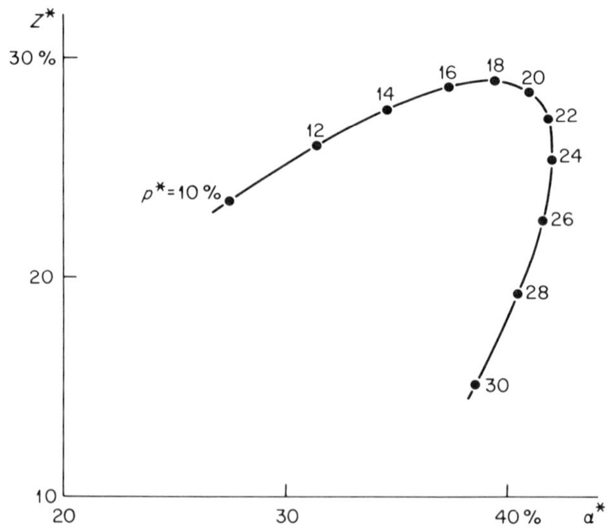

FIG. 12.9. Increase in profit vs increase in profit margin.

4. A differential price increase

A more general strategy may be formulated by management to maintain a relationship between p_1^* and p_2^*. Two common methods for price differentials are:

(a) $p_1^* = kp_2^*$
(b) $p_1^* - p_2^* = k$

In (a) the ratio between the two unit price changes is constant while in (b) there is a fixed percentage difference between them. The analysis with respect to any of the six objectives listed above may proceed for either of these methods for any specified value of k, and the strategy of $p_1^* = p_2^*$ then becomes a special case when $k = 1$ in (a) and $k = 0$ in (b).

ALLOCATION OF OVERHEADS

A crucial element in the analysis of performance in the multi-product case is the allocation of central fixed overhead costs to individual products or activities. A reference to this problem is made in Chapter 5 in connection with the allocation of overheads to divisions, plants or departments within an enterprise, since in addition to the divisional or departmental fixed costs there are central overhead costs incurred at the headquarters offices (which may include interest charges for borrowings arranged at the center) and these have to be covered as well.

There are many possible formulae for the distribution of central overheads, but the following conventions are perhaps most prominent. For the sake of convenience it is assumed here that the total fixed cost F is not subject to change (i.e. $F^* = 0$) and that our concern is with the two-product case.

(1) A fixed allocation of F

The assumption here is that the elements that make up the total fixed costs F can be identified and allocated to the individual products, so that

$$F = F_1 + F_2$$

where F_1 and F_2 are not affected by changes in volumes, revenues or profits.

(2) A partial allocation of F

The allocation of the total fixed costs may not always be possible. The products may share some production and warehousing facilities, as well as administrative and managerial resources, so that while some cost elements may be readily identifiable as to which product they belong to, others may not. The allocation may, therefore, take the form

$$F = F_0 + F_1 + F_2 \tag{12.25}$$

where F_0 remains an unallocated amount, a formulation which is reminiscent of one of the cost functions in Chapter 5. This method implies that while all the basic relationships for the individual products stated earlier apply, namely

$$Z_1 = R_1 - C_1$$

and

$$\alpha_1 = \frac{Z_1}{R_1}$$

etc., for the total product-mix the expression

$$C = C_1 + C_2$$

must be replaced by

$$C = C_1 + C_2 + F_0$$

since the fixed cost F_0 is not accounted for in C_1 and C_2, and similarly the total profit becomes

$$Z = Z_1 + Z_2 - F_0$$

Thus, Equation (12.1) no longer applies here, since now

$$\alpha = \frac{Z}{R} = \frac{Z_1 + Z_2 - F_0}{R}$$

$$= \rho_1 \alpha_1 + \rho_2 \alpha_2 - g_0 \tag{12.26}$$

so that Equation (12.1) is derived from (12.26) as a special case when $F_0 = 0$.

(3) An equal allocation of *F*

In the absence of reliable data that allow an allocation according to Equation (10.25), an equal allocation of F amongst the individual products is sometimes adopted. An example is shown in Table 12.1, where the total fixed costs of 24 000 are divided equally between the two products. Generally, this method of arbitrary allocation is unsatisfactory, because it fails to take account of the relative importance or contribution of the individual products, but when it is adopted the allocated fixed costs F_1 and F_2 are treated as in method (1) and the methodology discussed earlier in this chapter applies.

(4) Allocation of *F* in proportion to "contribution"

If the adoption of method (1) is not practical (for lack of information to allow cost elements to be identified with the individual products) and if method (3) is not acceptable for the reasons mentioned earlier, an alternative that takes account of the relative level of activity generated by each product needs to be formulated. There are several ways in which a measure of activity can be defined, leading to the allocation of F in proportion to

(i) profits, so that

$$\frac{F_1}{F_2} = \frac{Z_1}{Z_2} \quad \text{or} \quad F_1 = \frac{Z_1}{Z} F \tag{12.27}$$

(ii) variable costs, so that

$$\frac{F_1}{F_2} = \frac{S_1}{S_2} \quad \text{or } F_1 = \frac{S_1}{S} F \tag{12.28}$$

(iii) revenues, so that

$$\frac{F_1}{F_2} = \frac{R_1}{R_2} \quad \text{or} \quad F_1 = \rho_1 F \tag{12.29}$$

(iv) contributions to overheads, i.e.

$$\frac{F_1}{F_2} = \frac{R_1 - S_1}{R_2 - S_2} \quad \text{or } F_1 = \frac{R_1 - S_1}{R - S} F \tag{12.30}$$

where throughout we have

$$F_1 + F_2 = F$$

The first formula is too cumbersome, since the definitions of Z_1 and Z_2 in turn depend on the allocated F_1 and F_2, namely $Z_1 = R_1 - F_1 - S_1$ and a similar expression applies to Z_2. When these are substituted into the formula it becomes

$$\frac{F_1}{F_2} = \frac{R_1 - F_1 - S_1}{R_2 - F_2 - S_2}$$

which is not a particularly convenient instrument to implement. The second formula is straightforward enough, provided the variable costs S_1 and S_2 can be easily identified and segregated from each other. The third formula, where the allocation of F is carried out in proportion to the revenues R_1 and R_2, is in many cases the simplest to apply. If this formula is adopted, then

$$F_1 = \rho_1 F$$

and

$$F_2 = \rho_2 F$$

This means that although $F^* = 0$, changes in revenues affect F_1 and F_2, i.e.

$$F_1^* = \rho_1^*$$
$$F_2^* = \rho_2^*$$

In the example, in Table 12.1, $F = 24\,000$, so that this method of allocation would result in

$$F_1 = \rho_1 F = 0.58 \times 24\,000 = 13\,920$$
$$F_2 = \rho_2 F = 0.42 \times 24\,000 = 10\,080$$

and it is not difficult to verify that an allocation in proportion to variable costs in formula (ii) would lead to somewhat different results. The fourth formula has the attraction that it incorporates (ii) and (iii), so that after substituting the definitions of β_1, β and ρ_1, we get

$$F_1 = \frac{\beta_1 \rho_1}{\beta} F$$

with a similar expression for F_2. Clearly, this formula is not as simple as that given by (iii) and the resultant F_1 is different for both.

(5) Allocation of F_0 in proportion to "contribution"

This method is a combination of methods (1) and (4): first, the fixed cost elements that can be identified as belonging to each of the products are set aside, as in method (2), leaving an unallocated amount F_0. The next step is to allocate F_0 according to one of the formulae in method (4). In this way, the arbitrary overall allocation advocated by some of the methods is avoided and the selection of an allocation formula is confined to F_0.

(6) Unallocated F

An opposite view to that taken in method (3) is not to allocate F at all. The operating costs then consist solely of the variable costs S_1 and S_2 respectively and the notional profits are

$$Z_1 = R_1 - S_1$$
$$Z_2 = R_2 - S_2$$

while

$$C = S_1 + S_2 + F$$
$$Z = Z_1 + Z_2 - F$$

and

$$\alpha = \frac{Z}{R} = \frac{Z_1 + Z_2 - F}{R_1 + R_2}$$

so that Equation (12.26) is replaced by

$$\alpha = \rho_1 \alpha_1 + \rho_2 \alpha_2 - g$$

where $g = F/R$.

This method is similar to (2), except that now $F_1 = F_2 = 0$ and $F = F_0$. The derivation of Z^* leads to an expression similar to Equation (9.9). Starting with the definition of overall profit

$$Z = R - S - F$$

we obtain (for $F^* = 0$)

$$Z^* = \frac{R}{Z} R^* - \frac{S}{Z} R^*$$

Substituting

$$\frac{R}{Z} = \frac{1}{\alpha}$$

$$\frac{S}{Z} = \frac{S}{R}\frac{R}{Z} = \frac{1-\beta}{\alpha}$$

and we get

$$Z^* = \frac{1}{\alpha}[R^* - (1-\beta)S^*] \tag{12.31}$$

so that (for $\alpha > 0$)

$$Z^* > 0 \qquad \text{if } R^* > (1-\beta)S^* \tag{12.32}$$

As for the profit margin, since

$$\alpha^* = \frac{Z^* - R^*}{1 + R^*}$$

it follows that

$$\alpha^* > 0 \qquad \text{if } Z^* > R^*$$
$$\text{or } (1-\alpha)R^* > (1-\beta)S^* \tag{12.33}$$

This is by no means a comprehensive list of overhead allocation methods. The accounting literature abounds with many suggestions, some deemed more "theoretically appropriate" or more "practical" than others. They vary greatly in complexity, from simple formulae (such as those described above) to sophisticated procedures, such as the use of the dual variables derived from an optimum solution of a linear programming model†. It is generally agreed that there is no "true" or "best" allocation method and that the adoption of a particular method is a matter of convention. Naturally, availability of data, convenience, and simplicity of application are often overriding considerations. However, what it is important to realize is that the allocation of overheads does not affect the result for the overall profit Z or for the net margin α, though it may have a profound effect on the performance of individual products.

THE SINGLE-PRODUCT TWO-MARKET CASE

A manufacturing company often distributes its products to several markets. These may be distinct geographically, when the markets are defined as different territories (such as home and export markets), or represent market

† See, for example, Salkin, G. R. and Kornbluth, J. S. H. (1973): "Linear Programming in Financial Planning" (Haymarket Publications, London).

segmentation by outlets or types of customers (for example, retail outlets versus mail order business). The following assumptions are made:

- the product is the same irrespective of which market it is intended for
- there are two independent markets or outlets with volumes V_1 and V_2 respectively
- the unit prices for the two markets are p_1 and p_2 respectively
- the overall fixed cost F is not allocated to the two markets
- the marginal unit cost s is the same for the two markets
- $s^* = F^* = 0$

The case where $p_1^* \neq p_2^*$

For the purpose of this analysis, the problem may be treated as if we had two products, except that $s_1 = s_2 = s$, with two price elasticities of demand ε_1 and ε_2. The total cost is then

$$C = S_1 + S_2 + F$$

hence

$$C^* = f(1)\, V_1^* + f(2)\, V_2^* = -f(1)\, \varepsilon_1\, p_1^* - f(2)\, \varepsilon_2\, p_2^*$$

where the cost proportions are defined in this case as

$$f(1) = \frac{S_1}{C}$$

$$f(2) = \frac{S_2}{C}$$

$$f = \frac{F}{C}$$

so that $f(1) + f(2) + f = 1$. The total revenue is

$$R = R_1 + R_2 = p_1\, V_1 + p_2\, V_2$$

hence

$$\begin{aligned} R^* &= \rho_1\, R_1^* + \rho_2\, R_2^* \\ &= \rho_1\, p_1^*\, (1 - \varepsilon_1 - \varepsilon_1\, p_1^*) + \rho_2\, p_2^*\, (1 - \varepsilon_2 - \varepsilon_2\, p_2^*) \end{aligned}$$

The profit is $Z = R - C = R_1 + R_2 - S_1 - S_2 - F$

and the net profit margin

$$\alpha = \frac{Z}{R} = \frac{Z_1 - S_1}{R} + \frac{R_2 - S_2}{R} - \frac{F}{R}.$$

Substituting the definitions of β_1 and β_2 in Equation (12.2) and $g = F/R$, we derive the expected expression

$$\alpha = \rho_1 \beta_1 + \rho_2 \beta_2 - g = \beta - g$$

All the other relationships for the two-product case also hold here. If pricing policy considerations do not require that p_1^* and p_2^* be the same, then the alternative objectives of maximizing profit or profit margin may be pursued as follows:

(1) *Maximum* Z is obtained by maximizing Z_1 and Z_2, and since the products are assumed independent of each other we need to find

$$p_1^* \text{ to maximize } Z_1^*$$

and

$$p_2^* \text{ to maximize } Z_2^*$$

where, as F is unallocated here,

$$Z_1 = R_1 - S_1$$

or, after substituting β_1,

$$Z_1^* = \frac{1}{\beta_1} [R_1^* - (1 - \beta_1) S_1^*] \tag{12.34}$$

and a further substitution of R_1^* and S_1^* yields

$$Z_1^* = \frac{p_1^*}{\beta_1} (1 - \varepsilon_1 p_1^* - \varepsilon_1 \beta_1) \tag{12.35}$$

Assuming $\beta_1 > 0$, the value of Z_1 increases for $p_1^* > 0$ if the expression in the brackets is positive, i.e.

$$Z_1^* > 0 \qquad \text{if } 0 < p_1^* < \frac{1}{\varepsilon_1} - \beta_1 \tag{12.36}$$

Similarly, for a reduction in p_1 we have $Z_1^* > 0$ if the inequalities in (12.36) read the other way. These conditions are as stated in (9.34). As shown in Chapter 9, the value of p_1^* expected to maximize Z_1^* lies half-way between 0 and the bound stated in condition (12.36).

(2) *Maximum* α is obtained when

$$\alpha = \frac{Z}{R} = \frac{R - C}{R} = 1 - \frac{C}{R}$$

is maximized or when C/R is minimized. Now

$$\left(\frac{C}{R}\right)^* = \frac{C^* - R^*}{1 + R^*}$$

so that C/R can be reduced if $R^* > C^*$, for which a sufficient condition is

$$R_1^* > S_1^* = V_1^* \text{ and } R_2^* > S_2^* = V_2^*$$

i.e.

$$p_1^* (1 - \varepsilon_1 - \varepsilon_1 p_1^*) > -\varepsilon_1 p_1^*$$

or

$$0 < p_1^* < \frac{1}{\varepsilon_1} \tag{12.37}$$

and when the unit price is to be reduced, the inequalities need to read the other way. A similar expression applies for product 2 to ensure that $R_2^* > S_2^*$.

A uniform price change

If price changes are constrained to $p_1^* = p_2^* = p^*$ then it is convenient to start with expression (12.32), which gives the condition for overall profit to rise, namely

$$Z^* > 0 \qquad \text{if } R^* > (1 - \beta) S^*$$

Here

$$\begin{aligned} R^* &= \rho_1 R_1^* + \rho_2 R_2^* \\ &= \rho_1 p^* (1 - \varepsilon_1 - \varepsilon_1 p^*) + \rho_2 p^* (1 - \varepsilon_2 - \varepsilon_2 p^*) \\ &= p^* [1 - (1 + p^*)(\rho_1 \varepsilon_1 + \rho_2 \varepsilon_2)] \end{aligned}$$

As

$$R_1^* = p^* [1 - (1 + p^*) \varepsilon_1]$$

and

$$R_2^* = p^* [1 - (1 + p^*) \varepsilon_2]$$

by analogy we may define

$$R^* = p^* [1 - (1 + p^*) \varepsilon_R]$$

where

$$\varepsilon_R = \rho_1 \varepsilon_1 + \rho_2 \varepsilon_2 \tag{12.38a}$$

ε_R being the "overall elasticity" from the point of view of total revenue. Similarly,

$$S_1^* = V_1^* = -\varepsilon_1 p^*$$

$$S_2^* = V_2^* = -\varepsilon_2 p^*$$

so that

$$S^* = \frac{S_1}{S} S_1^* + \frac{S_2}{S} S_2^*$$
$$= -p^* \left(\frac{S_1}{S} \varepsilon_1 + \frac{S_2}{S} \varepsilon_2 \right)$$
$$= -\varepsilon_S p^*$$

where

$$\varepsilon_S = \frac{S_1}{S} \varepsilon_1 + \frac{S_2}{S} \varepsilon_2 \tag{12.38b}$$

is the perceived "overall elasticity" affecting the total value S.
Substituting these expressions for R^* and S^* we get the condition that, if the price is to rise, then

$$Z^* > 0 \qquad \text{if } 0 < p^* < \frac{1}{\varepsilon_R} - 1 + (1 - \beta) \frac{\varepsilon_S}{\varepsilon_R} \tag{12.39}$$

and for a reduction in price the inequalities in this condition read the other way. Similarly, the condition for an increase in the overall profit margin can be developed.

Example

Given the data in Table 12.6, find the conditions for increasing profit and for maximizing profit when $p_1^* \neq p_2^*$ and $p_1^* = p_2^* = p^*$ respectively.

(*a*) $p_1^* \neq p_2^*$. As F is unallocated here, we denote Z_1 and Z_2 as suggested earlier, namely

$$Z_1 = R_1 - S_1 = 40$$
$$Z_2 = R_2 - S_2 = 20$$

Hence

$$Z = Z_1 + Z_2 - F = 15$$

TABLE 12.6
An example.

	F	S	R	Z	ρ	α	β	g	ε
Product 1		60	100		0.625		40.0		1.2
Product 2		40	60		0.375		33.3		0.9
Total	45	100	160	15	1.00	9.4	37.5	0.281	

From Equation (12.35)

$$Z_1^* = \frac{p_1^*}{\beta_1}(1 - \varepsilon_1 p_1^* - \varepsilon_1 \beta_1)$$
$$= p_1^* (1.3 - 3.0 p_1^*)$$

Hence,

$$Z_1^* > 0 \qquad \text{if } 0 < p_1^* < 0.433$$

whereas negative p_1^* leads to a negative value of Z_1^*. Similarly, for product 2

$$Z_2^* = p_2^* (2.1 - 2.7 p_2^*)$$

and

$$Z_2^* > 0 \qquad \text{if } 0 < p_2^* < 0.778$$

the optimal values being

$$p_1^* = 21.7\% \text{ and } p_2^* = 38.9\%$$

leading to

$$Z_1^* = 14.1\% \text{ and } Z_2^* = 40.8\%$$

and the new expected profit level at these price increases is

$$Z = 40(1 + Z_1^*) + 20(1 + Z_2^*) - F$$
$$= 45.64 + 28.16 - 45 = 28.8$$

i.e. 92% above the base profit level.

(*b*) $p_1^* = p_2^* = p^*$. From $Z = Z_1 + Z_2 - F$ we have

$$Z^* = \zeta_1 Z_1^* + \zeta_2 Z_2^*$$

where $\zeta_1 = \frac{Z_1}{Z} = \frac{40}{15} = 2.667$

$$\zeta_2 = \frac{Z_2}{Z} = \frac{20}{15} = 1.333$$

Substituting Z_1^* and Z_2^* from the above we get

$$Z^* = 2.667 p^* (1.3 - 3.0 p^*) + 1.333 p^* (2.1 - 2.7 p^*)$$
$$= 1.333 p^* (4.7 - 8.7 p^*)$$

Alternatively, Equation (12.31) may be used, namely

$$Z^* = \frac{1}{\alpha}[R^* - (1 - \beta) S^*]$$

where

$$\begin{aligned} R^* &= \rho_1 R_1^* + \rho_2 R_2^* \\ &= 0.625\, p^* (1 - \varepsilon_1 - \varepsilon_1 p^*) + 0.375\, p^* (1 - \varepsilon_2 - \varepsilon_2 p^*) \\ &= -p^* (0.088 + 1.088\, p^*) \end{aligned}$$

and

$$\begin{aligned} S^* = (S_1 + S_2)^* &= \frac{S_1}{S} S_1^* + \frac{S_2}{S} S_2^* \\ &= 0.6\, S_1^* + 0.4\, S_2^* \\ &= -0.6\, \varepsilon_1\, p^* - 0.4\, \varepsilon_2\, p^* \\ &= -1.08\, p^* \end{aligned}$$

Substituting these expressions, as well as the values of α and β, we get

$$Z^* = 1.333\, p^* (4.7 - 8.7\, p^*)$$

as derived above. Hence profit is expected to rise if

$$0 < p^* < 0.540$$

and this result may be derived directly from the condition (12.39). Maximum profit is expected at $p^* = 27.0\%$ where

$$Z^* = 0.846$$

i.e. somewhat below the expected increase in (a), the profit level here being

$$Z = 15\,(1 + Z^*) = 27.7$$

The reason for this result being inferior to that in case (a) is that in (a) there is an added degree of freedom, since the two price changes may be considered independently of each other, whereas in (b) the constraint $p_1^* = p_2^*$ forces Z_1 and Z_2 below their respective optimum levels.

CROSS ELASTICITIES

The analysis so far has been based on the assumption that the products are totally independent of each other in the market place, and that a price change for one product affects only its own volume. However, if the products are distributed in the same market, price comparisons may well affect customer choice, either because of a degree of product substitution that may exist or because of the effect on the image of the company and its brand names. This interdependence may be formulated in the two-product case in the following way:

- A relative change p_1^* causes a relative change in volume V_{11}^* for product 1 and also a change V_{21}^* for product 2;
- A relative change p_2^* causes a change V_{22}^* for product 2 and V_{12}^* for product 1.

This notation is adopted in order to clarify which product is affected (the first digit in the suffix) and for what cause (given by the second digit), so that V_{21}^* (for example) is the relative change in volume for product 2 due to the unit price change for product 1.

Similarly, the elasticities are defined, as summarized in Table 12.7. ε_{11} and ε_{22} are the *direct elasticities*, reflecting the effect of each price change on its own product, while ε_{21} and ε_{12} are the *cross elasticities*. The proposed relationships between volume change and price change, also shown in Table 12.7, follow the form adopted in previous chapters for the single product case. A reasonable assumption to adopt is that the total effect on the volume of each product is the sum of the individual effects, i.e.

$$\left.\begin{aligned} V_1^* &= V_{11}^* + V_{12}^* = -\varepsilon_{11}\, p_1^* - \varepsilon_{12}\, p_2^* \\ V_2^* &= V_{22}^* + V_{21}^* = -\varepsilon_{22}\, p_2^* - \varepsilon_{21}\, p_1^* \end{aligned}\right\} \qquad (12.40)$$

Four important points need to be made in this context:

(1) This formulation makes no distinction between a price increase and a price decrease, but in practice (as stated in Chapter 8) the value of the elasticity coefficients may well depend on whether p^* is positive or negative, as well as on its magnitude. When such information is available, it needs to be incorporated in the analysis.

(2) Following the discussion in Chapter 8, the direct elasticity coefficients ε_{11} and ε_{22} are assumed to be positive, i.e. a price increase is expected to cause a

TABLE 12.7
Effect of price changes.

(a) Direct and cross elasticities

	Effect on Product 1	Effect on Product 2
Price change		
Product 1	ε_{11}	ε_{21}
Product 2	ε_{12}	ε_{22}

(b) Relative volume changes

Price change		
Product 1	$V_{11}^* = -\varepsilon_{11} p_1^*$	$V_{21}^* = -\varepsilon_{21} p_1^*$
Product 2	$V_{12}^* = -\varepsilon_{12} p_2^*$	$V_{22}^* = -\varepsilon_{22} p_2^*$

decline in volume and a price decrease to a rise in volume. Such an assumption is not warranted for the cross elasticity coefficients, as is illustrated by the following examples:

(a) The two products share the same brand name but aim at different segments of the market and with no evidence of substitution between them. If the price for one product increases, ill-informed customers may gain the impression that prices in general have risen and demand for both products may be adversely affected.
(b) One product is an expensive version of the other and substitution does take place. If the price of the cheaper model is increased, causing the price differential between the two models to narrow, volume for the cheaper model will decline but some of the lost volume may be diverted towards the expensive model and its volume may thereby increase.

(3) For various practical purposes it may be convenient to retain the concept of an "overall" elasticity coefficient, defined in the conventional way by

$$\begin{aligned} V_1^* &= -\varepsilon_1 \, p_1^* \\ V_2^* &= -\varepsilon_2 \, p_2^* \end{aligned} \tag{12.41}$$

In conjunction with Equation (12.35) this means that the overall elasticities are

$$\left.\begin{aligned} \varepsilon_1 &= \varepsilon_{11} + \varepsilon_{12} \, p_2^*/p_1^* \\ \varepsilon_2 &= \varepsilon_{22} + \varepsilon_{21} \, p_1^*/p_2^* \end{aligned}\right\} \tag{12.42}$$

so that even when the individual elasticity coefficients in Table 12.7 are independent of the magnitude of the price changes, the overall elasticities in (12.42) do depend on the ratio p_2^*/p_1^*.

(4) A word of caution needs to be voiced here about the practical difficulties of ascertaining the values of the four elasticities in Table 12.7. Studies aimed at determining direct elasticities can be formidable enough; extending them to cover cross elasticities needs to rely on a sophisticated data bank and a market research methodology that rarely yield results in sufficient detail or clarity. In the two-product case the matrix of elasticities is a modest 2×2, as shown in Table 12.7, but for n products it becomes $n \times n$ in size and the attendant effort needed to determine and validate the values of numerous coefficients becomes increasingly difficult to justify. In cases where the cross elasticities are relatively small in their absolute values, compared with direct elasticities, naturally more effort needs to be directed towards improving the information about the latter rather than the former.

Example

Take the data for the two-product case in Table 12.6 and add values for cross elasticities as given in Table 12.8.
Hence

$$V_1^* = -1.2\,p_1^* + 0.2\,p_2^*$$
$$V_2^* = -0.9\,p_2^* - 0.5\,p_1^*$$

TABLE 12.8
Example for data for elasticities.

	Effect on Product 1	Effect on Product 2
Price change for 1	$\varepsilon_{11} = 1.2$	$\varepsilon_{21} = 0.5$
Price change for 2	$\varepsilon_{12} = -0.2$	$\varepsilon_{22} = 0.9$

Revenues

$$R_1^* = p_1^* + V_1^*\,(1 + p_1^*)$$
$$R_2^* = p_2^* + V_2^*\,(1 + p_2^*)$$
$$R^* = 0.625\,R_1^* + 0.375\,R_2^*$$

Profits

$$Z_1^* = \frac{1}{0.4}\,(R_1^* - 0.6\,S_1^*) \qquad \text{From equation (12.34)}$$
$$= 2.5\,R_1^* - 1.5\,V_1^*$$
$$Z_2^* = 3\,R_2^* - 2\,V_2^*$$
$$Z^* = 2.667\,Z_1^* + 1.333\,Z_2^*$$

Results for R^* and Z^* for various combinations of p_1^* and p_2^* are given in Table 12.9, from which it is clear that for any given value of p_2^* the total profit is expected to rise with p_1^*, reaching a maximum in the range of $p_1^* = 15$ to 20%, and then falls. A similar pattern is observed when p_2^* is varied while p_1^* is kept constant. The optimum solution for Z^* lies at

$$p_1^* = 15.6\% \qquad \text{and } p_2^* = 44.8 \text{ or } 44.9\%$$

where

$$R^* = -6.6\% \qquad \text{while } Z^* = 96.6\%$$

Three comments need to be made in connection with these results:

(1) It is not surprising that a relatively large increase in p_2 is called for (compared with that for p_1), since the direct elasticity ε_{22} is below 1 and the

TABLE 12.9

Results for R^* and Z^* (top and bottom figures respectively in each cell) for combinations of p_1^* and p_2^* (all in %).

p_2^*	p_1^* =	−10	−5	0	5	10	15	20	25	30
−10	R^* =	0.4	−0.6	−2.0	−3.7	−5.8	−8.2	−11.1	−14.3	−17.9
	Z^* =	−73.6	−53.3	−36.9	−24.6	−16.3	−11.8	−11.6	−15.3	−22.9
−5		1.4	0.5	−0.9	−2.6	−4.7	−7.2	−10.1	−13.3	−16.9
		−53.9	−33.7	−17.6	−5.4	2.8	6.9	7.1	3.3	−4.6
0		2.4	1.4	0	−1.8	−3.9	−6.4	−9.3	−12.5	−16.1
		−36.0	−16.0	0	12.0	20.0	24.0	24.0	20.0	12.0
5		3.1	2.1	0.7	−1.0	−3.2	−5.7	−8.6	−11.9	−15.5
		−19.9	−0.1	15.8	27.6	35.4	39.3	39.1	34.9	26.8
10		3.7	2.7	1.3	−0.5	−2.7	−5.2	−8.1	−11.4	−15.0
		−5.6	14.1	29.7	41.4	49.1	52.7	52.4	48.1	39.7
15		4.1	3.1	1.7	−0.1	−2.3	−4.8	−7.8	−11.1	−14.7
		6.9	26.4	41.9	53.4	60.9	64.4	63.9	59.4	50.9
20		4.4	3.3	1.9	0.1	−2.1	−4.7	−7.7	−10.9	−14.6
		17.6	36.9	52.3	63.6	70.9	74.3	73.6	68.9	60.3
25		4.5	3.4	2.0	0.1	−2.1	−4.7	−7.6	−10.9	−14.6
		26.5	45.7	60.8	72.0	79.2	82.3	81.5	76.7	67.8
30		4.4	3.3	1.8	0	−1.2	−4.8	−7.8	−11.1	−14.9
		33.6	52.6	67.6	78.6	85.6	88.6	87.6	82.6	73.6
35		4.1	3.0	1.6	−0.3	−2.5	−5.2	−8.1	−11.5	−15.2
		38.9	57.7	72.6	83.4	90.2	93.1	91.9	86.7	77.6
40		3.7	2.6	1.1	−0.8	−3.0	−5.7	−8.7	−12.0	−15.8
		42.4	61.1	75.7	86.4	93.1	95.7	94.4	89.1	79.7
45		3.1	2.0	0.5	−1.4	−3.7	−6.3	−9.3	−12.7	−16.5
		44.1	62.6	77.1	87.6	94.1	96.6	95.1	89.6	80.1
50		2.4	1.2	−0.3	−2.2	−4.5	−7.2	−10.2	−13.6	−17.4
		44.0	62.3	76.7	87.0	93.3	95.7	94.0	88.3	78.7

cross elasticity ε_{12} is negative. This means that a rise in unit price for product 2 has a muted effect on the volume for product 2 while positively stimulating demand for product 1. Under such circumstances, profit is expected to rise sharply as p_2 increases, and it is only when p_2^* exceeds a relatively high value (in our case 45 %) that the expected increase in profit is halted.

(2) As stated in Chapter 8, it is not likely that in practice the elasticity coefficients remain constant over a wide range of price changes. In our case the results are based on the assumption that $\varepsilon_{22} = 0.9$, even when the unit price p_2 is increased by 50 %, and in reality such an assumption is rarely

likely to remain valid. In practice, however, the value of the results to the decision maker is that they point firmly to the desirable direction of change. More modest price increases than those suggested by the optimal solution can still lead to an expected substantial improvement, for example the combinations of a 5% price increase for product 1 coupled with a 10% increase in product 2 suggest in Table 12.9 that profit would rise by 41% with hardly any loss of revenue. The decision maker may, therefore, prefer this course of action as a first stage, then reassess the situation after a period of consolidation and collection of fresh data regarding elasticities.

(3) In addition to the presentation of tabulated results as in Table 12.9, it is often useful to produce iso-Z^* and iso-R^* contour maps, which for our example are combined in Fig. 12.10 for price increases in the range 0 to 20%.

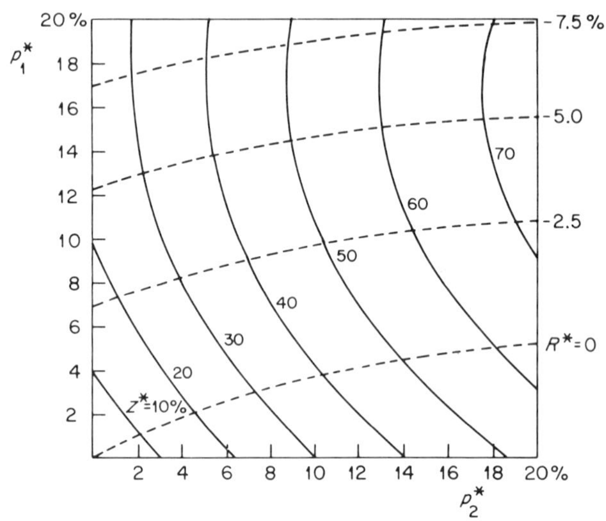

FIG. 12.10. Iso-Z^* and iso-R^* map for combinations of p_1^* and p_2^*.

CONCLUSION

If products in the product-mix are completely independent of each other, and if cost data for each can be determined, then each may be treated as a single product, using the methodologies discussed in earlier chapters. Interdependence of products occurs when

(1) products share certain costs, such as in production, inspection, warehousing, marketing, distribution and overall administration, so that while

the total costs for each of these functions can be readily ascertained, the proportion of the costs to be borne by each product is not evident from the cost structure;

(2) a change in design specifications, production and marketing of one product may affect customer appeal of others in the product range (these effects are not essentially dissimilar to those caused by actions of competitors in the market place and are therefore not discussed in this Chapter):

(3) a change in the pricing policy for one product may have an effect on others in the range.

In addition, there are many possible objectives that management is often pressed to consider, such as

- improve the profit level achieved for each product as well as for the aggregate product-mix
- improve the return on capital for individual products and for the aggregate
- similarly, for revenue, as well as net and gross profit margins

The problems posed by (1) above are relevant to the evaluation of performance criteria for individual products and many methods are available for allocation of central overheads to these products. Needless to say, the allocation method adopted may have a profound effect on the individual cost structure and hence on the perceived performance of an individual product, which may be rated as excellent when one allocation method is used but poor under another. The allocation method does not, however, affect the measures of overall performance of the product-mix.

As shown in earlier chapters, improvements in some performance measures do not always go hand-in-hand with improvements in others. Similarly, an improvement in the performance of an individual product (with respect to any given criterion) is not necessarily compatible with an improvement of the aggregate. It is often observed that relentless efforts are made to improve the standing of a selected product with the conviction that the results will be beneficial for the enterprise as a whole. Such a premise is not always justified and it is wise to corroborate its validity in every case.

A major factor that affects performance is pricing. A useful way of demonstrating the effect of price changes on the net profit margin is by the $\alpha - g$ chart, as shown in the examples in Figs 12.2–12.8, where the results for the individual products are contrasted with those of the aggregate. Such results raise the question as to whether individual unit price changes are preferable to a uniform price change across the board.

Finally, it is important to recognize the possible implication of (3) above. A formal procedure for handling the effect of price changes on volumes is suggested in Table 12.7, where a distinction is made between direct elasticities of demand and cross elasticities, the total effect on individual volumes being postulated in Equation (12.40). The analysis of performance may then proceed for each product and for the aggregate for any given method of allocation of overheads. The assumption throughout, though, is that the values for the direct and cross elasticities can be ascertained, and while this may be possible for a narrow product range and for relatively small price movements, serious difficulties are bound to arise when attempts are made to construct large scale elasticity matrices.

EXERCISES

1. If $\rho_1 = 0.6$ and $\zeta_1 = 0.5$ while $\alpha_1 = 9\%$ and $\alpha_2 = 13\%$, find the effect on the total revenue, profit and profit margin in the following cases:

   (a) R_1 increases by 20%, and C_1 by 15%
   (b) R_2 increases by 20%, and C_2 by 15%
   (c) both R_1 and R_2 increase by 20%, as do C_1 and C_2.
2. If $f = \eta_1 f_{(1)} + \eta_2 f_{(2)}$
   $\alpha = \rho_1 \alpha_2 + \rho_2 \alpha_2$
   $g = \rho_1 g_{(1)} + \rho_2 g_{(2)}$
   prove that
   $$g = f(1-\alpha)$$
3. Prove that

   (a) $$\rho_1^* = \frac{\rho_2 (R_1^* - R_2^*)}{1 + \rho_1 R_1^* + \rho_2 R_2^*}$$

   (b) $$\rho_2^* - \rho_1^* = \frac{R_2^* - R_1^*}{1 + R^*}$$

   (c) $$\frac{\rho_2^* + \rho_1^*}{\rho_2^* - \rho_1^*} = \rho_1 - \rho_2$$
4. Extend Table 12.2 to cover the range of price increases up to 40% when $p_1^* = p_2^* = p^*$. Compute α_1, α_2, α, $g_{(1)}$, $g_{(2)}$, Z_1^*, Z_2^* and Z.
5. For the example in Table 12.1, if $p_1^* = p_2^* = p^*$, find the value of p^* expected to maximize α_1.
6. Find the values of p_1^* and p_2^* that yield the intersection points I, II and III in Fig. 12.8.
7. Check whether the results in Table 12.4 are compatible with Equation (12.1).
8. Compute the results summarized in Table 12.5 for the six stated objectives.
9. If in Table 12.1 it is possible to determine that of the figures $F = 24\,000$ the

sums of 8000 and 3000 can be identified with products 1 and 2 respectively, calculate the allocated F_1 and F_2 according to the four formulae suggested in methods (4) and (5).

10. For the example in Table 12.1 the fixed costs are divided into

$$F = F_0 + F_1 + F_2$$

where $F_1 = 10\,000$ and $F_2 = 6000$. If only a uniform relative price change is contemplated, find the value of p^* that will maximize total profit Z when method (2) is selected for the allocation of fixed costs. What would the effect be if method (5) was selected instead?

11. For the example in Table 12.6 allocate the fixed costs F in proportion to the individual product revenues and proceed to find the optimal values of p_1^* and p_2^* to maximize the total profit Z when $p_1^* \neq p_2^*$ and $p_1^* = p_2^*$ respectively. Compare your results with those derived in the text.

12. (a) For the data in Tables 12.6 and 12.8, tabulate the corresponding values of p_1^* that in combination with $p_2^* = 0, 2, 4, \ldots, 20\%$ would be expected to yield the following respective values for R^*:

$$R^* = 0, \; -2.5\%, \; -5\%, \; -7.5\%, \; -10\%$$

(b) Is it possible in this example for revenue to rise when unit prices are increased?

13
Planning for Decision Making

Themes in this chapter

Use of incremental calculus for corporate planning
Incremental sensitivity tables
Possible errors of extrapolation from such tables
Deterministic appraisal
Most likely, optimistic and pessimistic scenarios
Risk simulation
Formats for monthly performance reports and for corporate planning
How to examine whether multiple objectives or constraints can be met
Sharing the benefits of
- **increase in profit**
- **increase in added value**
- **decrease in costs**

APPLICATIONS OF INCREMENTAL CALCULUS

As intimated in the Preface, the main purpose of the modeling methodology presented in this book is to help managers evaluate the consequences of alternative decisions on key performance criteria. There are four situations in which the methodology can be usefully employed:

(a) ***Examination of past performance***, to help ascribe possible cause and effect relationships, in order to identify reasons for changes in performance and thereby gain a better understanding than hitherto of the way the enterprise operates.
(b) ***Prediction of changes*** that are likely to occur in the future, changes which are outside the control of management (such as cost of materials and market demand) and which are likely to affect cost and revenue levels.
(c) ***Analysis of managerial decisions***, in relation to pricing and cost reduction exercises aimed either at nullifying the detrimental effects of external factors and/or at improving performance by an appropriate mixture of actions to affect revenue and cost.

(d) ***Helping in negotiations***, between managers and/or employees to determine the contribution of labor productivity to overall performance and to find solutions that satisfy various aspirations, which often conflict with each other.

These four types of application are the essence of a *corporate planning* function. First, it must incorporate a data bank for the values of variables in the past and a diagnostic element to provide the basis for determining functional relationships between variables. Common examples encountered by analysts are questions such as:

- What are the fixed costs? Are they really fixed, or are they partly affected (as suggested by evidence of past performance) by volume?
- Is the variable unit cost constant irrespective of volume, or is it a function of volume, and if so can the functional relationship be stated?
- What cost elements should be treated separately from the conventional fixed and variable cost components, and what do their values depend on?
- Are there economies or diseconomies of scale, and are they affected by variables other than volume?
- What evidence can be derived from past performance, or from data relating to competitors, about price and promotion elasticities of demand and about the mechanization elasticity? In the case of a multi-product output, can estimates be made both for direct and indirect elasticities?
- What measures of performance appear to dominate the thinking of the incumbent management and how do they compare with measures adopted by competitors?
- To what extent have these measures been compatible with each other and when have they been in conflict?
- Why have past overall performance goals not been achieved?
- Have surrogate objectives, defined for middle management or for subsystems within the business, helped or hindered the attainment of overall performance goals?

These questions help in the construction of models for the past, linking the major variables, highlighting correlations and explaining how performance criteria are affected by changes in the system. It is by understanding what has happened in the past that managers can hope to control performance in the future.

AN INCREMENTAL SENSITIVITY TABLE

Analysts and managers often seek to determine effects of changes at the margin: what would happen if wage rates increase by 1 %, or volume increases by 1%, or unit price increases by 1%, etc? Such information may be summarized in an *incremental sensitivity table*, an example of which is shown in Table 13.1, mainly with the factor input cost model in mind. The values in Table 13.1 help to highlight which changes are most significant for key variables and criteria, so that attention can be devoted to them in the monitoring and control function of the enterprise. It should be stressed that while the variables in Table 13.1 are obviously useful in many circumstances, the choice of variables must be governed by the needs of management in any given set of circumstances. (For an example of how the incremental sensitivity table was used in a case study of a steel plant see [3, pp. 135–6].)

TABLE 13.1
An incremental sensitivity table.

	Effect on (in %)											
+1% Change in	c_W	c_M	c_E	c	C	R	Z	r	α	β	f	g
Labor input												
No. of employees												
Regular hours												
Overtime hours												
Total hours												
Wage rate												
Labor productivity												
Total labor cost												
Materials input												
Volume/unit of output												
Price factors												
Total materials cost												
Other expenses												
Depreciation												
Borrowings												
Interest rate (one point increase)												
Output												
Unit Price												

An example

To illustrate the construction of an incremental sensitivity table, consider the following example:

Let
$$C = S + F + J + D$$
where
$$f_1 = 0.4, f_2 = 0.3, f_3 = 0.2, f_4 = 0.1$$

i.e.

$$f = f_2 + f_3 + f_4 = 0.6$$

Other given data:

$$\alpha = 12.5\%$$
$$\varepsilon = 1.2$$
$$\sigma = 0.2$$
$$I_W^* = 0.8\, R^*$$
$$\lambda = 0.5$$
$$i = 0.25$$

hence

$$\beta = \alpha + f - \alpha f = 65.0\%$$

Also

$$B^* = \frac{I_W^*}{\lambda}$$

$$J^* = j^* + B^*(1 + j^*)$$

$$I^* = (1 - i) I_W^* = 0.6\, R^*$$

An incremental sensitivity table is given in Table 13.2, which consists of two parts: the first postulates a 1% rise in the variables F, s, J, D, V_0 and p, each on its own, while the second part assumes a 1% fall in each. Since the given data suggest that the working capital changes with revenue (but not with total cost), a change in each of the variables F, s and J has no effect on V, I_W (and hence on I) and R.

A change in D, V_0 or p affects the volume V, assuming (as before) that

$V^* = V_0^*$ for an intrinsic change in demand
$V^* = \sigma D^*$ for a change in D
$V^* = -\varepsilon p^*$ for a change in p

so that a change in R takes place, as well as a change in I_W (hence in I) and this results in $r^* \neq Z^*$.

Comments of the ±1% sensitivity tables

A close examination of the table suggests the following comments in connection with this example:

(1) A change in F, s or J has an effect on unit cost according to the cost proportions, and the effect of a change in V_0 is of the same order of magnitude. A change in D has a negligible effect on unit cost, while a change in price has the largest effect here.

(2) The absolute magnitude of the effects of changes in D and p on revenue is roughly the same (the effects are in opposite directions) and are significantly smaller than the effect of V_0^*.

TABLE 13.2
Example for an incremental sensitivity table.

	Effect on (in %)							Resultant values of		
	V	I_W	C	c	R	Z	r	α (%)	β (%)	f
No change								12.5	65.0	0.600
1% rise in										
F			0.30	0.30		−2.10	−2.10	12.24	65.0	0.601
s			0.40	0.40		−2.80	−2.80	12.15	64.7	0.598
J			0.20	0.20		−1.40	−1.40	12.33	65.0	0.601
D	0.20	0.16	0.24	0.04	0.20	−0.08	−0.23	12.47	65.0	0.600
V_0	1.00	0.80	0.72	−0.28	1.00	2.96	2.35	12.74	65.0	0.599
p	−1.20	−0.17	−0.55	0.66	−0.21	2.14	2.27	12.79	65.3	0.603
1% fall in										
F			−0.30	−0.30		2.10	2.10	12.76	65.0	0.599
s			−0.40	−0.40		2.80	2.80	12.85	65.4	0.602
J			−0.20	−0.20		1.40	1.40	12.68	65.0	0.599
D	−0.20	−0.16	−0.24	−0.04	−0.20	0.11	0.23	12.54	65.0	0.600
V_0	−1.00	−0.80	−0.72	0.28	−1.00	−2.96	−2.37	12.25	65.0	0.601
p	1.20	0.15	0.54	−0.65	0.19	−2.28	−2.39	12.19	64.6	0.597

Note: The blank spaces signify no change.

(3) Changes in F, s and J are magnified, when their effect on Z and r is computed, while D^* has only a minor effect. V_0^* and p^* are also significant in this respect.

(4) The resultant values of β and f are little changed and are far less affected than α.

(5) With the exception of the results for p^*, the results in the two halves of the table seem symmetrical, i.e. a 1 % rise in a given variable has the same but opposite effect of a 1 % fall. Some of this symmetry is genuine, but some is the outcome of rounding errors. Also, as stated in Chapter 9, a 10 % rise in a given variable does not always lead to a tenfold result of a 1 % change. Thus, the properties of *symmetry* and *linear scaling* do not apply to all the selected performance measures, and this means that extrapolation from the ± 1 % incremental sensitivity table may sometimes involve undesirable errors.

(6) This conclusion is demonstrated in Table 13.3, where selected results for C^*, c^*, R^* and Z^* for ± 1 % changes are compared with ± 10 % changes in F, V_0 and p respectively. Symmetry and linear scaling apply when changes in F take place. They also apply for changes in V_0, except for the results for c^*. In the case of changes in p, evidently neither symmetry nor linear scaling applies. It is interesting to note that in this example (for which the elasticity ε is greater than 1) a change in V_0 has a greater effect on total cost than a change in p, but a lesser effect on unit cost. As for the resultant profit, a ± 1 % change in V_0 is somewhat more effective than a ± 1 % change in p, significantly more effective at the ± 10 % change level, but there is little difference between the outcomes for a change of -10 %. The reason for this result is that while Z^* is a linear function of V_0^* it is a quadratic function of p^*, as shown in Fig. 13.1, and this explains why scaling up the

TABLE 13.3
Comparison of the effect of changes for symmetry and linear scaling.

	C^*		c^*		R^*		Z^*	
	+	−	+	−	+	−	+	−
± 1 % change in								
F	0.30	−0.30	0.30	−0.30			−2.10	2.10
V_0	0.72	−0.72	−0.28	0.28	1.00	−1.00	2.96	−2.96
p	−0.55	0.54	0.66	−0.65	−0.21	0.19	2.14	−2.28
± 10 % change in								
F	3.00	−3.00	3.00	−3.00			−21.00	21.00
V_0	7.20	−7.20	−2.55	3.11	10.00	−10.00	29.60	−29.60
p	−5.82	5.06	7.02	−6.20	−3.20	0.08	15.17	−28.99

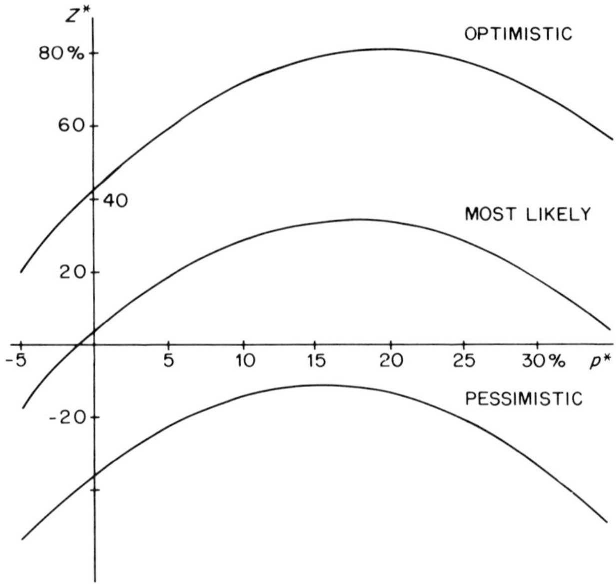

FIG. 13.1. An example of profit change for three scenarios.

figures in the $\pm 1\%$ incremental sensitivity table is valid for some changes but not for others.

ANALYSIS OF POSSIBLE SCENARIOS

Deterministic appraisal

While the incremental sensitivity table is undoubtedly of interest to managers in showing the effects of marginal changes, another problem should be borne in mind: managers are usually curious to find the possible outcome of *a combination of changes*, and not just changes in individual variables. The following combination of circumstances is an example of a set of changes specified by management for consideration:

an increase in wage rates of 5%
an increase in volume of 8%
an increase in fixed costs of 10%
an increase in the number of employees of 2%
an increase in unit price of 6%

This set of figures constitutes a *possible scenario*, for which the consequences need to be ascertained. The effect of change in each of the listed variables cannot be directly derived by extrapolation from each of the individual effects from the 1 % sensitivity table because of the inaccuracies involved in scaling, alluded to above.

Furthermore, the cumulative effect of several changes is not necessarily additive, namely it is not the sum of the effects of the individual changes. To illustrate this point, consider the results in Table 13.4, where the effect of a 1 % increase in V_0^* or p^* respectively is compared with a 1 % increase in both. The first two rows in Table 13.4 are taken from Table 13.2 and we find that only for V^* the additivity property holds (because of the definition $V^* = V_0 - \varepsilon p^*$), i.e. that the sum of the first two rows is equal to the effect of the simultaneous changes $V_0^* = 1.0$ % and $p^* = 1.0$ %. Similarly, the additivity property does not hold (again, with the exception of V^*) in the second part of Table 13.4 (where V_0 and p are each subjected to a fall of 1 %), nor in the third part (where the changes are +1.0 % and −1.0 % respectively).

The fact that the additivity property does not hold means that, even if we were to replace the 1 % sensitivity table by tables corresponding to 5 %, 8 %, etc., as postulated by the scenario specified earlier, the total effect on, say, profit is not the effect on profit of a 5 % change in wage rates plus the effect of an 8 % change in volume, plus etc. The interaction between the individual changes may cause a significant deviation from the arithmetic sum of the individual

TABLE 13.4
An example to test the property of additivity.

	V^*	I_w^*	C^*	c^*	R^*	Z^*	r^*
$V_0^* = 1.0\%$	1.00	0.80	0.72	−0.28	1.00	2.96	2.35
$p^* = 1.0\%$	−1.20	−0.17	−0.55	0.66	−0.21	2.14	2.27
Total	−0.20	0.63	0.17	0.38	0.79	5.10	4.62
$V_0^* = p^* = 1.0\%$	−0.20	0.64	0.05	0.25	0.80	6.05	5.55
$V_0^* = -1.0\%$	−1.00	−0.80	−0.72	0.28	−1.00	−2.96	−2.37
$p^* = -1.0\%$	1.20	0.15	0.54	−0.65	0.19	−2.28	−2.39
Total	0.20	0.65	−0.18	−0.37	−0.81	−5.24	−4.76
$V_0^* = p^* = -1.0\%$	0.20	−0.64	−0.05	−0.25	−0.80	−6.08	−5.62
$V_0^* = 1.0\%$	1.00	0.80	0.72	−0.28	1.00	2.96	2.35
$p^* = -1.0\%$	1.20	0.15	0.54	−0.65	0.19	−2.28	−2.39
Total	2.20	0.95	1.26	−0.93	1.19	0.68	−0.04
$V^* = 1.0\%$, $p^* = -1.0\%$	2.20	0.94	1.07	−1.11	1.18	1.94	1.23

effects, as shown above (and as demonstrated by examples of the multiplication rule in Chapter 4). Admittedly, when the individual changes are small, an additive procedure based on the 1 % sensitivity table may prove an adequate approximation for some of the key measures, but as the additive property does not apply as a general rule, the potential errors (shown in Table 13.4) are not insignificant.

Thus, the set of changes listed in the above scenario needs to be fed into the appropriate models to derive the overall outcome. Since all the values of the changes are given in deterministic terms, the analysis takes the form of a *deterministic appraisal* of the postulated scenario. All the models developed in the preceding chapters are deterministic in nature, since they are based on the assumption that the causal changes are finite and known (or assumed to be known).

In practice the manager is interested in many variants of the basic scenario: what would happen if the basic scenario generally applies, but the increase in volume turns out to be 10 % instead of 8 %? Similarly, what would happen if the other actual changes deviate from the forecast values listed in the basic scenario?

For example, let us assume that the variables listed in the scenario may be divided into two categories, as shown in Table 13.5. We visualize in this case that the outcome of wage negotiations is uncertain, so that a forecast is given of a range of increased rates of 2–8 %, with 5 % being the most likely outcome. Similarly, the level of future demand is unknown, but market intelligence

TABLE 13.5
An example of a postulated scenario.

(1) Variables outside the immediate and complete control of management:

		Best (most likely) estimate	Range	Optimistic outcome	Pessimistic outcome
increase in wage rates	w	5%	2–8%	2%	8%
increase in volume	V_0	8%	4–12%	12%	4%
increase in fixed costs	F	10%	8–12%	8%	12%

increase in the number of employees is assumed to depend on the increase in volume V_0^* (and not on the total volume increment V^*), with the following reference points:

volume increase V_0^* (%)	4	8	12
increase in number of employees N^* (%)	0	2	4

(2) Decision variable
unit price—to be determined

suggests an increase of 4–12%, and impending increases in rent and energy costs are likely to increase the fixed costs by 8–12%.

We further visualize that management will be successful in reducing the level of manning, partly as a consequence of introducing more efficient working methods, and partly due to employees' efforts, so that an increase of 4% in intrinsic demand can be achieved with the same workforce and for every 1% further increase in this volume the number of employees would have to increase by 0.5%. The remaining variable is that of the unit price, which—unlike the basis scenario—is unspecified, and alternative price changes may, therefore, be examined for the envisaged set of changes listed.

To proceed with the analysis, further information is required about the cost structure, the price elasticity of demand and the prevailing profit margin. Let us assume that in this case

$$C = S + F$$

where

$$S = W + s_1 V$$

i.e. the variable costs consist of the wage component W and a further component dependent on volume. Thus

$$C = W + s_1 V + F$$

where the cost proportions are given as

$$\frac{W}{C} = 0.3$$

$$\frac{s_1 V}{C} = 0.3$$

$$\frac{F}{C} = 0.4$$

Hence, assuming $s_1^* = 0$,

$$C^* = 0.3\,W^* + 0.3\,V^* + 0.4\,F^*$$

and

$$c^* = \frac{C^* - V^*}{1 + V^*}$$

If it is assumed that the change in volume is given by

$$V^* = V_0^* - \varepsilon p^*$$

where the first term denotes the intrinsic change in market demand and the second expresses the effect of a price increase p^*, and if

$$C_0^* = 0.3\,W^* + 0.3\,V_0^* + 0.4\,F^*$$

where C_0^* is the expected total cost increment for $p^* = 0$, then

$$C^* = C_0^* - 0.3\,\varepsilon p^*$$

The change in profit is

$$Z^* = \frac{1}{\alpha}[R^* - (1-\alpha)C^*]$$

$$= \frac{1}{\alpha}[p^*(1+V^*) + V^* - (1-\alpha)(C_0^* - 0.3\,\varepsilon p^*)]$$

Assuming that in this example

$$\varepsilon = 1.0$$
$$\alpha = 0.1$$

then

$$Z^* = 10\,R^* - 9\,C^*$$

or, after substituting V^*

$$Z^* = p^*(2.7 + 10\,V_0^*) - 10\,p^{*2} + 10\,V_0^* - 9\,C_0^*$$

The most likely scenario

A deterministic appraisal, based on the most likely estimates given in Table 13.5, involves the following values:

$$F^* = 0.1$$
$$w^* = 0.05$$
$$N^* = 0.02$$

The expected increase in the wage bill is, then,

$$W^* = w^* + N^* + w^*N^* = 0.071 \;\; (7.1\,\%)$$

and C_0^* is

$$C_0^* = 0.3 \times 0.071 + 0.3 \times 0.08 + 0.4 \times 0.1 = 0.085$$

and if c_0^* denotes the change in unit cost for $p^* = 0$, then

$$c_0^* = \frac{0.085 - 0.08}{1.08} = 0.005$$

i.e. if no action is taken to adjust the price level, the total cost is expected to increase by 8.5 % and the unit cost by 0.5 %. Substituting the value of c_0^* we get

$$Z^* = p^*\;(2.7 + 10\,V_0^*) - 10\,p^{*2} + 10\,V_0^* - 0.7677$$

V_0^* is estimated at 0.08, hence

$$Z^* = 3.5\,p^* - 10\,p^{*2} + 0.0323$$

from which the following conclusions may be drawn:

(1) If the unit price remains unchanged, the profit level is expected to rise by a modest 3.2%.
(2) If $p^* = 0.05$, as contemplated in the basic scenario, then the total cost is expected to rise by 7.0%, the unit cost by 3.9% and the profit by 18.2%.
(3) The unit cost remains unchanged, or is expected to decline, when $C^* \leqslant V^*$ i.e.

$$C_0^* - 0.3\,p^* \leqslant V_0^* - p^*$$

or

$$p^* \leqslant 1.429\;(V_0^* - C_0^*)$$

so that if $V_0^* < C_0^*$ then a price increase must lead to an increase in unit cost. In our case, substituting C_0^* and V_0^* for the deterministic case and we get

$$p^* \leqslant -0.008$$

i.e. the unit price remains unchanged if the price is reduced by just under 1%.
(4) Profit is maximized at

$$\frac{dZ^*}{dp^*} = 3.5 - 20\,p^* = 0$$

namely at

$$p^*_{\text{opt}} = 0.175$$

when profit is expected to rise by 31.9%.

The results of the deterministic appraisal for p^* as a decision variable are summarized in Table 13.6.

TABLE 13.6
Results for a deterministic appraisal based on the most likely estimates (all in %).

p^*	−5	0	5	10	15	17.5	20	25	30	35
V^*	13.0	8.0	3.0	−2.0	−7.0	−9.5	−12.0	−17.0	−22.0	−27.0
C^*	10.0	8.5	7.0	5.5	4.0	3.3	2.5	1.0	−0.5	−2.0
c^*	−2.6	0.5	3.9	7.7	11.9	14.1	16.5	21.7	27.6	34.3
R^*	7.4	8.0	8.2	7.8	7.0	6.3	5.6	3.8	1.4	−1.5
Z^*	−16.8	3.2	18.2	28.2	33.2	33.9	33.2	28.2	18.2	3.2
						↑ max				

Two other scenarios

Apart from the results based on the most likely estimates in Table 13.5, managers are often interested in the extreme scenarios related to the *optimistic* and *pessimistic* estimates. The former postulates the lowest increase in wage rates (2 %) coupled with a low increase in fixed costs (8 %) and a high increase in volume (12 %), whereas the pessimistic scenario envisages estimates at the other ends of the respective ranges. In this way the best and worst possible outcomes can be computed, as shown in Table 13.7.

A comparison of these results with those in Table 13.6 (Z^*, R^* and c^* are also presented in Figs 13.1–13.3) shows how sensitive the outcome for Z^* can be to the assumptions in the scenario. For example, a 5 % price increase, taken as the most likely estimate, is expected to lead to an 18.2 % increase in profit, whereas the corresponding figures for the best (optimistic) and worst

TABLE 13.7
Results for extreme scenarios (all in %).

p^*	−5	0	5	10	15	17.5	20	25	30	35
(a) Assuming optimistic estimates										
V^*	17.0	12.0	7.0	2.0	−3.0	−5.5	−8.0	−13.0	−18.0	−23.0
C^*	10.1	8.6	7.1	5.6	4.1	3.4	2.6	1.1	−0.4	−1.9
c^*	−5.9	−3.0	0.1	3.5	7.3	9.4	11.5	16.2	21.5	27.4
R^*	11.2	12.0	12.4	12.2	11.6	11.0	10.4	8.8	6.6	4.0
Z^*	20.4	42.4	59.4	71.4	78.4	80.0	80.4	77.4	69.4	56.4
(b) Assuming pessimistic estimates										
V^*	9.0	4.0	−1.0	−6.0	−11.0	−13.5	−16.0	−21.0	−26.0	−31.0
C^*	9.9	8.4	6.9	5.4	3.9	3.2	2.4	0.9	−0.6	−2.1
c^*	0.8	4.2	8.0	12.1	16.7	19.2	21.9	27.7	34.3	41.9
R^*	3.6	4.0	4.0	3.4	2.4	1.6	0.8	−1.3	−3.8	−6.9
Z^*	−53.6	−35.6	−22.6	−14.6	−11.6	−12.0	−13.6	−20.6	−32.6	−49.6

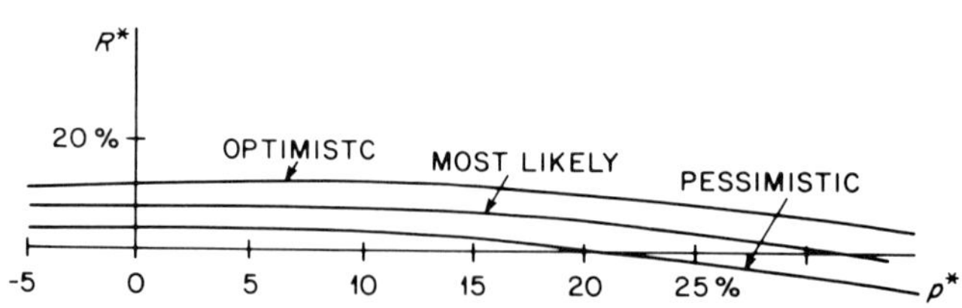

FIG. 13.2. Change in revenue for three scenarios.

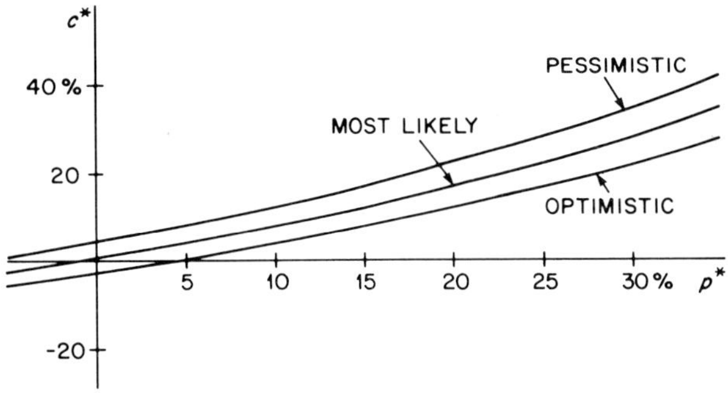

FIG. 13.3. Change in unit cost for three scenarios.

(pessimistic) cases are 59.4% and −22.6% respectively. Notice that although the most likely estimates in Table 13.5 lie in the middle of their respective ranges, the results in Table 13.6 are not always the average of the results for the two scenarios in Table 13.7. In contrast to the wide range of outcomes for Z^*, in Fig. 13.1, the three scenarios result in closer values for R^* and c^*, as shown in Fig. 13.2 and Fig. 13.3.

Risk simulation

The three scenarios—the most likely, the best and the worst—are of great interest to managers, but they may justifiably argue that the probability of the two extremes cases in Table 13.7 is rather small. To make full use of the information given in Table 13.5, we need to consider many scenarios which conform to the specified ranges of values. The procedure for doing so is based on the Monte Carlo simulation method, namely on drawing samples from distributions for the estimated variables, so that each set of samples constitutes a scenario for which the outcome is evaluated. The process is then repeated many times and a distribution (or a histogram) of the simulated outcomes can then be presented.

A comparison between the methods of deterministic appraisal and risk simulation is shown in Fig. 13.4 [3, p. 138]. Needless to say, these methods are not confined to the analysis of cost and profit, but can be used for any other measures of performance, such as cost components, added value, revenue, working capital, return on capital, productivity ratios, etc. An example of the application of risk simulation in a case study of steel plant [3, pp. 140–142] is given in Figs 13.5 and 13.6, showing histograms for changes in the total unit cost and return on capital.

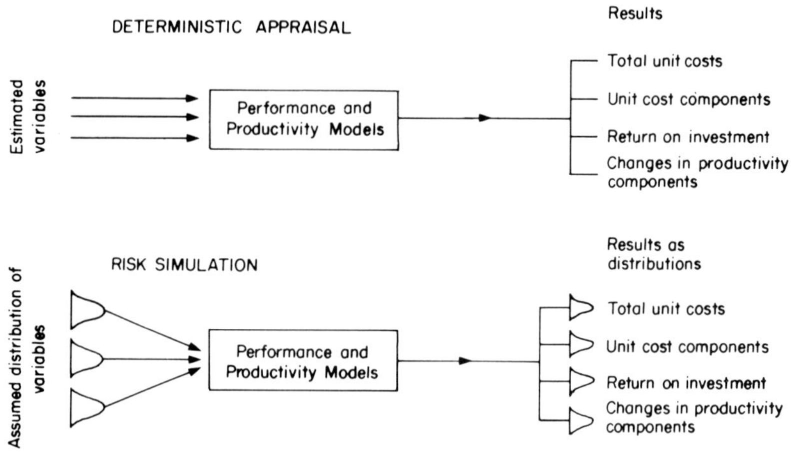

FIG. 13.4. The methods of deterministic appraisal and risk simulation [3, p. 138].

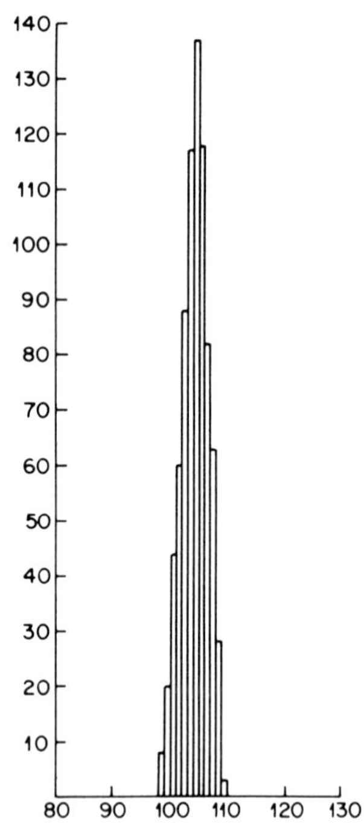

FIG. 13.5. Histogram for total costs from risk simulation (768 runs) [3, p. 141].

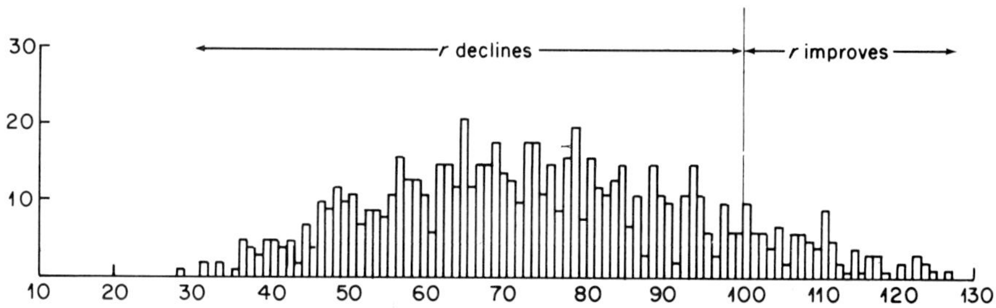

FIG. 13.6. Histogram for return on investment (r) from risk simulation (768 runs) for no price increase [3, p. 142].

The methodology of risk simulation raises two important procedural questions:

(1) *What distribution should be used for the changes in the various variables?* It is rare for sufficient data to be available for these distributions to be explicitly stated, so that assumptions have to be made on the basis of limited information, such as in Table 13.5. In our example, the best estimate for each variable lies in the middle of the range, in which case a symmetrical distribution is an appropriate assumption, but cases are sometimes encountered where skewed distributions need to be used. Experience with risk simulation suggests that the shape of the distributions is rarely a crucial assumption, so that normal distributions (with truncated tails beyond the limits of the given range), or even triangular distributions, are usually adequate.

(2) *What sampling method should be used?* The procedure described for our example envisages independent sampling from the three distributions for w^*, V^* and F^*, but circumstances may arise when the relative increments are interrelated. For example, a high value of F^* may be expected to arise only if a high value of V^* materializes, whereas a low positive V^* may require only a modest increase in fixed costs. If such relationships hold, then *biased* or *discriminate sampling* methods are called for, to ensure that simulation runs for incompatible data are not carried out (for a description of these methods see [2]).

Risk simulation is easily carried out on a computer (usually 1000 runs are perfectly adequate), and for relatively small models and simple sampling procedures results can even be obtained on suitable desk calculators.

Simulating the example in Table 13.5

For the purpose of simulating the example in Table 13.5, we assume that independent sampling may be used by drawing independent random samples from the distributions of w^*, V_0^* and F^*. The shape of these distributions is not specified in Table 13.5, but as the most likely estimates are in the middle of their respective ranges, it is reasonable to assume normal distributions here within the specified ranges. We may further assume that the range of values in each case covers six standard deviations, i.e. that the range accounts for 99.7 % of all the possible outcomes for the variable in question. This means that, as the difference between the upper and lower limit of the range for w^* is 6 %, one standard deviation for w^* is 1 %. Similarly, the standard deviations for V_0^* and F^* are 1.333 % and 0.667 % respectively. The risk simulation procedure is then carried out as follows:

(1) Generate a normal random deviate x_1 such that its value is between -3 and $+3$.
(2) Generate a second random deviate x_2, where $-3 \leqslant x_2 \leqslant 3$.
(3) Generate a third random deviate x_3, where $-3 \leqslant x_3 \leqslant 3$.
(4) Compute the random variable $w^* = (5+x_1)/100$ (from Table 13.5 we assume that the average value of the distribution for w^* is 0.05).
(5) Compute $V_0^* = (8+1.333\ x_2)/100$.
(6) Compute $F^* = (10+0.667\ x_3)/100$.
(7) Compute $N^* = 0.5\ (V_0^*-0.04)$; this assumed deterministic relationship is based on the data in Table 13.5.
(8) Compute $W^* = w^*+N^*+w^*N^*$.
(9) Compute $C_0^* = 0.3\ (W^*+V_0^*)+0.4\ F^*$.
(10) Compute $V^* = V_0^*-p^*$.
(11) Compute $C^* = C_0^*-0.3\ p^*$.
(12) Compute $c^* = \dfrac{C^*-V^*}{1+V^*}$.
(13) Compute $R^* = p^*+V^*+p^*V^*$.
(14) Compute $Z^* = 10\ R^*-9\ C^*$.
(15) Go back to step 1 and repeat the process, say, 1000 times.
(16) Record histograms for, say, C^*, c^*, R^* and Z^*.

Results

Histograms for c^*, R^* and Z^* are shown in Fig. 13.7 for $p^* = 0$ and in Fig. 13.8 for $p^* = 10\%$ (both for 1000 simulation runs). A summary of the risk simulation results, compared with the deterministic appraisals, is given in Table 13.8. For the risk simulation, the mean and standard deviation are recorded in each case, as well as the minimum and maximum values of the simulated outcomes. The results for the deterministic appraisal relate to the three scenarios discussed earlier. The following observations may be made:

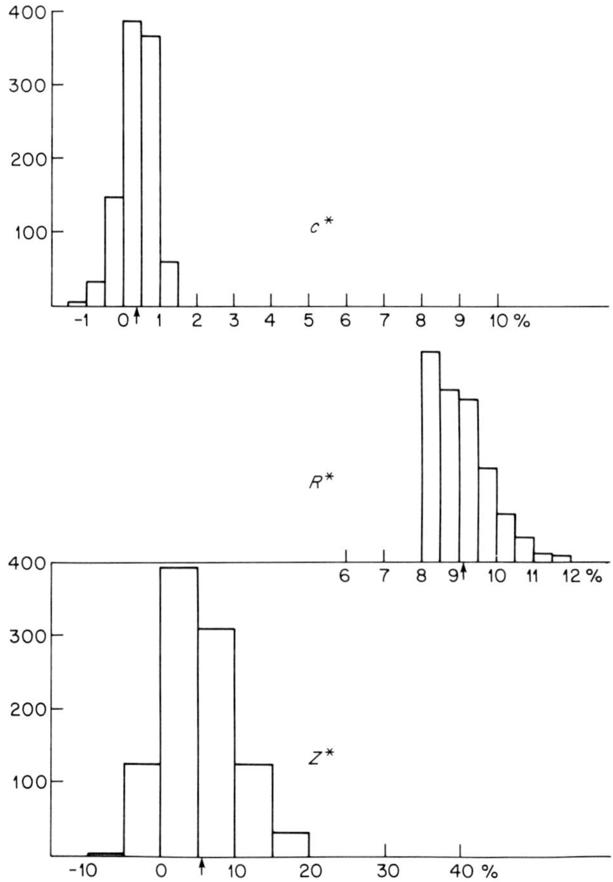

FIG. 13.7. Histograms for $p^* = 0$ (data in Table 13.5; arrow points to mean value).

(1) The histograms for c^* and R^* are far narrower than those for Z^*, again demonstrating the greater sensitivity of the latter to changes in the system.
(2) The histogram for c^* for $p^* = 10\%$ is similar to that for $p^* = 0$, except that it is shifted to the right, so that the mean rises from 0.4 %–7.5 %, while the standard deviation is subject to a moderate increase. Similar observations may be made about the histograms for R^* and Z^* respectively.
(3) All the histograms are somewhat skewed, particularly those relating to R^*.
(4) The mean values of the histograms derived by risk simulation are fairly close to the results of the deterministic appraisal for the most likely case.
(5) The optimistic and pessimistic scenarios are well outside the ranges of the

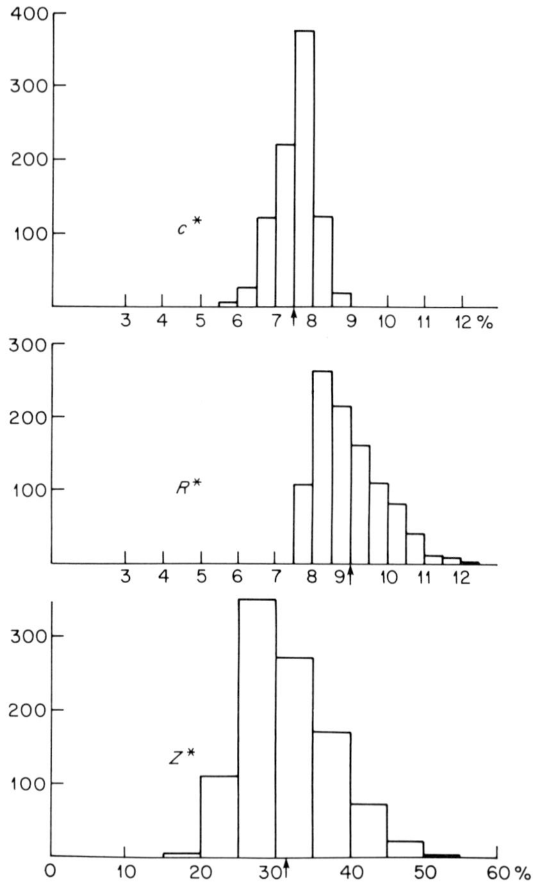

FIG. 13.8. Histograms for $p^* = 10\%$ (data in Table 13.5; arrow points to mean value).

histograms, and this serves to emphasize the very low probability of these scenarios.

A FORMAT FOR CORPORATE PLANNING

The analysis of various scenarios, whether confined to deterministic appraisal or embellished by risk simulation, provides management with basic information for use in determining future budgets. Another important piece of information is a comparison of current performance with current budget and

TABLE 13.8
Results for risk simulation and deterministic appraisal (in %).

	Risk simulation				Deterministic appraisal		
	Min	Mean	Max	s.d.	Pessimistic	Most likely	Optimistic
$p^* = 0$							
change in unit cost c^*	−1.3	0.4	1.4	0.44	4.2	0.5	−3.0
change in revenue R^*	8.0	9.1	12.0	0.79	4.0	8.0	12.0
change in profit Z^*	−5.6	5.4	24.7	5.04	−35.6	3.2	42.4
$p^* = 10$							
change in unit cost c^*	5.6	7.5	8.8	0.55	12.1	7.7	3.5
change in revenue R^*	7.8	9.0	12.0	0.87	3.4	7.8	12.2
change in profit Z^*	18.2	31.5	52.4	5.9	−14.6	28.2	71.4

TABLE 13.9
Monthly performance report.

	Last year actual			This year			Budget		
	This month	To-date	Total	Actual this month	Actual to-date	Forecast total	This month	To-date	Total
Sales (R)									
Volume index (V)									
Cost of sales (S)									
Gross profit ($R-S$)									
Added value (A)									
Overheads									
Factory									
Selling and distribution									
Administration									
Other									
Total (F)									
Promotion and advertising (D)									
Interest charges (J)									
Total cost ($C = S+F+J+D$)									
Profit before interest and tax ($R-C+J$)									
Profit before tax ($Z = R-C$)									
Other data									
Average number of employees (N)									
Employees' remuneration (W)									
Closing stock value									
Closing working capital (I_W)									

Fixed investment (I_F)
Total investment (I)
Capital expenditure
Depreciation
Net cash flow

Profit ratios
Gross margin (β)
Net margin (α)
Return on capital (r)

Cost proportions of
Cost of sales (S/C)
Fixed overheads (F/C)
Interest (J/C)
Promotion (D/C)

Revenue ratios (proportion of sales)
Cost of sales (S/R)
Fixed overheads (F/R)
Employees' remuneration (W/R)
Stock value to sales
Working capital (I_W/R)
Total investment (I/R)

Ratios for employees
Sales per employee (R/N)
Sales to remuneration (R/W)
Profit per employee (Z/N)
Added value per employee (A/N)
Remuneration to added value (W/A)
Output per employee (V/N)

with the past. Such information is derived from a *performance report*, usually prepared monthly, such as in the framework shown in Table 13.9.

The details to be included in a performance report must take account of the type of operations the enterprise is engaged in and its environment. The format in Table 13.9 is an example of data commonly sought by senior executives, but it obviously needs to be modified to suit the circumstances for which it is to be applied. For the current year, figures are given for:

- the month to which the report pertains
- the cumulative value from the beginning of the financial year
- the latest forecast for the total financial year

These figures are compared with the corresponding values for last year and with the budget figures (the totals for last year and for the budget of the current remain the same for all monthly reports in a given year). Thus, the nine columns in the report provide a good summary of current operations, show where progress has been made, compared with last year, and highlight deviations from the budget. In an enterprise which comprises several subsidiaries or separate operating divisions, a performance report for each subsidiary or division is called for, as well as a consolidated account for the enterprise as a whole.

Corporate plans and forward budgets follow the same format as in Table 13.9, except that the columns read differently, such as:

For a one year plan—twelve monthly columns (one for each month), total for the first six months and for the last six months for last year and for this year respectively, as well as totals for the whole of last year and for the whole of this year.

For a three or five year plan—annual figures for last year, this year and each year in the planned horizon.

MULTIPLE DECISION VARIABLES AND MULTIPLE OBJECTIVES

As pointed out in earlier chapters, while one measure, such as return on capital, is often singled out as the prime indicator of corporate performance, an enterprise is inherently a multi-objective system, hence the plethora of yardsticks and ratios used by financial analysts in performance reports (see Chapter 2). Inevitably, some objectives are in conflict, so that a domain of feasible decisions needs to be circumscribed to focus managerial attention on the available alternatives. The idea of a feasible domain is not that it can meet fully the aspirations of all the protagonists, who favor (or have an interest in) particular performance criteria, but that it can satisfy them within prescribed limits. Several examples of how our modeling methodology can be of help are described below.

Profit versus revenue

In Chapter 9 an example is given to demonstrate the conflict between profit and revenue, leading to a definition of a tradeoff between the two (see Figs 9.24 and 9.25). A more elaborate example by Eilon and Cosmetatos [1] is described briefly below.

The cost function is given by

$$C = sV + F + D$$

and there are two decision variables, namely p^* and D^*. The total change in volume is

$$V^* = V_0^* + V_p^* + V_D^*$$

where V_0^* is estimated at 2%, V_p^* is the effect of a change in price and V_D^* is the effect of a change in the promotion expenditure. Other known data and assumptions are:

(1) $\alpha = 0.20$

(2) The fixed costs are adjusted for an increase or decrease in output, so that

$$f_1 = \begin{cases} 0.6 \text{ if output is to rise} \\ 0.3 \text{ if output is to fall} \end{cases}$$

$$f_2 = \begin{cases} 0.3 \text{ if output is to rise} \\ 0.5 \text{ if output is to fall} \end{cases}$$

$$f_3 = 0.1$$

(3) Changes in volume cause a short-term change in the unit variable cost. Details are shown in Table 13.10. The relationships between p^* and V_p^* and that between D^* and V_D^* are also given in Table 13.10.

(4) The range of values contemplated for the decision variables is −5% to 5% for p^* and 0 to 10% for D^*

The result is shown in Fig. 13.9, where Z^* is plotted against R^* with p^* and D^* as decision variables [1]. Each curve corresponds to a given value of D^* and the values of p^* are marked on the curves. The solid line represents the dominating set of solutions and suggests that $D^* = 10\%$ should be adopted, while p^* may take any value between 0 and 3% (solutions for a price reduction are not shown in the diagram). The curve shows the tradeoff between an increase in profit and an increase in revenue, and management can now debate

TABLE 13.10
Data for an example [1] (all in %).

(a) Effect on s^*

V^* =	−10	−8	−6	−4	−2	0	2	4	6	8	10	12
s^* =	0.6	−0.1	−0.5	−0.6	−0.4	0	0.6	1.4	2.6	4.3	6.5	9.4

(b) Effect of p^*

p^* =	−5	−4	−3	−2	−1	0	1	2	3	4	5
V_p^* =	5.5	4.2	3.0	1.9	0.9	0	−1.4	−2.9	−4.6	−6.7	−9.7

(c) Effect of D^*

D^* =	0	2	4	6	8	10
V_D^* =	0	0.4	1.3	2.5	3.3	3.7

the merits of various solutions on this curve, in order to determine which combination of Z^* and R^* should be aimed for.

An added concern to management is bound to be the effect of the decision variables on the unit cost, the results for which are shown in Table 13.11. If it is

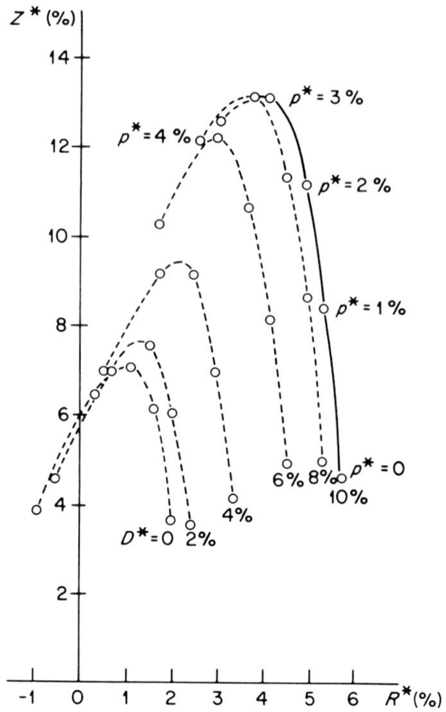

FIG. 13.9. An example of change in revenue vs change in profit ($p^* \geqslant 0$) [1].

TABLE 13.11
Results for c^* [1] (all in %).

$D^* = 0$		2	4	6	8	10
$p^* = 0$	-0.4	-0.3	-0.2	-0.1	0	0.2
1	-0.1	0	0	0	0.1	0.3
2	0.5	0.5	0.3	0.2	0.3	0.4
3	1.4	1.4	1.1	0.7	0.6	0.8
4	2.7	2.7	2.3	1.8	1.6	1.5
			$c^* > 1.0\%$			

thought undesirable to allow the unit cost to rise by more than 1%, then the feasible domain for p^* and D^* is constrained, as shown in Table 13.11, to $p^* \leqslant 2\%$ and to $p^* = 3\%$ provided $D^* \geqslant 4.5\%$.

Unit cost versus employee remuneration

In Chapter 6 the conflict between increasing wage rates and reducing unit labor cost is discussed. A further example to illustrate the modeling approach is given by the following case.

Suppose that the total cost is given by the function

$$C = sV + F + W$$

the cost proportions being

$$f_1 = \frac{sV}{C} = 0.4$$

$$f_2 = \frac{F}{C} = 0.2$$

$$f_W = \frac{W}{C} = 0.4$$

It is assumed that $s^* = F^* = 0$. In negotiations with the labor union, it transpires that the union is expecting a minimum increase in rates of 4%, in exchange for higher productivity, provided the total number of employees does not decrease by more than 1%. Management, in anticipation of increased volume, does not wish the wage bill increase to exceed 4% and the unit cost to increase by 1%. Furthermore, management would not like f_W to exceed its current value.

These constraints may be summarized as follows:

Management constraints	Union constraints
$W^* \leqslant 0.04$	$w^* \geqslant 0.04$
$c^* \leqslant 0.01$	$N^* \geqslant -0.01$
New $f_W \leqslant 0.4$	

This is not the place to argue whether the constraints are reasonable, and whether they would not be modified under pressure, or as a result of skillful negotiations by one of the parties concerned. The modeling exercise is not a substitute for negotiations, nor does it purport to train either side how "to get the best deal" or how "to pull a fast one". The purpose is merely to investigate whether there are circumstances for which the constraints can be met, and if so to determine the feasible domain for further consideration.

The basic equations and constraints for this case are:

$$C^* = 0.4\, V^* + 0.4\, W^*$$

$$c^* = \frac{C^* - V^*}{1 + V^*} \leqslant 0.01$$

$$W^* = w^* + N^* + w^* N^* \leqslant 0.04$$

$$w^* \geqslant 0.04$$

$$N^* \geqslant -0.01$$

$$\text{new } f_W \leqslant 0.4$$

Figure 13.10 shows the relationship between w^* and N^* with an iso-W^* line for $W^* = 4\,\%$. The area bounded by the lines $N^* = -1\,\%$, $w^* = 4\,\%$ and $W^* = 4\,\%$ meets the constraints stated above for these variables, and this means that a solution that satisfies these three requirements is feasible. In fact, the minimum demands stipulated by the union at $w^* = 4\,\%$ and $N^* = -1\,\%$ can be met with $W^* = 3\,\%$, as shown in Fig. 13.10.

Now, the constraint that c^* should not exceed $1\,\%$ leads to

$$\begin{aligned} C^* &= V^* + c^*\,(1 + V^*) \\ &\leqslant V^* + 0.01\,(1 + V^*) \end{aligned}$$

Substituting the function for C^*

$$0.4\, V^* + 0.4\, W^* \leqslant V^* + 0.01\,(1 + V^*)$$

or

$$W^* \leqslant 0.025 + 1.525\, V^*$$

This linear relationship, with an equality sign, is shown in Fig. 13.11, where the domain below the line satisfies the constraint $c^* < 1\,\%$. The horizontal line $4\,\%$

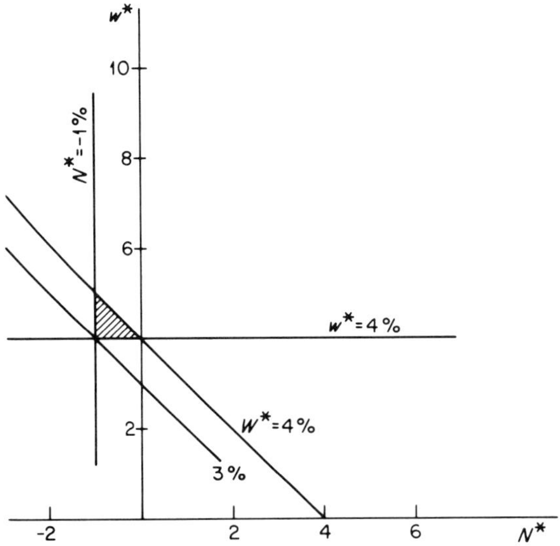

FIG. 13.10. The feasible domain for constraints on w^*, N^* and W^*.

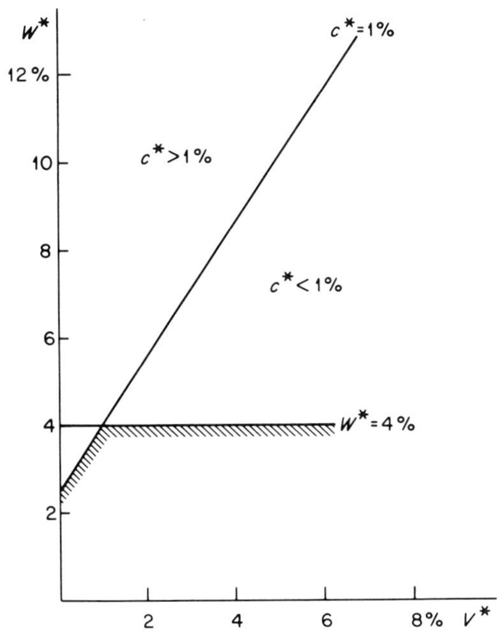

FIG. 13.11. Constraints for c^* and W^*.

represents the upper limit specified by management. It is clear that up to $V^* = 1\%$ the c^* constraint is more demanding, whereas for $V^* > 1\%$ the constraint for c^* will allow the minimum of W^* to increase well above the stated 4%.

To meet both the constraints for W^* and c^* we have:

for $V^* = 0$, $W^* = 0.025$ then $C^* = c^* = 0.01$

for $V^* = 0.01$, $W^* = 0.04$ then $C^* = 0.02$ and $c^* = 0.01$

for $V^* > 0.01$, $W^* = 0.04$ then $C^* = 0.4\,V + 0.016$ and $c^* = \dfrac{0.016 - 0.6\,V^*}{1 + V^*}$

The results for c^* are tabulated in Table 13.12.

As for the new value of f_W it can be calculated from

$$\text{new } f_W = \text{old } f_W \frac{1 + W^*}{1 + C^*} = 0.4 \frac{1 + W^*}{1 + C^*}$$

The computed results are also given in Table 13.12, from which it is clear that the constraint that the value of f_W should not exceed 0.4 cannot be met for low volume increases. It is only when V^* is at least 6% that the constraint for f_W can be satisfied.

The implications for labor productivity can be derived as follows:

$$\pi = \frac{V}{N} = \frac{Vw}{W}$$

Hence

$$\pi^* = \frac{V^* + w^* + V^* w^* - W^*}{1 + W^*}$$

Thus, for $w^* = W^*$ (which is valid for $V^* \geqslant 0.01$)

$$\pi^* = V^*$$

and even when $w^* \neq W^*$ we still have π^* as an increasing linear function of V^*. If this level of increase in productivity cannot be achieved, then one or several of the constraints stated above cannot be met.

The remaining constraint is for N^*. From $W = wN$

$$N^* = \frac{W^* - w^*}{1 + w^*}$$

In our case $w^* = W^*$, so that $N^* = 0$ and the constraint is therefore met. However, if the union were to press for $w^* = 6\%$ (while $W^* = 4\%$), then the resultant value for N^* would be -1.9%, i.e. a reduction in manpower almost double the figure stipulated in the set of constraints.

TABLE 13.12

Results for an example with constraints on c^* and W^*

V^*	0	1	2	3	4	5	6	7	8	9	10
W^*	2.5	4.0	4.0	4.0	4.0	4.0	4.0	4.0	4.0	4.0	4.0
c^*	1.0	1.0	0.4	−0.2	−0.8	−1.3	−1.9	−2.4	−3.0	−3.5	−4.0
new f_W	0.406	0.408	0.406	0.405	0.403	0.402	0.400	0.398	0.397	0.395	0.394

c^* dominates ⟵|⟶ W^* dominates

Wage rate, added value and profit

Another example, that concerns constraints associated with added value, is given in the following case. The total cost function is

$$C = M + W + E$$

where

$$f_M = 0.6$$
$$f_W = 0.2$$
$$f_E = 0.2$$

and $E^* = 0$, while the cost of materials M is proportional to volume (i.e. $M^* = V^*$). Volume is affected by price and it is estimated that

$$V^* = -1.2\, p^*$$

and the current profit margin is $\alpha = 0.12$. Improved manufacturing methods (assumed here to involve no capital cost) are estimated to reduce the labor force by 2%, but the employees demand an increase of 5% in wage rates as compensation. The objectives of management are:

(1) an increase in profit of at least 20% ($Z^* \geqslant 0.2$)
(2) an increase in added value of at least 10% ($A^* \geqslant 0.1$)
(3) an increase in added value per employee of at least 5% ($(A/N)^* \geqslant 0.05$)
(4) no increase in the ratio of remuneration to added value ($(W/A)^* \leqslant 0$)

The decision variable is p^*, but a price increase above 12% is likely to have severe repercussions on demand and is therefore undesirable. The question is: can all these constraints be met?

Assuming that $w^* = 0.05$ and $N^* = -0.02$, we find that

$$W^* = w^* + N^* + w^* N^* = 0.029$$

Now,

$$C^* = f_M M^* + f_W W^* + f_E E^*$$

Substituting the cost proportions, M^* and W^* (with $E^* = 0$)

$$C^* = 0.6\, V^* + 0.0058 = -0.72\, p^* + 0.0058$$

Hence

$$Z^* = \frac{1}{\alpha}\left[R^* - (1-\alpha)C^*\right]$$

$$= \frac{1}{0.12}(R^* - 0.88\, C^*)$$

where

$$R^* = -0.2\, p^*(1 + 6\, p^*)$$

The change in added value is derived as follows

$$A = W + E + Z$$

$$A^* = \frac{W}{A} W^* + \frac{Z}{A} Z^*$$

but

$$\frac{A}{C} = \frac{W}{C} + \frac{E}{C} + \frac{Z}{C} = f_W + f_E + \frac{\alpha}{1-\alpha} = 0.5364$$

so that

$$\frac{W}{A} = \frac{W}{C}\frac{C}{A} = \frac{f_W}{0.5364} = 0.3729$$

and

$$\frac{Z}{A} = \frac{Z}{C}\frac{C}{A} = \frac{\alpha}{1-\alpha}\frac{1}{0.5364} = 0.2542$$

hence

$$A^* = 0.3729\, W^* + 0.2542\, Z^*$$

$$= 0.0108 + 0.2542\, Z^*$$

If the constraint $A^* \geqslant 0.1$ is to be observed, then

$$0.0108 + 0.2542\, Z^* \geqslant 0.1$$

or

$$Z^* \geqslant 0.351$$

so that the constraint for $A^* \geqslant 0.1$ is far more demanding than the one stated earlier for Z^* to the effect that $Z^* \geqslant 0.2$.

The last constraints (3) and (4) in the set of management objectives are

$$\left(\frac{A}{N}\right)^* = \frac{A^* - N^*}{1 + N^*} \geqslant 0.05$$

or

$$A^* \geqslant 0.05\,(1 + N^*) + N^* = 0.029$$

and

$$\left(\frac{W}{A}\right)^* = \frac{W^* - A^*}{1 + A^*} \leqslant 0$$

or

$$A^* \geqslant W^* = 0.029$$

so that both constraints imply that added value should rise by more than 2.9 %, a much less demanding requirement than $Z^* \geqslant 0.2$ and $A^* \geqslant 0.1$.

Results for p^* as a decision variable are shown in Table 13.13 for the range of p^* between -4 % and 20 % (i.e. well beyond the constraint of 12 %) from which

TABLE 13.13
Results for an example with constraints on A^* and Z^*.

p^*	−4	−2	0	2	4	6	8	10	12	14	16	18	20
Z^*	−20.3	−11.9	−4.3	2.6	8.6	13.8	18.3	21.9	24.7	26.7	28.0	28.4	28.0
A^*	−4.1	−1.9	0	1.7	3.3	4.6	5.7	6.6	7.4	7.9	8.2	8.3	8.2
$\left(\frac{A}{N}\right)^*$	−2.1	0.1	2.0	3.8	5.4	6.7	7.9	8.8	9.6	10.1	10.4	10.5	10.4
$\left(\frac{W}{A}\right)^*$	7.3	4.9	2.9	1.1	−0.4	−1.6	−2.7	−3.5	−4.2	−4.6	−4.9	−5.0	−4.9

we conclude that:

$Z^* \geqslant 20\%$ when $p^* \geqslant 8.9\%$
$A^* \geqslant 10\%$ cannot be achieved, the highest value for A^* being 8.3% at $p^* = 18\%$, if such a value were permissible

$$\left(\frac{A}{N}\right)^* \geqslant 5\% \text{ and } \left(\frac{W}{A}\right)^* \leqslant 0 \text{ at } p^* \geqslant 3.5\%$$

It is, thus, clear that all the constraints in this case can be met within the range specified for p^*, except for $A^* \geqslant 10\%$. At the maximum permissible price increase we expect $A^* = 7.4\%$. If the constraint for A^* may be ignored, then a tradeoff can be readily determined between w^* and Z^*, as shown in Fig. 13.12 for various values of p^*. It is inferred from this diagram that for every 1% rise in wage rate the value of Z^* drops by 1.44%.

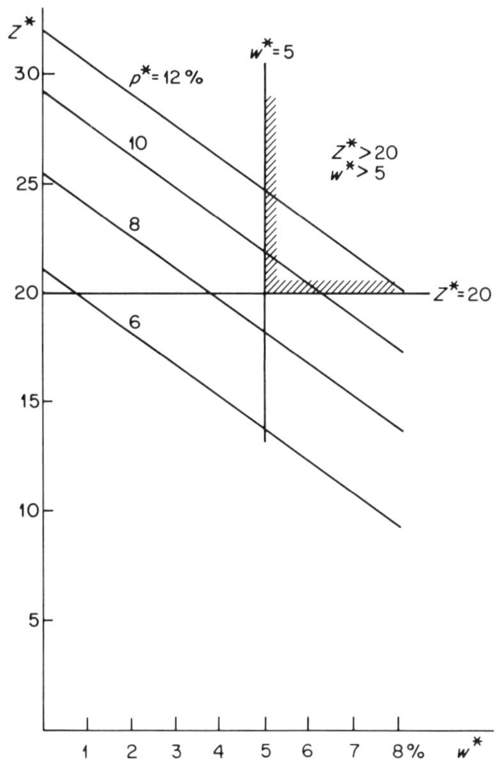

FIG. 13.12. Constraints for C^* and w^*.

SHARING THE BENEFITS

Extra benefits

The last two examples illustrate how management and labor unions often set about defining their objectives or constraints independently of each other. And yet, the outcomes are obviously interrelated. The case described in Fig. 13.12 highlights the conflict between profit and wage rate, showing how one can be improved at the expense of the other for any given level of p^*. The answer in that case is to manipulate the decision variable p^* in such a way as to satisfy the minimum requirements of both sides.

In fact, if $p^* = 12\%$ is to be implemented, there is an opportunity for greater payoffs than the stated minima, as shown in Fig. 13.13, where the triangle corresponds to the feasible domain in Fig. 13.12 (except that now the coordinates have the same scale). The oblique line I-II stands for $p^* = 12\%$ and any point on this line represents a particular share of the extra benefits (i.e. over and above the respective minimum values). The two extreme points are:

Point I: $Z^* = 25\%$, $w^* = 5\%$; here all the extra benefit goes into profit
Point II: $Z^* = 20\%$, $w^* = 8.3\%$; here all the extra benefit goes into the wage rate.

Management would normally prefer point I while the employees would prefer point II.

The problem is reminiscent of a two-person non-zero-sum game, where each player can assure himself of a minimum payoff, but getting more than the minimum requires cooperation by both sides to find a solution perceived as

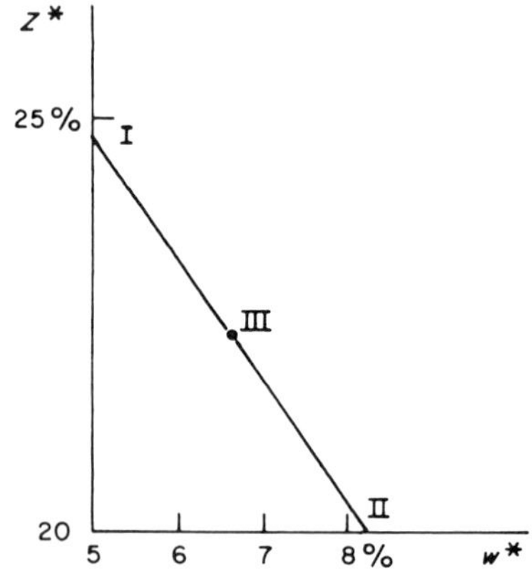

FIG. 13.13. A median solution for sharing extra benefits.

equitable to both. There are, of course, many ways in which the problem can be resolved, such as [4, 5]:

(1) The absolute benefit should be the same for each side, i.e. if the minimum values for Z^* and w^* are denoted as $Z^*_{\min}$ and $w^*_{\min}$ respectively, then this solution would require that

$$Z^* - Z^*_{\min} = w^* - w^*_{\min}$$

(2) The respective relative benefits should be the same, namely

$$\frac{Z^*}{Z^*_{\min}} = \frac{w^*}{w^*_{\min}}$$

(3) Each side gets a benefit which is half way between his minimum and the possible maximum. Thus, management is allocated $Z^* = 22.5\%$ which is the average of 20 % and 25 %, while the employees get 6.65 %, which is half way between 5 % and 8.3 %. This is called the *median solution* and is shown in Fig. 13.13 at point III (which is midway between I and II). This method is also applicable when no minimum values for Z^* and w^* are specified, in which case the feasible range of solutions is represented by the line $p^* = 12\%$ across the positive quadrant in Fig. 13.12.

Other sharing methods

Clearly, numerous formulae can be devised to give a reward to employees when the performance of the enterprise improves. We mention here only a few:

1. Profit sharing by a fixed relationship between w^* and Z^*

The relationship may take the form

$$w^* = kZ^*$$

where k is a constant. The median solution when $Z^*_{\min} = w^*_{\min} = 0$ is essentially a formula of this type. In some cases Z^* is divided into several bands (e.g. 0–20 %, 20–40 %, etc.) with different values of k for each band.

2. Absolute profit sharing

The additional profit made is

$$\delta Z = ZZ^*$$

and it is divided so that

$$\delta Z = G + U$$

where $G =$ the amount retained by management for distribution of dividends, for capital expenditure and reserves,
$U =$ the amount paid to employees to augment current remuneration.
A possible formula for this sharing scheme could be

$$U = k\,\delta Z$$

3. Sharing added value

Since added value is

$$A = W + E + Z$$

management is anxious to control the level of remuneration in order to allow sufficient money to be available for meeting the other expenses E and for the residual profit Z. Possible formulae are:

$$W = k\,A$$

or

$$W^* = k\,A^*$$

4. Sharing the benefit of cost reduction

It is sometimes argued that when increased profit cannot be directly related to the effort made by the whole workforce, it may be more appropriate to reward employees for the reduction in total cost. If volume does not change, then a change in cost δC (assuming that it is negative) can lead to a benefit, such as

$$U = -k\delta C$$

or

$$w^* = -kC^*$$

But when a change in volume does take place then a more appropriate measure is the unit cost, so that the total saving is estimated by

$$-V(1 + V^*)\delta c$$

where $V(1 + V^*)$ is the new volume and δc the reduction in unit cost (so that if δc is negative, positive savings are made). A share of this saving may then be distributed to the employees.

Use of models for such schemes

The sharing methods described above are merely illustrations of the problem of how to reward employees for benefits accrued to the enterprise. Any scheme considered for adoption needs to be examined in some detail as to the possible consequences to various performance criteria when circumstances change. We have seen several examples of how the modeling approach of incremental calculus can be used for the purpose, though the details with which any model needs to be developed is a matter of judgement, depending on prevailing conditions in the corporate environment.

CONCLUSION

Modeling based on incremental calculus can be invaluable, both for analysis of past performance, as well as for planning purposes. Performance reports, prepared on a monthly, quarterly or annual basis, vary in the degree of detail which the top management of an enterprise thinks is appropriate. An example of a format, found to be suitable in many circumstances, is shown in Table 13.9, and a similar layout is convenient for forward budget projections.

An important ingredient in the corporate planning function is the ability to evaluate the possible consequences of changing circumstances. This is where the modeling approach presented in this volume can be applied (for example in the form of incremental sensitivity tables to examine changes at the margin), particularly when a combination of several possible changes is envisaged. The method used is that of *scenario analysis*, where each scenario corresponds to a given set of specified changes.

If the postulated changes are deterministic, the analysis is a deterministic appraisal. However, it is often found that estimates of future changes take the form of forecast ranges, such as in the example in Table 13.5. In such cases, the analysis can be carried out by one or both of the following methods:

(1) A series of *deterministic appraisals* of chosen scenarios, including (or confined to) three of particular interest:
   (a) The *most likely scenario*, based on the most likely estimates.
   (b) The *optimistic scenario*, assuming that all the changes will be beneficial from the point of view of performance outcome.
   (c) The *pessimistic scenario*, assuming that the worst will happen.

(2) *Risk simulation*, based on repeated sampling from distributions of the variables subject to change, thereby simulating many possible scenarios, so that histograms of selected outcomes can be presented.

Another issue of concern to planners is how to deal with multiple objectives and/or constraints. Conflicts often ensue and it is instructive to find what tradeoff functions pertain between conflicting criteria. In addition, the modeling approach helps to ascertain whether a given set of constraints and/or objectives can be fully satisfied, and if so to proceed to define the feasible domain within which alternative courses of action can be evaluated.

Examples of this approach are given in this chapter to show how objectives and constraints relating to changes in profit, revenue, wage rates, total wage bill, unit cost, added value, added value per employee, ratio of wages to added value, etc., can be accommodated in suitably constructed models. Thus, the methodology offers opportunities of providing assistance in the process of negotiations, for example between management and employees, or between managers who espouse different objectives. It can also be useful in evaluating the possible consequences of various benefit schemes, such as profit sharing.

REFERENCES

[1] EILON, S. and COSMETATOS, G. P. (1980). Models for collective decision making in industry. *Europ. J. Operational Res.* **4**, 374–379.

[2] EILON, S. and FOWKES, T. R. (1973). Sampling procedures for risk simulation. *Operational Res. Q.* **24**, 242–252 (also In "Applications of Management Science in Banking and Finance", Chapter 4, Gower Press, London, 1972).

[3] EILON, S., GOLD, B. and SOESAN, J. (1976). "Applied Productivity Analysis for Industry". Pergamon Press, Oxford.

[4] EILON, S. (1973). Further thoughts on the minimax principle. *Omega* **1**, 505–511.

[5] EILON, S. (1975). A median solution for a two-person non-zero-sum game. *Omega* **3**, 107–111.

EXERCISES

1. Calculate the results in Table 13.2.
2. Tabulate Z^* for changes from -10% to 30% in V_0 and p respectively for the data on which Table 13.6 is based.
3. Compute the results shown in Table 13.7.
4. In the example in Table 13.5 it is inferred from the data that

$$N^* = 0.5\,(V_0^* - 0.04)$$

   In other words, the change in the number of employees is dependent solely on the change in the intrinsic demand V_0^*. A more realistic assumption might be to relate N^* to the total change in volume V^*. If the relationship

$$N^* = 0.5\,(V^* - 0.04)$$

   is then assumed, compute the outcomes for C^*, c^*, R^* and Z^* when $p^* = 0$, 5, 10 and 17.5% for the most likely scenario in Table 13.5 and compare them with Table 13.6.
5. Figures 13.7 and 13.8 give results for a risk simulation for the data given in Table 13.5 for $p^* = 0$ and 10% respectively. Carry out a risk simulation for the same data for $p^* = 5\%$ and 17.5% respectively.
6. Repeat the risk simulation in the previous exercise for

$$p^* = 0,\ 5,\ 10\%$$

when the assumption for N^* takes the form

$$N^* = 0.5\,(V^* - 0.04)$$

7. The results in Table 13.12 are based on the constraints:

$$\begin{aligned} c^* &\leqslant 0.01 \\ W^* &\leqslant 0.04 \\ w^* &\geqslant 0.04 \\ N^* &\geqslant -0.01 \end{aligned}$$

If the constraint for W^* is ignored, calculate the maximum values allowable for W^* and N^*, as well as the corresponding labor productivity improvement π^*, for $w^* = 4$, 6 and 8% respectively and when V^* is between 0 and 10%.

8. For the example in Fig. 13.12, calculate the value of Z^* for $p^* = 5\%$ and 10% respectively, when w^* ranges from 0 to 8%.

9. If the cost function takes the form

$$C = sV + F$$

where $f_1 = 0.6$, and if demand is expected to increase by 4%, find whether the following constraints, which are ranked in the order of priority, can be met:

$$\begin{aligned} Z^* &\geqslant 0.2 \\ \alpha^* &\geqslant 0 \\ c^* &\leqslant 0.03 \\ R^* &\geqslant 0. \end{aligned}$$

The decision variable is p^*, which may be considered within the range -10% to 20% and which leads to

$$V_p^* = -1.8\,p^*$$

It is also known that $\alpha = 8\%$.

10. Determine when the following constraints can be met:

$$W^* \geqslant kA^*$$

$$\left(\frac{W}{A}\right)^* \leqslant 0$$

$$\left(\frac{Z}{A}\right)^* \geqslant 0$$

where

$$\frac{Z}{A} = 0.3$$

Notation

Where symbols are used infrequently in this volume, their meaning is defined when they are introduced in the text, and hence not all these symbols are included in the list shown here. Note that some symbols are used for several purposes; hence, the appropriate definitions relate to the context in which they are used, and these definitions are given in the text.

A = added value = $R - M = W + E + Z$
B = borrowings
C = total cost
D = advertising costs
E = other expenses
F = fixed cost
G = amount of added profit retained after profit sharing
H = total hours worked = NT
I = total investment = $I_F + I_W$
I_F = fixed investment
I_W = working capital
J = interest charges on borrowings = jB
K = cash and equity finance
K = constant
M = cost of materials input
N = average number of employees
P (or $V_{\max}$) = capacity
Q = physical materials input
R = revenue = pV
S = variable cost
T = average number of hours per employee
T_b = number of basic hours per employee
T_o = number of overtime hours per employee
U = amount of added profit distributed to employees
V = physical output volume
V_0^* = relative change in volume due to intrinsic changes in the market
V_D^* = relative change in volume due to a change in advertising = σD^*
V_p^* = relative change in volume due to a change in price = $-\varepsilon p^*$

W = total wages, salaries and benefits
W_b = wages for basic hours
W_o = wages for overtime hours
X = cost of x units of a given factor input
Y = index number compared with 100
Y_{nb} = index number in year n compared with the base year b
Z = profit before tax $= R - C = (p - c)\,V$

a = a constant
b = a constant; or refers to the base period; or refers to basic hours
c = average cost per unit of output $= C/V$
c_W = cost per unit due to labor
c_M = cost per unit due to materials
c_E = cost per unit due to other expenses
c_F = cost per unit due to fixed costs
c_S = cost per unit due to variable costs

$$e = \text{capacity utilization} = \frac{\text{output}}{\text{capacity}} = V/P$$

= usually $e(3)$

$$e(1) = \text{technical efficiency} = \frac{\text{minimum resources}}{\text{actual resources}}$$

$$e(2) = \text{cost efficiency} = \frac{\text{minimum cost}}{\text{actual cost}}$$

$$e(3) = \text{capacity utilization} = \frac{\text{actual output}}{\text{maximum output}}$$

$$e(4) = \text{revenue efficiency} = \frac{\text{actual revenue}}{\text{maximum revenue}}$$

f_1, f_2, etc. = cost proportions
$f_1 = s/c$
$f_2 = F/C$ sometimes abbreviated to f (where $f_3 = f_4 = 0$)
$f_3 = J/C$
$f_4 = D/C$
f_W = cost proportion of labor $= W/C$
f_M = cost proportion of materials $= M/C$
f_E = cost proportion of other costs $= E/C$
$f = 1 - f_1$
g = fixed cost as proportion of revenue
$= F/R$
h_1, h_2 = working capital requirement coefficients

i = proportion of total investment allocated to fixed assets $= \dfrac{I_F}{I}$

j = interest rate
k = ratio of capacity to total investment $= P/I$
k = a coefficient, or the ratio of a term in a sum to the total sum (such as $k_1 = x_1/(x_1 + x_2)$.
m = average cost per unit of materials input
n = refers to period n
p = price per unit
r = return on investment $= Z/I$
s = marginal cost per unit of output
$u_1 = (1 + j^*) f_3 h_1/\lambda$
$u_2 = (1 + j^*) f_3 h_2/\lambda$

$$v = (u_1 + u_2)\frac{V^*}{1 + V^*}$$

w = average remuneration per employee
w_b = wage rate per basic hour
w_o = wage rate per overtime hour
w_H = average wage rate per hour
x = a variable
x_1 = a variable, or the value of the variable x in period 1
x^* = relative incremental change $= \dfrac{\delta x}{x}$

y = a function of x
y_{nb} = index number for period n compared with period b taken as 1.0
z = average profit per unit of output $= Z/V = p - c$

α = net profit margin $= \dfrac{Z}{R} = \dfrac{p-c}{p}$

β = gross profit margin $= \dfrac{p-s}{p}$

$$\beta_0 = \frac{(1-\alpha) f_4}{\sigma}$$

$$\gamma = \frac{f_4}{\sigma} - f$$

δ = increment, e.g. δx = increment of x
ε = relative change elasticity of demand

ε_a = arc elasticity of demand
ε_p = point elasticity of demand
ζ_i = profit proportion for product i
η_i = proportion of the total cost attributable to product $i = C_i/C$
θ = provision of capacity by fixed investment $= P/I_F$
λ = proportion of borrowings in the capital employed $= B/I_W$
μ = mechanization elasticity $= -s^*/F^*$
ξ = factor price for a given factor input

$$\pi_x = \text{productivity of factor input } x = \frac{\text{Physical output}}{\text{Factor input } x}$$

π_N = productivity of labor (per employee)
π_H = productivity of labor (per hour)
π_I = productivity of investment
π_{I_F} = productivity of fixed investment $= e\theta$
ρ_i = revenue proportion for product i
σ = promotion elasticity $= V^*/D^*$
$\phi = p - s$ = marginal profit per unit

Glossary of Terms

Acid test or quick ratio—the ratio of liquid assets to current liabilities, used as a measure of liquidity.

Added value—the difference between revenue and purchased materials and services.

Average unit fixed cost—the ratio of fixed cost to physical output.

Average unit variable cost—the ratio of variable cost to physical output.

Base year or **Base period** or **baseline**—the period with which a comparison is made.

Break-even analysis—examination of profit performance with a view to determining the break-even point and the effect of deviating from it.

Break-even point—the level of physical output (or sales) or revenue at which revenue equals cost, so that no profit is obtained.

Budget—*see* Corporate planning.

Capacity—the maximum, or nominal, amount of output of a manufacturing (or conversion) facility is capable of producing, under specified working conditions.

Capacity utilization—*see* Efficiency

Capital employed or **Total investment** or **Capital** for short—the total financial resources of an enterprise available for the conduct of its operations. Total investment consists of: *Fixed investment* or *Fixed assets*—machinery, plant, building and land owned by the enterprise; *Working capital*—current assets minus current liabilities.

Corporate planning—a projection of future activities, and the level of resources to be devoted to various sectors of the business, for the purpose of achieving certain predetermined goals and/or putting the resources of the enterprise to good use. A corporate plan may be constructed for one year, three years, or longer, the first year of the plan expressing the intent of senior management for immediate implementation. The one year plan is accompanied by a *budget*, which is a plan cast in financial terms, outlining monetary expectations of revenues, costs and profit, their major constituent parts, as well as selected performance ratios.

Cost—the value of input resources expended over a given time period.

Cost function—the relationship between cost and output.

Cost proportion—the ratio of a given cost element to total cost.

Current assets—short-term assets, including cash, debtors and stocks.

Current liabilities—debts and claims to be settled in the short term, including creditors and taxes.

Current ratio—the ratio of current assets to current liabilities, used as a measure of liquidity.

Demand function—the relationship between demand for a product and factors that affect customer spending, for example the relationship between price and volume.

Depreciation—the decrease in the value of fixed assets due to the passage of time, wear and obsolescence.

Deterministic appraisal—an analysis of a given scenario of assumed known changes in certain variables which affect the performance of a system.

Earnings per share—the ratio of profit after tax to the number of shares.

Effectiveness—the extent to which a system manages to complete given tasks or achieve pre-specified objectives.

Efficiency—a measure of utilization of available resources, defined as a ratio, such as: *Capacity efficiency* or *Capacity utilization*—actual output to maximum output, for a given input; *Cost efficiency*—minimum costs to actual costs, for a given output; *Revenue efficiency*—actual revenue to maximum revenue, for a given output; *Technical efficiency*—minimum resources to actual resources, for a given output.

Elasticity—the postulated response of the value of one variable to a given percentage change in another. Examples: *Mechanization elasticity*—the percentage change in unit variable cost due to a given percentage change in fixed cost (resulting from capital investment in manufacturing facilities); *Price elasticity of demand*—the proportion change in demand due to a given proportion change in price; three alternative definitions are given in Chapter 8 for: *Point elasticity*, *Arc elasticity* and *Relative change Elasticity*; *Promotion elasticity*—the percentage change in demand due to a given percentage change in promotion expenditure.

Ex ante—before the event (i.e. planned).

Ex post—after the event (actual results).

Factor unit cost—the cost per unit of physical output due to a given input.

Fixed cost—the component of total cost which is unaffected by change in volume, sometimes called *Fixed overheads*.

Fixed investment—*see* Capital employed.

Gearing—the ratio between loans and shareholders funds.

Gross profit margin—the ratio of profit plus fixed costs (or overheads) to revenue.

Incremental calculus—analysis based on relative incremental changes in the value of variables.

Index number—the value of a variable in a given period compared with its value in a chosen base period, the latter usually taken as 100. Index numbers are often used to indicate changes in the levels of prices, costs, volume, and revenues.

Indirect cost– *see* Overheads.

Iso cost map—a map with contour lines, each being the locus of all points involving the same level of cost. Similarly, there are *Iso revenue* and *Iso profit* maps.

Labor performance ratio—the ratio between output (physical, revenue, profit or added value) to a measure of labor input (number of employees, manhours, total wages).

Labor productivity—the ratio of physical output per employee or per man-hour.

Liabilities—the total amount owed, including short and long term debts, and the amount due to shareholders.

Liquid assets—current assets minus stocks.

Marginal cost—the change in total cost when output is changed by one unit.

Monte Carlo technique—a simulation conducted for the purpose of determining the probabilities of various possible outcomes, based on repeating a large number of hypothetical experiments involving uncertain values of key variables. *See* Risk simulation.

Non-dimensional ratio—a ratio in which both the numerator and denominator are measured in the same units (for example, physical volume or financial values).

Operation cost—total cost, excluding tax and interest charges.

Overheads or **Indirect cost**—costs which are not readily assignable to individual products or divisions of an enterprise and are therefore conveniently treated as a fixed cost. *See* cost and fixed cost.

Product mix—the range of products offered for sale.

Productivity—the ratio of physical output to a given input.

Profit—the difference between revenue and cost. Note the distinction between: Profit before interest and tax (PBIT); Profit after interest but before tax (PBT); Profit after interest and tax (PAT).

Profit margin or **Net profit margin**–the ratio of profit before tax (sometimes before interest and tax) to revenue.

Profit sharing—a scheme that provides for a proportion of the profit to be distributed among employees that qualify for the purpose.

Quick ratio—*see* Acid test.

Reckoning—providing an analysis, which takes relevant factors into account, of past performance of a business and its future posture.
Relative increment—the ratio $\delta x/x$ where δx is the change in the value of a given variable x and its original value prior to the change.
Return on capital or **Return on investment**—the ratio of profit (usually PBT) to total capital employed.
Revenue or **Turnover**—sales income.
Revenue proportion—the ratio of a given cost element to revenue.
Risk simulation—an analysis based on the Monte Carlo technique, which takes successive samples from postulated distributions of specified variables to determine the value of a function dependent on these variables, and—by repeating the process many times—derives a distribution of the outcome.
Stock turnover—ratio of cost of sales to stock.

Total investment—*see* Capital employed
Transfer price—the price charged to one division of a company for goods or services supplied by another division, to which the price is credited.
Turnover—*see* Revenue

Unit cost—the average cost per unit of physical output, i.e. the ratio of total cost to physical output.
Unit price—the average price obtained per unit of physical output, i.e. the ratio of revenue to physical output.
Unit profit—the ratio of profit before tax to physical volume.
Utility—subjective assessment of the level of benefit or payoff accrued.

Variable cost or the **Cost of sales**—the difference between total cost and fixed cost (or overheads).

Work in progress—semi-finished goods.
Working capital—*see* Capital employed.

Index

Q

R

S

T

U

V

W